T0343370

NATURAL GAS HYDRATES

NATURAL GAS HYDRATES

A Guide for Engineers

Third Edition

JOHN CARROLL
Gas Liquids Engineering, Calgary, Canada

Amsterdam • Boston • Heidelberg • London • New York • Oxford
Paris • San Diego • San Francisco • Singapore • Sydney • Tokyo

Gulf Professional Publishing is an imprint of Elsevier

Gulf Professional Publishing is an imprint of Elsevier
225 Wyman Street, Waltham, MA 02451, USA
The Boulevard, Langford Lane, Kidlington, Oxford, OX5 1GB, UK

Library of Congress Cataloging-in-Publication Data
Application Submitted

British Library Cataloguing in Publication Data
A catalogue record for this book is available from the British Library

ISBN: 978-0-12-800074-8

For information on all Gulf Professional Publishing publications visit our web site at store.elsevier.com

This book has been manufactured using Print on Demand technology. Each copy is produced to order and is limited to black ink. The online version of this book will show color figures where appropriate.

CONTENTS

ACKNOWLEDGMENT

There are many people who I must thank. Without their help and support this book would not have been possible.

First, I would like to thank my employer Gas Liquids Engineering Ltd., Calgary, Alberta, Canada, and in particular the principals of the company Douglas MacKenzie and James Maddocks. They allowed me the time to build the hydrates course upon which this book is based and for providing me the time to write the manuscript. I would also like to thank them for the other resources they provided. This book would have been impossible without them. I would also like to thank my colleague Peter Griffin, also from Gas Liquids Engineering, for his encouragement. With his help I have been able to present this material throughout the world.

Words cannot express my thanks to Alan E. Mather of the University of Alberta, Edmonton, Alberta, Canada. He was my patient supervisor during my time as a graduate student and he continues to be my mentor. The core of my knowledge of thermodynamics, and in particular how it relates to phase equilibrium, is as a result of his teaching. Over the years we have collaborated on many interesting projects. In addition, he proofread early versions of the manuscript, which was enormously valuable.

The book is the result of a one-day course on gas hydrates. I have received positive feedback from many of those who have attended. Some of their ideas have been added to the book. Thus I thank all of those who attended the course in the past.

I would be amiss if I did not also thank my loving wife, Ying Wu, for her endless support, encouragement, and love.

Finally, I would like to express my gratitude to the *Gas Processors Association* (*GPA*) and the *Gas Processors Suppliers Association* (*GPSA*), both of Tulsa, Oklahoma, for permission to reproduce several figures from the *GPSA Engineering Data Book* (11th ed.). Furthermore, over the years these associations have sponsored a significant amount of research into gas hydrates. This research has been valuable both to the author of this book and to others working in the field.

PREFACE TO THE THIRD EDITION

The objective of the third edition is the same as the first two—to give engineers in the field the concepts to understand hydrates. From this understanding they should be able to implement strategies to prevent them from forming and to combat them when they do. Gas hydrates continue to be a significant concern in the natural gas business. Companies spend millions of dollars attempting to mitigate problems that arise from these icelike solid materials.

With each new edition there are new discoveries to explore; new concepts to examine. Although the chapter structure remains unchanged from the second edition, there are several new topics included in almost every chapter. Most of these ideas come from people who have attended my one-day course on hydrates.

For the author, hydrates remain a continuing interest because of their unusual properties and new discoveries. This makes them an engaging research topic. But as a process engineer, they remain a concern in my daily work as they are for many other engineers.

Although the book is intended for engineers, others who have to deal with hydrates will find some value in the material presented.

PREFACE TO THE SECOND EDITION

The goal of the second edition is the same as the first—to provide practicing engineers the tools to deal with hydrates.

One of the reasons that the author finds hydrates so interesting is their unusual properties. Since the time of the first edition several new properties have come to light and are discussed in the second edition. These include the type o hydrate formed from mixtures of methane and ethane, hydrates of hydrogen, the role of isopropanol in hydrate formation, etc. All of these topics will be discussed.

Another addition to the book is discussion of a few other hydrate formers. Notably, the hydrates of ethylene and propylene are included.

More examples are taken from the literature and additional comparisons are made. A new section on the prediction of hydrate formation in sour gas is also included.

PREFACE TO THE FIRST EDITION

Gas hydrates are of particular interest to those working in the natural gas industry. Thus the main audience for this book are the engineers and scientists who work in this field. Provided in this book are the tools for predicting hydrate formation and details on how to combat them.

The genesis of this book was a one-day course presented to engineers who work in the natural gas business. In particular these companies produce, process, and transport natural gas. The book has been expanded from the original set of class notes. Much of the new material came from feedback from attendees.

Many people outside the field of natural gas have also attended the course and found some value in the material. These include oceanographers studying the hydrate deposits on the seabeds throughout the world. Astronomers investigating the possibility of hydrates on the planets of the solar system as well as other celestial bodies may also find some of the material in this book of some use. And those who are simply curious about these interesting compounds will find this book to be useful.

The structure of the book is a little unusual. The chapters are meant to be approximately independent, however, they do follow from the more simple introductory topics to the more advanced applications. Occasionally it is necessary to take a concept in a subsequent chapter in order to make a point in the current chapter. This is unfortunate, but it is also necessary.

The purpose of this book is to explain exactly what gas hydrates are, under what conditions they form, and what can be done to combat their formation. Another purpose of this book is to explore some of the myths associated with gas hydrates. The material is organized and presented in such a way that the average engineer can use the information in their day-to-day work.

In some sections of the book, especially those dealing with dehydration, pipeline heat loss calculations, and line heater design, the reader would benefit greatly if they have the ability to calculate the physical properties of natural gas. The topic of the properties of natural gas is not covered in this book.

CHAPTER 1

Introduction

This chapter is an attempt to introduce hydrates, without much background material. Many of the words and principles will be better defined in subsequent chapters of this book. However, they are needed here to present the basic introductory concepts. If you are a little confused as you read this chapter, hopefully things will become clearer as you progress through the book.

In its most general sense, a hydrate is a compound containing water. For example, there is a class of inorganic compounds called "solid hydrates." These are ionic solids where the ions are surrounded by water molecules and form crystalline solids. However, as used in this book, and commonly in the natural gas industry, hydrates are solid phase composed of a combination of certain small molecules and water.

So hydrates are crystalline solid compounds formed from water and small molecules—without water there are no hydrates and without the small molecules that stabilize the structure there are no hydrates. They are a subset of compounds known as clathrates or inclusion compounds. A clathrate compound is one where a molecule of one substance is enclosed in a structure built up from molecules of another substance. Here, water builds up the structure and the other molecule resides within. The size of the other molecule must be such that it can fit within the water structures. More details of the nature of these structures formed by water and the molecules within are presented in Chapter 2 of this book.

Even though the clathrates of water, the so-called hydrates, are the focus of this work, they are not the only clathrate compounds. For example, urea forms interesting inclusion compounds as well.

Although hydrates were probably encountered by others earlier, credit for their discovery is usually given to the famous English chemist, Sir Humphrey Davy. He reported of the hydrate of chlorine in the early nineteenth century. In particular, he noted (1) that the ice-like solid formed at temperatures greater than the freezing point of water, and (2) that the solid was composed of more than just water. When melted, the hydrate of chlorine released chlorine gas.

Davy's equally famous assistant, Michael Faraday, also studied the hydrate of chlorine. In 1823, Faraday reported the composition of the chlorine

Natural Gas Hydrates
ISBN 978-0-12-800074-8
http://dx.doi.org/10.1016/B978-0-12-800074-8.00001-6

hydrate. Although his result was inaccurate, it was the first time that the composition of a hydrate was measured.

Throughout the nineteenth century, hydrates remained basically an intellectual curiosity. Early efforts focused on finding which compounds formed hydrates and under what temperatures and pressures they would form. Many of the important hydrate formers were discovered during this era.

Amongst the nineteenth century hydrate researches who deserve mention are the French chemists Villard and de Forcrand. They measured the hydrate conditions for a wide range of substances, including hydrogen sulfide.

However, it would not be until the twentieth century that the industrial importance of gas hydrates would be established.

Over the years there have been many, many experimental studies of hydrate formation. These include the hydrates for single components, binary mixtures, as well as multicomponent mixtures. Some of these studies are discussed in the chapters that follow. If the reader has doubts about methods used in the work, they should consult the literature. They may not find the exact data for their situation, but they may find data that are useful for testing the models they chose to employ.

1.1 NATURAL GAS

Even though all terrestrial gases (air, volcanic emissions, swamp gas, etc.) are natural, the term "natural gas" is customarily reserved for the mineral gases found in subsurface rock reservoirs. These gases are often associated with crude oil. Natural gas is a mixture of hydrocarbons (such as methane, ethane, propane, etc.) and a few nonhydrocarbons (hydrogen sulfide, carbon dioxide, nitrogen, etc., and water).

The light hydrocarbons in natural gas have value as fuels and as feedstocks for petrochemical plants. As a fuel, they are used for heating and cooking in private homes, to generate electricity, and increasingly as fuel for motor vehicles. In the chemical plants, they are converted a host of consumer products; everything from industrial chemicals, such as methanol, to plastics, such as polyethylene.

The nonhydrocarbons tend to be less valuable. However, depending upon the market situation, hydrogen sulfide has some value as a precursor to sulfur. Sulfur in turn has several applications, the most important of which is probably the production of chemical fertilizer. Carbon dioxide and nitrogen have no heating value and thus are useless as fuels.

Natural gas that contains significant amounts of sulfur compounds, and hydrogen sulfide in particular, is referred to as "sour." In contrast, natural gas with only minute amounts of sulfur compounds is called "sweet." Unfortunately there is no strict-defining sulfur content that separates sour gas from sweet gas. As we have noted, sales gas typically contains less than about 15 ppm and is indeed sweet, but for other applications there are other definitions. For example, in terms of corrosion, the sweet gas may contain more sulfur compounds and not require special materials.

Strictly speaking, gas that contains carbon dioxide but no sulfur compounds is not sour. However, gas that contains carbon dioxide shares many characteristics with sour gas and is often handled in the same way. Probably the most significant differences between carbon dioxide and hydrogen sulfide are the physiological properties, and this is what really separates the two. Hydrogen sulfide is highly toxic, whereas carbon dioxide is essentially nontoxic, except at very high concentrations. Furthermore, hydrogen sulfide has an obnoxious odor, whereas carbon dioxide is odorless.

1.1.1 Sales Gas

An arrangement is made between the company producing the natural gas and the pipeline company for the quality of the gas the purchaser will accept. Limits are placed on the amounts of impurities, heating value, hydrocarbon dew point, and other conditions. This arrangement is what defines "sales gas."

Amongst the impurities that are limited in the sales gas is water. One of the reasons why water must be removed from natural gas is to help prevent hydrate formation.

In terms of water content, a typical sales gas specification would be less than approximately 10 lb of water per million standard cubic feet of gas (10 lb/MMCF). In the United States, the value is usually 7 lb/MMCF, whereas in Canada, it is 4 lb/MMCF, and other jurisdictions have other values. For those who prefer SI units, 10 lb/MMCF is equal to 0.16 grams per standard cubic meter (0.16 g/Sm^3) or 160 milligrams per standard cubic meter (160 mg/Sm^3). More discussion of units and standard conditions is presented later in this chapter.

There are several other restrictions on the composition of sales gas. For example, there is a limit on the amount of hydrogen sulfide present (typically on the order of about 10 parts per million or 10 ppm) and the amount of carbon dioxide (typically around 2 mole percent). These too vary from jurisdiction to jurisdiction and contract to contract.

1.1.2 Hydrates

In combination with water, many of the components commonly found in natural gas form hydrates. One of the problems in the production, processing, and transportation of natural gas and liquids derived from natural gas is the formation of hydrates. Hydrates cost the natural gas industry millions of dollars annually. In fact, individual incidents can cost $1,000,000 or more depending upon the damage inflicted. There is also a human price to be paid because of hydrates. Sadly, there have been deaths either directly or indirectly associated with hydrates and their mishandling.

However, the importance of natural gas hydrates was not apparent in the early era of the gas business. In the early era of the natural gas business, gas was produced and delivered at relatively low pressure. Thus, hydrates were never encountered. In the twentieth century, with the expansion of the natural gas industry, the production, processing, and distribution of gas became high-pressure operations. Under pressure, it was discovered that pipelines and processing equipment were becoming plugged with what appeared to be ice, except that the conditions were too warm for ice to form. It was not until the 1930s that Hammerschmidt (1934) clearly demonstrated that the "ice" was actually gas hydrates, and that the hydrates were a mixture of water and the components of natural gas.

In the petroleum industry, the term "hydrate" is reserved for substances that are usually gaseous at room temperature. These include methane, ethane, carbon dioxide, and hydrogen sulfide. This leads to the term "gas hydrates" and also leads to one of the popular misconceptions regarding these compounds. It is commonly believed that nonaqueous liquids do not form hydrates. However, liquids may also form hydrates. An example of a compound that is liquid at room conditions, yet forms a hydrate, is dichlorodifluoromethane (Freon 12). But we are getting ahead of ourselves. More details about what compounds form hydrates will be given in Chapter 2.

1.2 THE WATER MOLECULE

Many of the usual properties of water (and yes, if you aren't aware of it, water does have some unusual properties) can be explained by the structure of the water molecule and the consequences of this structure.

Of particular interest to us is the fact that the structure of the water molecule leads to the possibility of hydrate formation. In the next sections, it will be demonstrated that water does indeed have some unusual properties.

1.2.1 The Normal Boiling Point of Water

As an example of the unusual properties of water, consider the boiling point. We will use some simple chemistry to demonstrate that the boiling point of water is unusually high. The boiling points used in this discussion are taken from Dean (1973).

The periodic table of elements is not just a nice way to display the elements. The original design of the table came from aligning elements with similar properties. Thus, elements in the rows of the tables have similar properties or at least properties that vary in a periodic, predictable manner. The 6A column in the table consists of oxygen, sulfur, selenium, and tellurium. We would expect these elements and their compounds to have similar properties, or at least to behave in a predictable pattern.

The hydrogen compounds of the column 6A elements are: water (hydrogen oxide), hydrogen sulfide, hydrogen selenide, and hydrogen telluride. All have the chemical formula H_2X, where X represents the group 6A element. If we look at the normal boiling points of H_2S, H_2Se, and H_2Te, we should be able to predict the boiling point of water. Figure 1.1

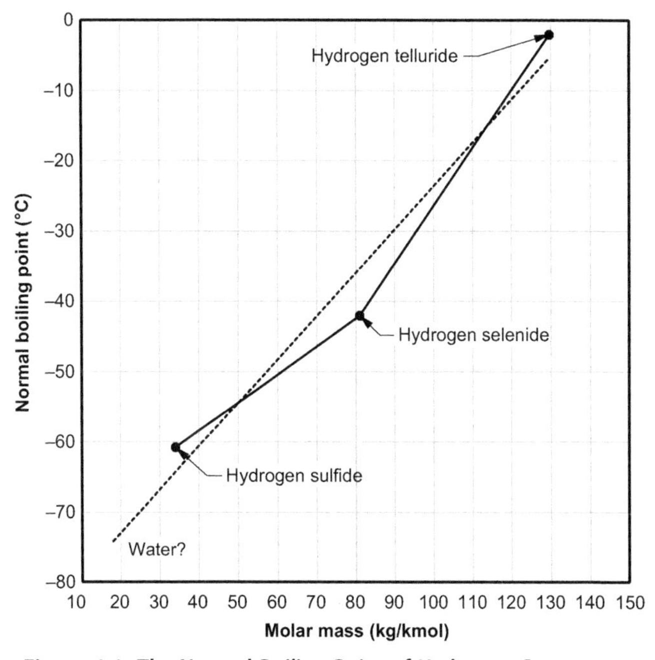

Figure 1.1 *The Normal Boiling Point of Hydrogen Component.*

shows a plot of the normal boiling points for these three compounds. Note that as the size of the molecule increases, so does the normal boiling point. Although it is not exactly linear, we can use a linear approximation to estimate the boiling point of water. This extrapolation yields an estimated boiling point of −74 °C! Since the boiling point of water is 100 °C, this is clearly a very poor estimation. There is probably something unusual about water.

It is worth noting that a similar plot could be constructed showing the melting points for these compounds. Again, the predicted melting point of water, based on the other substances, is much too low.

As a second example, consider the homologous series of normal alcohols. Figure 1.2 shows a plot of the normal boiling points of the alcohols as a function of their molar mass (molecular weight). In this case, the relation is nearly linear. Assume that water is the smallest member of this group of compounds, and extrapolate the correlation to estimate the boiling point. This yields 43 °C for the boiling point, which is significantly lower than the actual value.

Why does water have such an anomalously large boiling point?

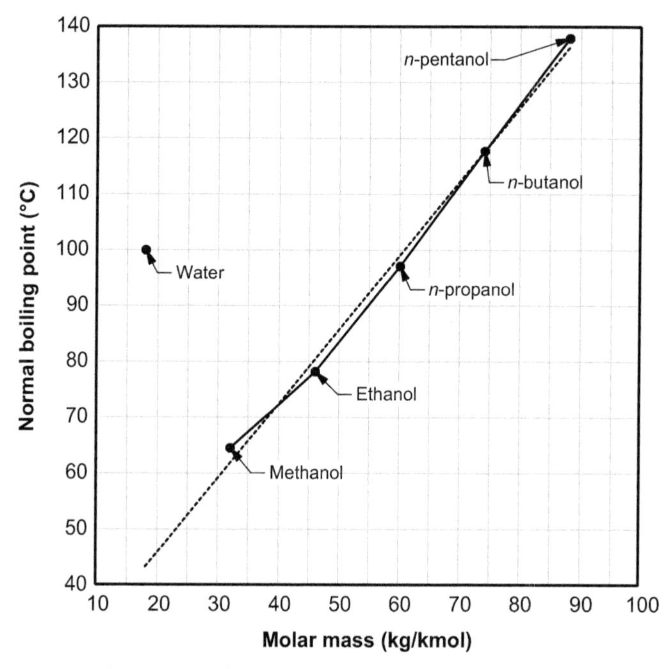

Figure 1.2 ***The Normal Boiling Point of Alcohols.***

1.2.2 Enthalpy of Vaporization

In Table 1.1, the enthalpies of vaporization of several components at their boiling point are listed. The table includes both polar and nonpolar substances. From this table, it can be seen that water has a fairly large enthalpy of vaporization, even in comparison to other polar substances.

It takes significantly more energy to boil 1 kg of water than it does to boil any of the hydrocarbons listed in the table—approximately five times as much energy.

Again, we must ask the question, why does water behave so anomalously?

1.2.3 Expansion upon Freezing

Another unusual property of water is that it expands upon freezing. In common terms, this means that ice floats on water. In engineering terms, the density of ice (917 kg/m^3 or 57.2 lb/ft^3) is less than that of liquid water (1000 kg/m^3 or 62.4 lb/ft^3) at the freezing point.

The reason for this expansion is that the water atoms arrange themselves in an ordered fashion and the molecules in the crystal occupy more space than those in the liquid water. The reason for this behavior is also due to the shape of the water molecule and something called the hydrogen bond.

The molecules in solid water form a hexagonal crystal. This is most obvious in snow, with its characteristic pattern structures (see for example Fig. 1.3).

Table 1.1 The Enthalpy of Vaporization of Several Substances at Their Normal Boiling Point

Compound	Nature	Enthalpy of Vaporization (kJ/kg)
Water	Polar	2257
Methanol	Polar	1100
Ethanol	Polar	855
Acetone	Polar	521
Ethylene glycol	Polar	800
Ammonia	Polar	1369
Methane	Nonpolar	510
n-Pentane	Nonpolar	357
n-Octane	Nonpolar	306
Benzene	Nonpolar	394
o-Xylene	Nonpolar	347
Cyclohexane	Nonpolar	358

Data taken from Dean (1973), pp. 9–85 to 9–95.

Figure 1.3 ***Micro Photographs of Snowflakes.*** *Courtesy of the National Oceanic and Atmospheric Administration (NOAA), Washington, DC—http://www.photolib.noaa.gov. Original photographs by Wilson Bentley (1865–1931), which were not copyright.*

1.2.4 The Shape of the Water Molecule and the Hydrogen Bond

Virtually all of the unusual properties of water noted earlier can be explained by the shape of the water molecule and the interactions that result from its shape.

The water molecule consists of a single atom of oxygen bonded to two hydrogen atoms, as depicted in Fig. 1.4. In the water molecule, the bond

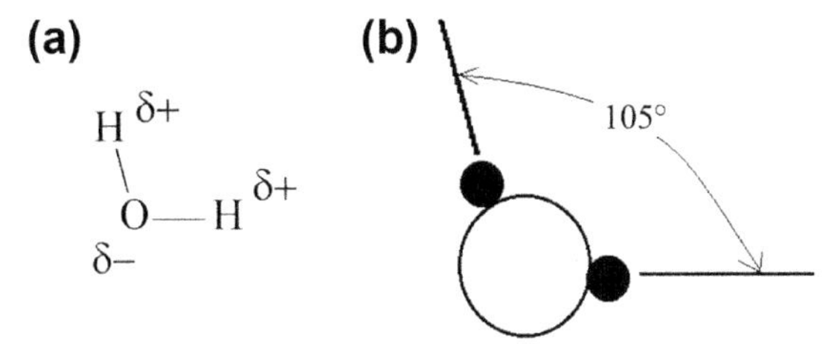

Figure 1.4 ***The Shape of the Water Molecule.*** (a) Stick representation showing induced charges which result in hydrogen bonding. (b) Ball model showing the angle between the hydrogen molecules.

between the oxygen and hydrogen atoms is a covalent bond. A covalent bond is essentially a shared pair of electrons. The angle between the two hydrogen atoms in the water molecule is about 105°.

What Fig. 1.4 does not show is that there are two pairs of unbonded electrons on the "back" of the oxygen molecule. These electrons induce negative charges on the oxygen molecule and a small positive charge on the hydrogen atoms. The induced electrostatic charges on the molecule (denoted δ+ for the positive charge and δ− for the negative) are shown in Fig. 1.4(a). Thus, the water molecules will tend to align with a hydrogen molecule lining up with an oxygen.

This aligning of the hydrogen and oxygen atoms is called a "hydrogen bond." The hydrogen bond is essentially an electrostatic attraction between the molecules. It should be noted that each water molecule has two pairs of unbonded electrons and thus has two hydrogen bonds—two water molecules "stick" to each water molecule. The hydrogen bond is only 1/10 or 1/20 as strong as a covalent bond, which is what hold the oxygen and hydrogen atoms together in the water molecule, but this is still strong enough to explain the properties discussed earlier.

The hydrogen bonds are particularly strong in water, although they are present in other substances, such as the alcohols discussed earlier. It is for this reason that the normal boiling points of the alcohols are significantly larger than their paraffin analogs.

When the water molecules line up, they form a hexagonal pattern. This is the hexagonal crystal structure discussed earlier. From elementary geometry, it is well know that the angle between the sides of a regular hexagon is 120°, which is greater than the 105° angle in the water molecule. This seeming paradox is overcome because the hexagonal pattern of the water molecules is not planar. The hexagonal pattern of the water molecules in the ice crystal is shown in Fig. 1.5. In this figure, the

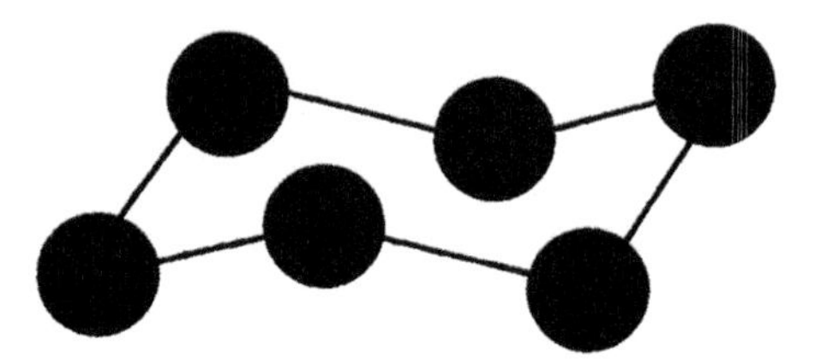

Figure 1.5 ***The Three-Dimensional Hexagonal Arrangement of the Water Molecules in an Ice Crystal.***

circles represent the water molecules and the lines represent the hydrogen bonds.

1.3 HYDRATES

It is a result of the hydrogen bond that water can form hydrates. The hydrogen bond causes the water molecules to align in regular orientations. The presence of certain compounds causes the aligned molecules to stabilize and a solid mixture precipitates.

The water molecules are referred to as the "host" molecules, and the other compounds, which stabilize the crystal, are called the "guest" molecules. In this book, the guest molecules are more often referred to as "formers." The hydrate crystals have complex, three-dimensional structures where the water molecules form a cage and the guest molecules are entrapped in the cages.

The stabilization resulting from the guest molecule is postulated to be due to van der Waals forces, which is the attraction between molecules that is not as a result of electrostatic attraction. As described earlier, the hydrogen bond is different from the van der Waals force because it is due to strong electrostatic attraction, although some classify the hydrogen bond as a van der Waals force.

Another interesting thing about gas hydrates is that there is no bonding between the guest and host molecules. The guest molecules are free to rotate inside the cages built up from the host molecules. This rotation has been measured by spectroscopic means. Therefore, these compounds are best described as a solid solution.

The formation of a hydrate requires the following three conditions:

1. The right combination of temperature and pressure. Hydrate formation is favored by low temperature and high pressure.
2. A hydrate former must be present. Hydrate formers include: methane, ethane, and carbon dioxide.
3. A sufficient amount of water—not too much, not too little.

Figure 1.6 gives a visual of the three criteria for hydrate formation. The three are interconnected—violate one and a hydrate does not form. Although this figure gives a quick visual image, it lacks the detailed provided by the discussion presented earlier. However, it provides a useful visual.

These three points will be examined in some detail in subsequent chapters, but they deserve a few comments at this point. As was noted, low temperature and high pressure favor hydrate formation. The exact temperature and pressure depends upon the composition of the gas. However, hydrates form at temperatures greater than 0 °C (32 °F), the freezing point of water. The nature of hydrate formers is discussed in detail in Chapter 2.

In order to prevent hydrate formation, one merely has to eliminate one of the three conditions stated above. Typically, we cannot remove the hydrate formers from the mixture. In the case of natural gas, it is the hydrate formers that are the desired product. So we attack hydrates by addressing the other two considerations.

Other phenomena that enhance hydrate formation include the following:

1. Turbulence

  a. High velocity

  Hydrate formation is favored in regions where the fluid velocity is high. This makes choke valves particularly susceptible to hydrate formation. First, there is usually a significant temperature drop when natural gas is choked through a valve due to the Joule–Thomson effect. Second, the velocity is high through the narrowing in the valve.

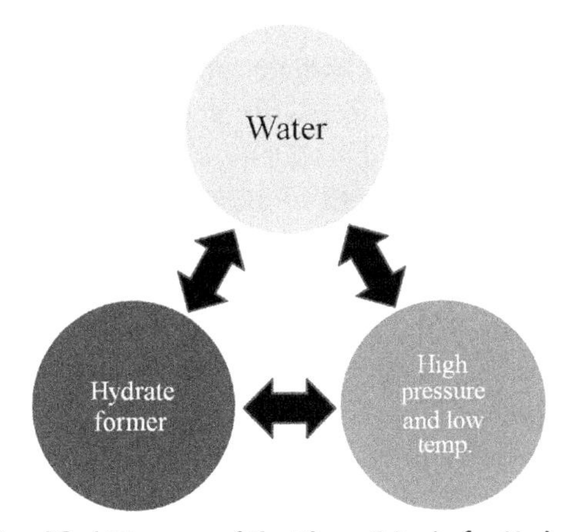

Figure 1.6 ***Simplified Diagram of the Three Criteria for Hydrate Formation.***

b. Agitation
Mixing in a pipeline, process vessel, heat exchanger, etc. enhances hydrate formation. The mixing may not be due to an actual mixer but perhaps to a tortuous routing of the line.

2. Nucleation sites
In general terms, a nucleation site is a point where a phase transition is favored, and in this case the formation of a solid from a fluid phase. An example of nucleation is the deep fryer used to make French fries in fast-food restaurants throughout the world. In the fryer, the oil is very hot but it does not undergo the full rolling-boil because there are no suitable nucleation sites. However, when the potatoes are placed into the oil, it boils vigorously. The French fries provide an excellent nucleation site.

Good nucleation sites for hydrate formation include an imperfection in the pipeline, a weld spot, or a pipeline fitting (elbow, tee, valve, etc.). Corrosion byproducts, silt, scale, dirt, and sand all make good nucleation sites as well.

3. Free-water
No, this is not a contradiction to earlier statements. Free-water is not necessary for hydrate formation, but the presence of free-water certainly enhances hydrate formation.

The presence of free water also assures that they is plenty of water present, which is more likely to form a plug.

In addition, the water–gas interface is a good nucleation site for hydrate formation.

The items in the above list enhance the formation of a hydrate, but are not necessary. Only the three conditions given earlier are necessary for hydrate formation.

Another important aspect of hydrate formation is the accumulation of the solid. The hydrate does not necessarily agglomerate in the same location as it is formed. In a pipeline, the hydrate can flow with the fluid phase, especially the liquid. It would tend to accumulate in the same location as the liquid does. Usually it is the accumulation of the hydrates that causes the problems. In a multiphase pipeline, it is the accumulation that blocks the line, plug, and damage equipment.

Often, pigging is sufficient to remove the hydrate from the pipeline. Pigging is the process of inserting a tool (called a "pig") into the line. Modern pigs have many functions, but the main one remains pipeline cleaning. The pig fits tightly into the line and scraps the inside of the pipe. It is transported along the line with the flow of the fluid, and by doing so it removes any solids (hydrate, wax, dirt, etc.) from inside the line. The pigging can also be used to remove accumulations of liquids.

However, the pigging must be scheduled such that the accumulation of hydrates does not become problematic. Usually, pigging is not used to clean hydrates from a line. Other measures are more commonly used to deal with hydrates, and these are detailed in subsequent chapters of this book.

Another benefit of pigging is the removal of salt, scale, etc. which is important for the proper operation of a pipeline. It also means that potential nucleation sites for hydrate formation are removed.

1.3.1 Temperature and Pressure

As was noted earlier, hydrate formation is favored by low temperature and high pressure. For each gas it is possible to generate a hydrate curve that maps the region in the pressure–temperature plane where hydrates can form. Much of the rest of this book is dedicated to the tools used to predict this locus. Again, without getting to far ahead of ourselves, some preliminary discussion of hydrate curves is appropriate.

Figure 1.7 shows a typical hydrate curve (labeled "hydrate curve"). The region to the left and above this curve (high pressure, low temperature) is where hydrates can form. In the region to the right and below the hydrate curve, hydrates can never form—in this region, the first criteria is violated. Therefore, if your process, pipeline, well, etc. operates in the region labeled "no hydrates," then hydrates are not a problem. On the other hand, if it is in the region labeled "hydrates region," then some remedial action is required to avoid hydrates.

It might seem as though we can treat the temperature and pressure as separate variables, but when discussing hydrates, they are linked. For example, you cannot say, "A hydrate will not form at 10 °C for the gas mixture shown in Fig. 1.7." You must qualify this with a pressure. So at 10 °C and 5 MPa the process is in the "hydrate region," whereas at 10 °C and 1 MPa the process is in the region where a hydrate will not form. Thus, we must talk about a combination of temperature and pressure, and not each variable on its own.

Furthermore, it is common to add a margin of safety, even to the best hydrate prediction methods. This margin can be 3–5 °C (5–10 °F), but typically 3 °C is used. The author typically uses 3 °C, but the reader may have their own margin or perhaps there is one specified by their company.

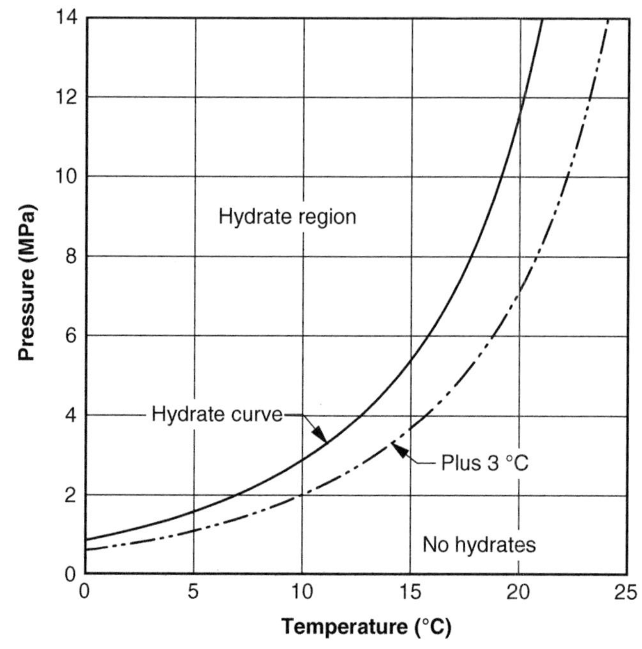

Figure 1.7 ***Pressure-Temperature Diagram Showing the Hydrate Region, the Region with No Hydrates, and a Safety Margin.***

A margin of safety is shown in Fig. 1.7 ("plus 3 °C") and the buffer zone between the estimated hydrate curve and the +3 °C curve is noted.

Throughout the remainder of the book, only the calculated hydrate curves will be shown, but the user should keep this safety margin in mind.

1.4 WATER AND NATURAL GAS

Water is often associated with natural gas. In the reservoir, water is always present. Thus, produced natural gas is always saturated with water. In addition, formation water is occasionally produced along with the gas. As the temperature and pressure change during the production of the gas, water can condense out. Methods for estimating the water content of natural gas are presented in Chapter 10.

In addition, water is often involved in the processing of natural gas. The process to sweeten natural gas (i.e., to remove hydrogen sulfide and carbon dioxide, the so-called "acid gases") often employs aqueous solutions. The most common of these processes involve aqueous solutions of alkanolamines. The sweet gas (i.e., the product of the sweetening process) from these processes is saturated with water.

There are several processes that are design to remove water from natural gas. These will be discussed in Chapter 6 of this book.

This association of water and natural gas mean that hydrates will be encountered in all aspects of the production and processing of natural gas.

A large portion of this book is dedicated to the prediction of the conditions at which hydrates will form. Armed with this knowledge, engineers working in the natural gas industry know whether or not hydrates will be a problem in their application.

Once it has been determined that hydrates are a problem, or potentially a problem, what can be done? Another large segment of this book addresses this topic.

1.4.1 Free-water

There is a myth in the natural gas industry that "free-water" (i.e., an aqueous phase) must be present in order to form a hydrate. In subsequent sections of this book, we will demonstrate that this is not correct. Free-water certainly increases the possibility that a hydrate will form, but it is not a necessity.

A strong argument demonstrating that free-water is not necessary for hydrate formation is presented in Chapter 9 on phase diagrams.

Another argument, the so-called "frost argument," asks the simple question: Is it necessary to have free-water in order to form ice? The answer is no. Frost forms without liquid water forming. The frost sublimes from the air onto my car on winter nights. The water goes directly from the air to the solid phase without a liquid being encountered. The air–water mixture is a gas, and the water is not present in the air in a liquid form. If you have an old-fashioned freezer, i.e., one that is not frost-free, just look inside. A layer of frost builds without liquid water ever forming. Hydrates can "frost" via the same mechanism.

One of the reasons why it is believed that free-water is required for hydrate formation is that hydrates formed without free-water may not be problematic. The inside of a pipe may "frost" with hydrates, but still function properly, or the amount of hydrate may be small and thus does not plug lines or damage processing equipment. Such "frost" hydrates can be easily cleaned using the pigging process discussed earlier.

The process of going directly from the gas to the solid is called sublimation, and it is not that rare. For example, carbon dioxide sublimes at atmospheric pressure. Solid CO_2, commonly called "dry ice," goes directly from the solid to the vapor without forming a liquid. At atmospheric pressure, CO_2 goes directly from the solid to the vapor at a temperature of about $-78\ ^\circ C$

(−108 °F). Another example of a solid that sublimes at atmospheric pressure is naphthalene, the main component of mothballs. The reason why you can smell mothballs is because the naphthalene goes directly from the solid to the vapor, and it is the vapor that you can smell. In reality, all pure substances sublime at pressures below their triple point pressure, and this includes pure water. So it should come as no surprise that hydrates, under the right set of conditions, can sublime directly from the gas phase to the solid phase.

1.5 HEAVY WATER

Deuterium is an isotope of hydrogen. In the simple hydrogen molecule there is one proton, one electron, and no neutrons, protons, electrons, and neutrons, being the elementary particles that make up the atom. Deuterium, on the other hand, is composed of one proton, one electron, and one neutron. Because of the additional particle, deuterium is "heavier" than normal hydrogen.

Water is composed of two hydrogen atoms and an oxygen atom. Heavy water, also called deuterium oxide, is composed of two deuterium atoms and an oxygen atom.

Now the question arises, does heavy water form a hydrate? The answer is yes. Heavy water still exhibits hydrogen bonding—the key to hydrate formation. However, it requires slightly more pressure to form hydrates in heavy water than in regular water (for example, see Chun et al. (1996)).

1.6 ADDITIONAL READING

Although this book is meant to provide the design engineer with tools for dealing with hydrates, it is not meant to be the definitive volume on the subject. The reader is referred to the works listed in the Bibliography for additional information on the subject of gas hydrates.

The author of this book maintains this website and will attempt to keep it updated. A website for this book, which includes the some of the programs mentioned, updates, and other material, is maintained at: http://members.shaw.ca/hydrate.

1.7 UNITS

Both American Engineering Units (ft-lb-sec) and SI Units (m-kg-s) are commonly employed in the natural gas business throughout the world. Depending upon the location, one system tends to dominate over the other, but both are in widespread use. Thus, in order to reach as wide an audience as possible, both sets of units are used in this book. Usually, the appropriate conversions are given.

In addition, in this book, the common abbreviations of the units are used, even though for the same quantity, the abbreviations often differ. For example, in American Engineering Units, "hr" and "sec" are the commonly used abbreviations for hour and second, whereas "h" and "s" are the standard in SI Units for the same quantities.

Another confusing aspect between the two systems of units is the use of the prefix M. In the SI system, M means "mega" or 10^6. In American Engineering Units, M means "1000" or 10^3 (based on the Roman numeral M, which equals 1000). Thus, in American Engineering Units, two M's (MM) means one million. Therefore, 1 MJ is "one million joules or 1 megajoule," whereas 1 MBtu is "1000 British thermal units" and MMBtu is "one million British thermal units."

Finally, the symbol MMCF as used in this book means "one million standard cubic feet." When the volume is expressed as actual cubic feet, then the unit will be followed by "[act]." In the SI systems, an S in front of the volumetric term indicates that it is at standard conditions. As used in this book, standard conditions are 60 °F and 14.696 psia or 15.6 °C and 101.325 kPa.

Chemists refer to "standard temperature and pressure" (STP), which they use as 0 °C and 1 atm. In Europe in particular, it is common to refer to "normal conditions," which are 20 °C and 1 atm. Furthermore, the International Union of Pure and Applied Chemistry (IUPAC, 2006) is encouraging the switch from 1 atm to 1 bar as the standard pressure, where 1 bar equals 0.9869 atm, which also equals 14.50 psia. However, this change does not seem to have been widely accepted in the engineering community.

The reader should be careful when converting temperature differences. For example, 10 °C is equal to 50 °F. However, a temperature difference of 10 °C is equal to a temperature difference of 18 °F. Also, a temperature difference of 10 °C equals a difference of 10 K and not 274.15.

Unit conversion tables and programs are available elsewhere and will not be repeated here.

1.8 QUANTIFYING ERROR

Throughout this book, comparisons will be made between data (experimental and field) and models. There, error is defined as the difference between a measured value and a calculated one, and the absolute error is the absolute difference between the measured value and the calculated one.

The following four quantities are applied to a set of data: (1) average error (AE), (2) absolute average error (AAE), (3) average relative error (ARE), and (4) average absolute relative error (AARE). These four quantities are defined in the following equations:

$$\mathrm{AE} = \frac{1}{N}\sum_{i=1}^{N}\left(x_{\mathrm{measure},i} - x_{\mathrm{calculated},i}\right) \tag{1.1}$$

$$\mathrm{AAE} = \frac{1}{N}\sum_{i=1}^{N}\left|x_{\mathrm{measure},i} - x_{\mathrm{calculated},i}\right| \tag{1.2}$$

$$\mathrm{ARE} = \frac{1}{N}\sum_{i=1}^{N}\left(x_{\mathrm{measure},i} - x_{\mathrm{calculated},i}\right)\Big/x_{\mathrm{measure},i} \tag{1.3}$$

$$\mathrm{AARE} = \frac{1}{N}\sum_{i=1}^{N}\left|x_{\mathrm{measure},i} - x_{\mathrm{calculated},i}\right|\Big/x_{\mathrm{measure},i} \tag{1.4}$$

where N is the number of points in a set of data, x_{measure} is a measured quantity, and $x_{\mathrm{calculated}}$ is a calculated value at the same conditions. With the error, both positive and negative values are possible, and these tend to cancel each other when the average is calculated. However, for the absolute error, only positive values result, and thus the AAE and ARE are always positive.

For a good correlation, the AE should be close to zero. Typically, the AAE will be greater than the AE. If the AE and AAE have the same value, this indicates a systematic deviation. That is, all of the calculated values are greater than the measured value or all of the calculated values are less than the measure value. However, both the error and the absolute error should be used together to judge the quality of a calculation.

Typically, the AE and the AAE will be used for temperatures and the relative errors for other quantities. Note that the AE and AAE have the same units as the quantity in question (i.e., for temperatures in Celsius and Fahrenheit degrees) where the relative errors are dimensionless and usually expressed in a percent.

The maximum and minimum errors might also be given. As their names describe, these are simply the minimum absolute error for a set of data.

Examples

Example 1.1

In our gathering system, the temperature is never less than 15 °C (59 °F) and the pressure is never greater than 8 MPa (1160 psia). The composition of the gas, in mole fraction, is given in the table below. The gas flows at a rate of 141×10^3 Sm^3/day (5 MMSCFD) and the initial water rate is 0.1 m^3/day (7 barrels per day, bpd). Should we be concerned about hydrates in the gathering system?

Nitrogen	0.55	Isobutane	0.54
CO_2	0.13	*n*-Butane	1.01
H_2S	0.04	Isopentane	0.29
Methane	86.83	*n*-Pentane	0.36
Ethane	6.92	Hexane	0.29
Propane	2.82	C_7+	0.22

Answer: Check list—the three criteria for hydrate formation:

1. *Water*: The information indicates that free-water is present. Furthermore, this is not a large excess of water.
2. *Hydrate Formers*: The gas mixture contains a number of hydrate formers: nitrogen, CO_2, H_2S, methane, ethane, propane, and isobutane. We will learn more about hydrate formers in Chapter 2.
3. The right combination of *temperature and pressure*: This is the subject of Chapters 3 and 4. By the end of those chapters, you will have the tools to decide whether or not this combination of temperature and pressure are in the hydrate forming region. For now: UNDECIDED. (see Example 9.8).

APPENDIX 1A PERIODIC TABLE OF THE ELEMENTS

1A	2A	3B	4B	5B	6B	7B		8B		1B	2B	3A	4A	5A	6A	7A	noble
								1 **H** 1.0079									2 **He** 4.0026
3 **Li** 6.941	4 **Be** 9.012											5 **B** 10.810	6 **C** 12.011	7 **N** 14.007	8 **O** 15.999	9 **F** 18.998	10 **Ne** 20.180
11 **Na** 22.990	12 **Mg** 24.305											13 **Al** 26.982	14 **Si** 28.086	15 **P** 30.974	16 **S** 32.066	17 **Cl** 35.453	18 **Ar** 39.948
19 **K** 39.098	20 **Ca** 40.08	21 **Sc** 44.956	22 **Ti** 47.867	23 **V** 59.942	24 **Cr** 51.996	25 **Mn** 54.938	26 **Fe** 55.845	27 **Co** 58.933	28 **Ni** 58.693	29 **Cu** 63.546	30 **Zn** 65.39	31 **Ga** 69.723	32 **Ge** 72.61	33 **As** 74.922	34 **Se** 78.96	35 **Br** 79.904	36 **Kr** 83.80
37 **Rb** 85.469	38 **Sr** 87.62	39 **Y** 88.906	40 **Zr** 91.224	41 **Nb** 92.206	42 **Mo** 95.94	43 **Tc** (98)	44 **Ru** 101.07	45 **Rh** 102.91	46 **Pd** 106.42	47 **Ag** 107.87	48 **Cd** 112.41	49 **In** 114.82	50 **Sn** 118.71	51 **Sb** 121.76	52 **Te** 127.60	53 **I** 126.90	54 **Xe** 131.29
55 **Cs** 132.91	56 **Ba** 137.33	57-71	72 **Hf** 178.49	73 **Ta** 180.95	74 **W** 183.84	75 **Re** 186.21	76 **Os** 190.23	77 **Ir** 192.22	78 **Pt** 195.08	79 **Au** 196.97	80 **Hg** 200.59	81 **Tl** 204.38	82 **Pb** 207.2	83 **Bi** 208.98	84 **Po** (209)	85 **At** (210)	86 **Rn** (222)
87 **Fr** (223)	88 **Ra** (226)	89-103	104 **Rf** (261)	105 **Db** (262)	106 **Sg** (263)	107 **Rh** (264)	108 **Hs** (265)	109 **Mt** (268)									

La series	57 **La** 138.91	58 **Ce** 140.12	59 **Pr** 140.91	60 **Nd** 144.24	61 **Pm** (145)	62 **Sm** 150.36	63 **Eu** 151.96	64 **Gd** 157.25	65 **Tb** 158.93	66 **Dy** 162.50	67 **Ho** 164.93	68 **Er** 167.26	69 **Tm** 168.93	70 **Yb** 173.04	71 **Lu** 174.97
Ac Series	89 **Ac** (227)	90 **Th** 232.04	91 **Pa** 231.04	92 **U** 238.04	93 **Np** (237)	94 **Pu** (244)	95 **Am** (243)	96 **Cm** (247)	97 **Bk** (247)	98 **Cf** (251)	99 **Es** (252)	100 **Fm** (257)	101 **Md** (258)	102 **No** (259)	103 **Lr** (262)

REFERENCES

Chun, M.-K., Yoon, J.-H.Y., Lee, H., 1996. Clathrate phase equilibrium for water + deuterium oxide + carbon dioxide and water + deuterium oxide + chlorodifluoromethane (R22) systems. J. Chem. Eng. Data 41, 1114–1116.

Dean, J.A. (Ed.), 1973. Lange's Handbook of Chemistry, 11th ed. McGraw-Hill, New York, NY.

Hammerschmidt, E.G., 1934. Formation of gas hydrates in natural gas transmission lines. Ind. Eng. Chem. 26, 851–855.

Hammerschmidt, E.G., 2006. IUPAC Compendium of Chemical Terminology. Electronic version. http://goldbook.iupac.org/S05910.html.

BIBLIOGRAPHY

Byk, S.S., Fomina, V.I., 1968. Gas hydrates. Russ. Chem. Rev. 37, 469–491.

Cox, J.L. (Ed.), 1983. Natural Gas Hydrates: Properties, Occurrence and Recovery. Butterworth Publishers, Boston, MA.

Deaton, W.M., Frost, E.M., July, 1946. Gas Hydrates and Their Relation to the Operation of Natural-Gas Pipelines. Monograph 8, U.S. Bureau of Mines.

Englezos, P., 1993. Clathrate hydrates. Ind. Eng. Chem. Res. 32, 1251–1274.

Holder, G.D., Zetts, S.P., Pradhan, N., 1988. Phase behavior in systems containing clathrate hydrates. Rev. Chem. Eng. 5, 1–70.

Koh, C.A., 2002. Towards a fundamental understanding of natural gas hydrates. Chem. Soc. Rev. 31, 157–167.

Makogon, Y.F., 1997. Hydrates of Hydrocarbons. PennWell Publishing Co., Tulsa, OK.

Pedersen, K.S., Fredenslund, Aa, Thomassen, P., 1989. Properties of Oils and Natural Gases. Gulf Publishing, Houston, TX (Chapter 15).

Sloan, E.D., 1989. Natural gas hydrate phase equilibria and kinetics: understanding the state-of-the-art. Rev. Inst. Fr. Pét. 45, 245–266.

Sloan, E.D., 1998. Clathrate Hydrates of Natural Gases, second ed. Marcel Dekker, New York, NY.

Sloan, E.D., Koh, C.A., 2008. Clathrate Hydrates of Natural Gases, third ed. CRC Press, Boca Raton, FL.

Sloan, E.D., 2003. Fundamental principles and applications of natural gas hydrates. Nature 426, 353–359.

CHAPTER 2

Hydrate Types and Formers

Hydrates of natural gas components and other similar compounds are classified by the arrangement of the water molecules in the crystal, and hence the crystal structure. The water molecules align, because of the hydrogen bonding, into three-dimensional sphere-like structures often referred to as a cage. A second molecule resides inside the cage and stabilizes the entire structure.

There are two types of hydrates commonly encountered in the petroleum business. These are called type I and type II, sometimes referred to as structures I and II. A third type of hydrate that also may be encountered is type H (also known as structure H), but it is less commonly encountered.

Table 2.1 provides a quick comparison among type I, type II, and type H hydrates. These hydrates will be reviewed in more detail throughout this chapter.

Figure 2.1 shows the types of polyhedral cages involved in type I and II hydrates. In these diagrams, the water molecule is on the corner of the polyhedral, and the edge of the polyhedral represents the hydrogen bond. The structures for the type H hydrate, being significantly more complex, are not described in such detail, except to note that the small cages are the regular dodecahedron. The information in Table 2.1 and Fig. 2.1 will become clearer as the reader covers the entire chapter.

2.1 TYPE I HYDRATES

The simplest of the hydrate structures is the type I. It is made from two types of cages: (1) dodecahedron, a 12-sided polyhedron where each face is a regular pentagon and (2) tetrakaidecahedron, a 14-sided polyhedron with 12 pentagonal faces and 2 hexagonal faces. The dodecahedral cages are smaller than the tetrakaidecahedral cages; thus, the dodecahedra are often referred to as small cages whereas the tetrakaidecahedral cages are referred to as large cages.

Type I hydrates consist of 46 water molecules. If a guest molecule occupies each of the cages then the theoretical formula for the hydrate is $X \cdot 5\,^3\!/_4\,H_2O$, where X is the hydrate former.

Natural Gas Hydrates
ISBN 978-0-12-800074-8
http://dx.doi.org/10.1016/B978-0-12-800074-8.00002-8

Table 2.1 Comparisons among Type I, Type II, and Type H Hydrates

	Type I	Type II	Type H
Water molecules per unit cell	46	136	34
Cages per unit cell			
Small	6	16	3
Medium	—	—	2
Large	2	8	1
Theoretical formula[1]			
All cages filled	$X \cdot 5\frac{3}{4} H_2O$	$X \cdot 5\frac{2}{3} H_2O$	$5X \cdot Y \cdot 34H_2O$
Mole fraction hydrate former	0.1481	0.1500	0.1500
Only large cages filled	$X \cdot 7\frac{2}{3} H_2O$	$X \cdot 17H_2O$	—
Mole fraction hydrate former	0.1154	0.0556	—
Cavity diameter (Å)			
Small	7.9	7.8	7.8
Medium	—	—	8.1
Large	8.6	9.5	11.2
Volume of unit cell (m^3)	1.728×10^{-27}	5.178×10^{-27}	
Typical formers	CH_4, C_2H_6, H_2S, CO_2	N_2, C_3H_8, i-C_4H_{10}	See text

[1] X is the hydrate former and Y is a type H former.

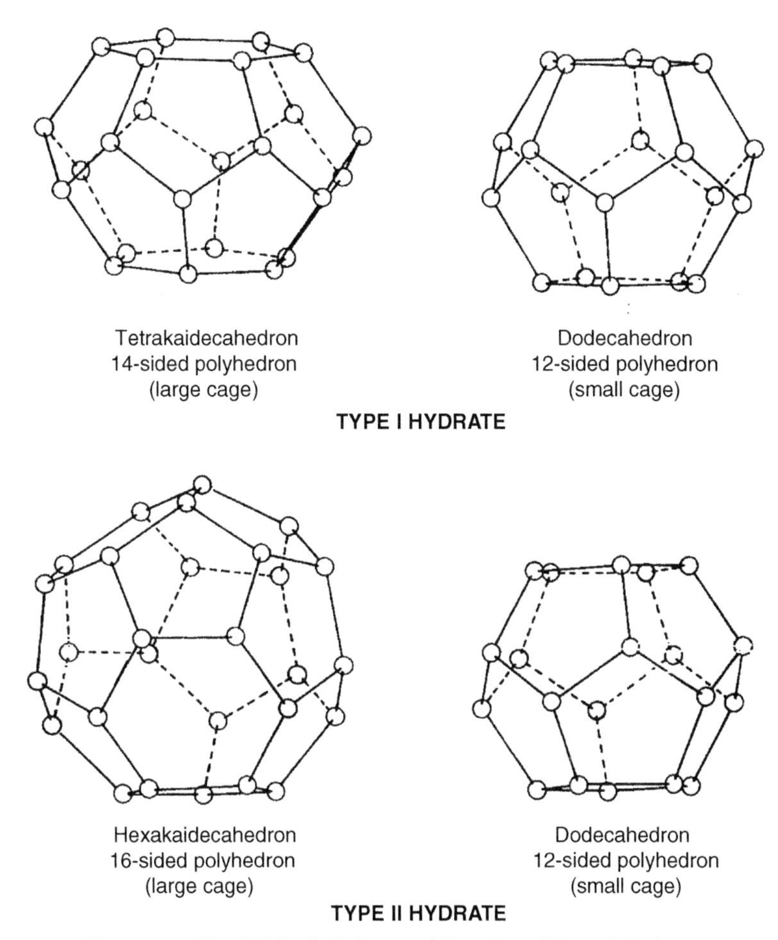

Figure 2.1 ***The Polyhedral Cages of Type I and Type II Hydrates.***

Often, in the literature, you will find oversimplifications for the hydrate crystal structure. For example, it is common that only the dodecahedron is given as the unit crystal structure. This is incorrect. The correct structures are given here.

One of the reasons why it took a long time to establish the crystal structure of hydrates is because hydrates are nonstoichiometric. That is, a stable hydrate can form without a guest molecule occupying all of the cages. The degree of saturation is a function of the temperature and the pressure. Therefore, the actual composition of the hydrate is not the theoretical composition given in the previous paragraph.

2.1.1 Type I Formers

Some of the common type I hydrate formers include methane, ethane, carbon dioxide, and hydrogen sulfide. In the hydrates of CH_4, CO_2, and H_2S, the guest molecules can occupy both the small and the large cages. On the other hand, the ethane molecule occupies only the large cages.

2.2 TYPE II HYDRATES

The structure of the type II hydrates is significantly more complicated than that of type I. The type II hydrates are also constructed from two types of cages. The unit structures of a type II hydrate are: (1) dodecahedron, a 12-sided polyhedron where each face is a regular pentagon, (2) hexakaidecahedron, a 16-sided polyhedron with 12 pentagonal faces and 4 hexagonal faces. The dodecahedral cages are smaller than the hexakaidecahedron cages.

The type II hydrate consists of 136 molecules of water. If a guest molecule occupies all of the cages, then the theoretical composition is $X \cdot 5\frac{2}{3}H_2O$, where X is the hydrate former. Alternatively, as is more commonly the case, if the guest occupies only the large cages, then the theoretical composition is $X \cdot 17H_2O$.

As with type I hydrates, the type II hydrates are not stoichiometric, so the compositions of the actual hydrates differ from the theoretical values.

2.2.1 Type II Formers

Among the common type II formers in natural gas are nitrogen, propane, and isobutane. It is interesting that nitrogen occupies both the large and small cages of the type II hydrate. On the other hand, propane and isobutane only occupy the large cages.

2.3 TYPE H HYDRATES

Type H hydrates are much less common than either type I or II. In order to form this type of hydrate, it requires a small molecule, such as methane, and a type H former. As such, type H hydrates are always double hydrates.

The type H hydrates are constructed from three types of cages: (1) dodecahedron, a 12-sided polyhedron where each face is a regular pentagon, (2) an irregular dodecahedron with 3 square faces, 6 pentagonal faces, and 3 hexagonal faces, and (3) an irregular icosahedron, a 20-sided polyhedron, with 12 pentagonal faces and 8 hexagonal faces.

The unit crystal is made up of three dodecahedral cages (small), two irregular dodecahedral cages (medium), and one icosahedral cage (large). It is composed of 34 water molecules.

Type H hydrates are always double hydrates. Small guest molecules, such as methane, occupy the small, medium, and some of the large cages of the structure whereas a larger molecule, such as those given below, occupies the large cage. The larger molecule, the so-called type H former, is of such a size that it only fits in the large cage of this structure.

Because two formers are required to form a type H hydrate, it is a little difficult to give the theoretical formula. However, if we assume that the small molecule, X, only enters the two smaller cages and we know that the large molecule, Y, only enters the large cages, then the theoretical formula is $Y \cdot 5X \cdot 34H_2O$.

2.3.1 Type H Formers

Type I and II hydrates can form in the presence of a single hydrate former, but type H requires two formers to be present, a small molecule, such as methane, and a larger type H forming molecule.

Type H formers include the following hydrocarbon species: 2-methylbutane, 2,2-dimethylbutane, 2,3-dimethylbutane, 2,2,3-trimethylbutane, 2,2-dimethylpentane, 3,3-dimethylpentane, methylcyclopentane, ethylcyclopentane, methylcyclohexane, cycloheptane, and cyclooctane. Most of these components are not commonly found in natural gas; or perhaps it is better to state that most analyses do not test for these components.

2.4 THE SIZE OF THE GUEST MOLECULE

Von Stackelberg (1949) discovered the relationship between the size of a molecule and the type of hydrate formed. He plotted a chart, reproduced here, after some modifications, as Fig. 2.2, showing the nature of the hydrate based on the size of the guest molecule.

At the top of the diagram are the small molecules and the size increases as one goes downward on the chart. Hydrogen and helium are the smallest molecules with diameters of only 2.7 and 2.3 Å, respectively (note $1 \text{ Å} = 1 \times 10^{-10}$ m). From the chart it can be seen that molecules with diameters less than about 3.8 Å do not form hydrates.

As molecules increase in size, that is as we move down the chart, we come upon the first hydrate formers, which include krypton and nitrogen. There is a region bounded by two rather broad hash marks and molecules

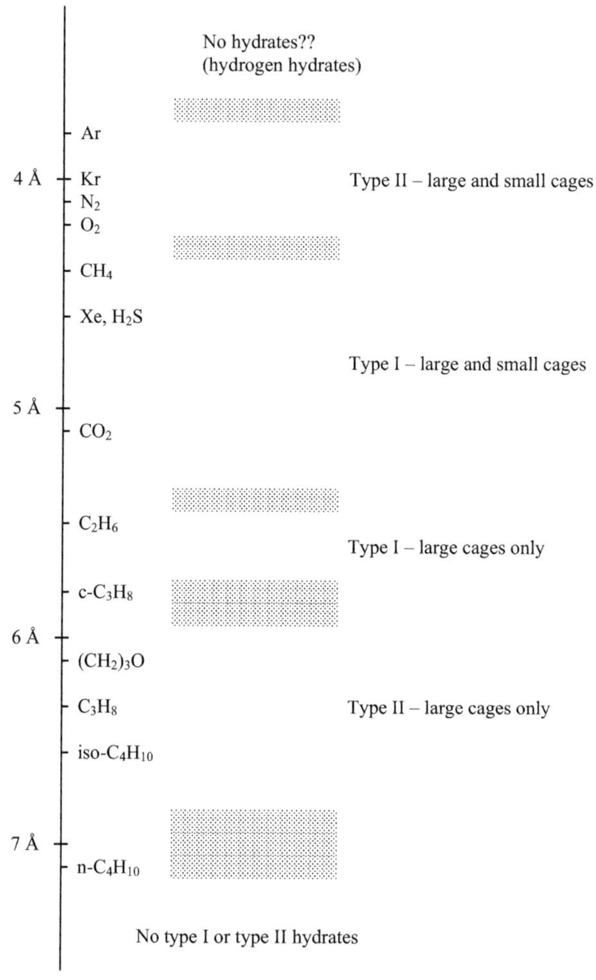

Figure 2.2 ***Comparison of Guest Size, Hydrate Type, and Cavities Occupied for Various Hydrate Formers.*** *Modified from original by von Stackelberg (1949).*

with sizes in this range (from about 3.8 to 4.2 Å) that form type II hydrates. These substances are sufficiently small such that they can occupy both the small and the large cages of this hydrate structure.

Even further down the chart is the next region (ranging from approximately 4.4–5.4 Å), which includes methane, hydrogen sulfide, and carbon dioxide. Molecules with sizes in this range are type I formers and the molecules are small enough to occupy both the small and the large cages.

Molecules that are larger still occupy the next region on the chart (from about 5.6 to 2.8 Å). The region is quite narrow and the only important substance in it is ethane. Compounds with sizes in this range form type I hydrates, but they only occupy the large cages. These molecules are too large to enter the small cages of a type I hydrate.

The next region, which represents even larger molecules (ranging from about 6.0 to 6.9 Å), contains propane and isobutane. These molecules are type II formers but only occupy the large cages of the type II structure. Molecules with sizes in this range are too large to enter the small cages of a type II hydrate.

Eventually, a limited is reached. Molecules larger than about 7 Å do not form either a type I or type II hydrate. Therefore, molecules such as pentane, hexane, and larger paraffin hydrocarbons are non-formers.

From the chart, we can see that cyclopropane (c-C_3H_8) and *n*-butane are in the hatched regions. These special components will be discussed in more detail later in this chapter.

Slightly larger molecules can form type H hydrates, but the maximum size for these compounds to form a hydrate is about 9 Å.

2.5 *N*-BUTANE

In the study of hydrates, *n*-butane is an interesting anomaly (Ng and Robinson, 1976). By itself, *n*-butane does not form a hydrate. However, its size is such that it can fit into the large cages of the type II hydrate lattice. Thus, in the presence of another hydrate former, *n*-butane can enter the cages.

It is based on the observed behavior of *n*-butane that the location of the last hatched section is established. *n*-Butane is a transition component. Molecules smaller than *n*-butane are hydrate formers, whereas molecules larger than *n*-butane do not form type I or type II hydrates.

2.6 OTHER HYDROCARBONS

Other types of hydrocarbons (alkenes and alkynes, for example) also form hydrates, provided they are not too large. Compounds such as ethylene, acetylene, propylene, and propyne are hydrate formers.

Other hydrocarbons are special cases and they will be discussed in subsequent sections of this chapter.

2.7 CYCLOPROPANE

Cyclopropane is another interesting case. It can form either a type I or II hydrate. The type of hydrate formed depends on the temperature and the pressure (Hafemann and Miller, 1969; Majid et al., 1969).

Regardless of the type of hydrate formed, cyclopropane is of such a size that it only enters the large cages.

This unusual behavior makes cyclopropane a transition component. Cyclopropane is the boundary between type I (large cage only) and type II (large cages only) formers.

2.8 2-BUTENE

The geometric isomers of 2-butene are also an interesting case. The two isomers of 2-butene are:

```
   H       H          H3C      H
    \     /             \     /
     C = C               C = C
    /     \             /     \
 H3C       C3H         H       C3H

  cis-2-Butene        trans-2-Butene
```

Like *n*-butane, they are too large to form hydrates on their own; however, in a mixture with methane, hydrates do form. However, the *trans* isomer behaves like *n*-butane, inasmuch as it will enter the hydrate lattice in the presence of hydrate formers. On the other hand, the cis isomer is too large to enter the hydrate lattice. Mixtures of methane + *trans*-2-butene form a type II hydrate with the *trans*-2-butene entering the large cages of the type II lattice. Mixtures of methane + *cis*-2-butene form a type I hydrate because this isomer is not capable of entering the lattice. For evidence of this behavior, see Holder and Kamath (1984).

2.9 HYDROGEN AND HELIUM

As was mentioned earlier, hydrogen and helium are small gases that are known to be non-formers of hydrates. Even when they are in a mixture with hydrate formers present, they do not enter the hydrate lattice in spite of their small size.

It is believed that van der Waals forces between the guest and the host molecules are the reason for the stabilization of the hydrate crystal. Furthermore, the van der Waals forces are believed to be a result of the electrons in a compound. Hydrogen (H_2) and helium have only two electrons per molecule and, thus, the van der Waals forces are weak. This is one explanation why these small molecules do not form hydrates.

2.9.1 Update

New evidence shows that under extreme pressure hydrogen does indeed form a hydrate (Mao and Mao, 2004). The pressures required are 200–300 MPa (29,000–43,500 psia) for temperatures in the range from 240 to 249 K (−33 to −24 °C, −28° to −11 °F). Such a hydrate was found to be type II, which is as expected based on the chart of von Stackelberg discussed earlier. However, an unusual feature is that two hydrogen atoms enter the individual hydrate cages.

Notwithstanding those observations, for almost all engineering applications hydrogen can be considered a non-former, which does not enter the hydrate cages.

2.10 CHEMICAL PROPERTIES OF POTENTIAL GUESTS

Another limiting factor is the chemical properties of the potential hydrate former. A molecule may be sufficiently small but it may not form a hydrate. Typically, if a gas is highly soluble in water, it will not form a hydrate.

It has been shown that CO_2 and H_2S, which have significant solubility in water, form hydrates. In addition, SO_2, which is quite soluble in water, is also a hydrate former. As a rule of thumb, these represent the upper limit in terms of solubility. Gases that are highly soluble in water, such as ammonia and hydrogen chloride, do not form hydrates even though their size might lead one to believe that they should.

Alternatively, if the molecule interferes with the hydrogen bonding, a hydrate will not form. Methanol, which is a small molecule, does not form a hydrate because it is hydrogen bonded and, hence, interferes with the hydrogen bonding among the water molecules. In addition, methanol is highly soluble in water. We will return to the subject of methanol and its importance in the formation of hydrates later on in this chapter.

2.11 LIQUID HYDRATE FORMERS

It is a commonly held belief that hydrates do not form in the presence of a nonaqueous liquid phase. That is, that hydrates will not form in the presence of condensate or oil. This is not true. Many experimental investigations demonstrate that liquid hydrocarbons can indeed form hydrates. As discussed briefly in Chapter 1, all that is required is the presence of a hydrate former, enough water present to form a hydrate, and the right combination of pressure and temperature. No mention was made of the phase of the former.

There is an unfortunate habit of referring to these compounds as "gas hydrates." This leaves the impression that they form only with gases.

2.12 HYDRATE FORMING CONDITIONS

Experimental investigations have been done for almost all of the common components in natural gas. These results are summarized in the book by Sloan (1998). It would be wise for the reader to search the literature for experimental data when there is any doubt about the hydrate forming conditions for a specific component or mixture. The book by Sloan (1998) is a good place to start.

2.12.1 Pressure–Temperature

The following tables show the pressure and temperature at which a hydrate will form for the common hydrate formers in natural gas. Table 2.2 gives the data for methane, Table 2.3 for ethane, Table 2.4 for propane, Table 2.5 for isobutane, Table 2.6 for hydrogen sulfide, and Table 2.7 for carbon dioxide.

Table 2.2 Hydrate Forming Conditions for Methane

Temperature (°C)	Pressure (MPa)	Phases	Composition (mol%)		
			Aqueous	Vapor	Hydrate
0.0	2.60	L_A–H–V	0.10	0.027	14.1
2.5	3.31	L_A–H–V	0.12	0.026	14.2
5.0	4.26	L_A–H–V	0.14	0.026	14.3
7.5	5.53	L_A–H–V	0.16	0.025	14.4
10.0	7.25	L_A–H–V	0.18	0.024	14.4
12.5	9.59	L_A–H–V	0.21	0.024	14.5
15.0	12.79	L_A–H–V	0.24	0.025	14.5
17.5	17.22	L_A–H–V	0.27	0.025	14.5
20.0	23.4	L_A–H–V	0.30	0.027	14.6
22.5	32.0	L_A–H–V	0.34	0.028	14.6
25.0	44.1	L_A–H–V	0.37	0.029	14.7
27.5	61.3	L_A–H–V	0.41	0.029	14.7
30.0	85.9	L_A–H–V	0.45	0.029	14.7

Composition for aqueous phase and for the hydrate is the mole percent of the hydrate former (CH_4). For the vapor, the composition is the mole percent water.

Table 2.3 Hydrate Forming Conditions for Ethane

Temperature (°C)	Pressure (MPa)	Phases	Composition (mol%) Aqueous	Nonaqueous	Hydrate
0.0	0.53	L_A–H–V	0.037	0.126	11.5
2.0	0.61	L_A–H–V	0.041	0.117	11.5
4.0	0.77	L_A–H–V	0.047	0.107	11.5
6.0	0.99	L_A–H–V	0.054	0.096	11.5
8.0	1.28	L_A–H–V	0.062	0.086	11.5
10.0	1.68	L_A–H–V	0.072	0.075	11.5
12.0	2.23	L_A–H–V	0.083	0.065	11.5
14.0	3.10	L_A–H–V	0.096	0.052	11.5
14.6	3.39	L_A–L_H–V–H	0.098	0.049-V 0.025-L_H	11.5
15.0	4.35	L_A–L_H–H	0.098	0.025	11.5
16.0	10.7	L_A–L_H–H	0.103	0.023	11.5
16.7	15.0	L_A–L_H–H	0.105	0.022	11.5
17.5	20.0	L_A–L_H–H	0.106	0.022	11.5

Composition for the aqueous phase and for the hydrate is the mole percent of the hydrate former (C_2H_6). For the nonaqueous phase (either the vapor or a second liquid), the composition is the mole percent water. The phase designated L_H is a C_2H_6-rich liquid.

Table 2.4 Hydrate Forming Conditions for Propane

Temperature (°C)	Pressure (MPa)	Phases	Composition (mol%) Aqueous	Nonaqueous	Hydrate
0.0	0.17	L_A–H–V	0.012	0.36	5.55
1.0	0.21	L_A–H–V	0.014	0.31	5.55
2.0	0.26	L_A–H–V	0.017	0.27	5.55
3.0	0.32	L_A–H–V	0.019	0.23	5.55
4.0	0.41	L_A–H–V	0.023	0.19	5.55
5.0	0.51	L_A–H–V	0.027	0.17	5.55
5.6	0.55	L_A–L_H–V–H	0.028	0.158-V 0.0094-L_H	5.55
5.6	1.0	L_A–L_H–H	0.028	0.0093	5.55
5.6	5.0	L_A–L_H–H	0.028	0.0088	5.55
5.7	10.0	L_A–L_H–H	0.028	0.0083	5.55
5.7	15.0	L_A–L_H–H	0.028	0.0079	5.55
5.7	20.0	L_A–L_H–H	0.028	0.0074	5.55

Composition for aqueous phase and for the hydrate is the mole percent of the hydrate former (C_3H_8). For the nonaqueous phase (either the vapor or a second liquid), the composition is the mole percent water. The phase designated L_H is a C_3H_8-rich liquid.

Table 2.5 Hydrate Forming Conditions for Isobutane

Temperature (°C)	Pressure (MPa)	Phases	Composition (mol%)		
			Aqueous	Nonaqueous	Hydrate
0.0	0.11	L_A–H–V	0.0058	0.55	5.55
0.5	0.12	L_A–H–V	0.0062	0.52	5.55
1.0	0.14	L_A–H–V	0.0070	0.46	5.55
1.5	0.15	L_A–H–V	0.0082	0.39	5.55
1.8	0.17	L_A–L_H–V–H	0.0081	0.40-V 0.0067-L_H	5.55
2.0	0.5	L_A–L_H–H	0.0080	0.0067	5.55
2.3	1.0	L_A–L_H–H	0.0079	0.0068	5.55
2.3	2.0	L_A–L_H–H	0.0079	0.0067	5.55
2.4	4.0	L_A–L_H–H	0.0079	0.0066	5.55
2.4	6.0	L_A–L_H–H	0.0079	0.0064	5.55
2.5	8.0	L_A–L_H–H	0.0079	0.0063	5.55
2.5	10.0	L_A–L_H–H	0.0079	0.0063	5.55
2.7	15.0	L_A–L_H–H	0.0077	0.0060	5.55
2.8	20.0	L_A–L_H–H	0.0076	0.0057	5.55

Composition for aqueous phase and for the hydrate is the mole percent of the hydrate former (i-C_4H_{10}). For the nonaqueous phase (either the vapor or a second liquid), the composition is the mole percent water. The phase designated L_H is an i-C_4H_{10}-rich liquid. Values for the compositions of the non-hydrate phases are preliminary.

Table 2.6 Hydrate Forming Conditions for Hydrogen Sulfide

Temperature (°C)	Pressure (MPa)	Phases	Composition (mol%)		
			Aqueous	Nonaqueous	Hydrate
0.0	0.10	L_A–H–V	0.37	0.62	14.2
5.0	0.17	L_A–H–V	0.52	0.54	14.3
10.0	0.28	L_A–H–V	0.74	0.46	14.4
15.0	0.47	L_A–H–V	1.08	0.38	14.5
20.0	0.80	L_A–H–V	1.58	0.32	14.6
25.0	1.33	L_A–H–V	2.28	0.28	14.6
27.5	1.79	L_A–H–V	2.84	0.25	14.7
29.4	2.24	L_A–L_S–V–H	3.35	0.24-V 1.62-L_S	14.7
30.0	8.41	L_A–L_S–H	3.48	1.69	14.7
31.0	19.49	L_A–L_S–H	3.46	1.77	14.7
32.0	30.57	L_A–L_S–H	3.41	1.83	14.7
33.0	41.65	L_A–L_S–H	3.36	1.88	14.7

Composition for aqueous phase and for the hydrate is the mole percent of the hydrate former (H_2S). For the nonaqueous phase (either the vapor or a second liquid), the composition is the mole percent water. The phase designated L_S is an H_2S-rich liquid.

Table 2.7 Hydrate Forming Conditions for Carbon Dioxide

Temperature (°C)	Pressure (MPa)	Phases	Composition (mol%)		
			Aqueous	Nonaqueous	Hydrate
0.0	1.27	L_A–H–V	1.46	0.058	13.8
2.0	1.52	L_A–H–V	1.67	0.056	13.9
4.0	1.94	L_A–H–V	1.92	0.053	13.9
6.0	2.51	L_A–H–V	2.21	0.051	14.1
8.0	3.30	L_A–H–V	2.54	0.049	14.2
9.8	4.50	L_A–L_C–V–H	2.93	0.051-V 0.21-L_C	14.2
10.0	7.5	L_A–L_C–H	2.97	0.22	14.5
10.3	10.0	L_A–L_C–H	3.0	0.24	14.7
10.8	15.0	L_A–L_C–H	3.1	0.25	14.7
11.3	20.0	L_A–L_C–H	3.1	0.27	14.7

Composition for aqueous phase and for the hydrate is the mole percent of the hydrate former (CO_2). For the nonaqueous phase (either the vapor or a second liquid), the composition is the mole percent water. The phase designated L_C is a CO_2-rich liquid.

The table for methane has been somewhat arbitrarily limited to 30 °C (86 °F). At this temperature, the hydrate formation pressure is 85.9 MPa (12,460 psia) and this pressure is probably a limit to that encountered in normal petroleum operations. It is worth noting that experimental data show that the methane hydrate does form at temperatures greater than 30 °C. In fact, the hydrate can be observed at the highest pressures measured in the laboratory.

The pressure–temperature loci for these hydrate formers are shown in Fig. 2.3 in SI units and in Fig. 2.4 in American engineering units. These are the same values as those presented in the tables mentioned earlier.

In every case, the three-phase loci involving two liquid phases are very steep. Small changes in the temperature have a dramatic effect on the pressure. This is clearly seen in Fig. 2.3. Methane does not have such a locus.

2.12.2 Composition

The compositions for the various phases: aqueous liquid, hydrate, non-aqueous liquid, and vapor, are also given in these tables.

Note that for ethane, propane, and isobutane, the compositions of the hydrate does not appear to be a function of the temperature or the pressure (i.e., they are constant). The reason for this is that these molecules only enter the large cages of their respective hydrate, and in the large cages there is a

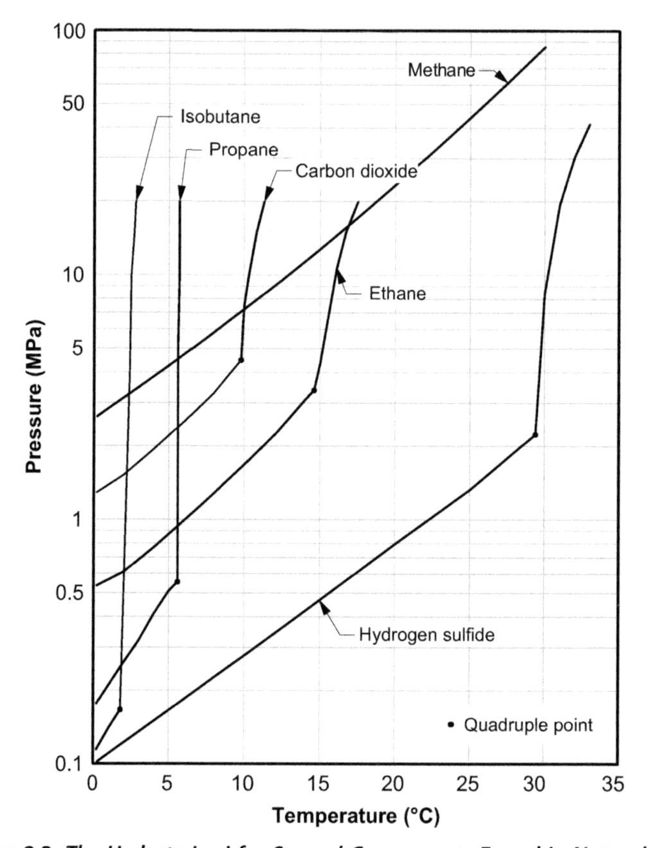

Figure 2.3 ***The Hydrate Loci for Several Components Found in Natural Gas.***

high degree of occupancy. The values given in the tables are essentially the 100% saturation values.

The compositions for the aqueous phase, the vapor, and, if it exists, the nonaqueous liquid were calculated using AQUAlibrium 2.[1] The compositions of the hydrate are estimated from a combination of experimental data, computer software, and the crystal structure of the hydrate.

The water content of the nonaqueous phases is given in different units in Appendix A.

The reader should not read too much into the magnitudes of the compositions of the nonaqueous phases. For example, it appears as though

[1] AQUAlibrium software was developed by John Carroll and is currently marketed by FlowPhase (www.flowphase.com). AQUAlibrium is discussed briefly in Chapter 10.

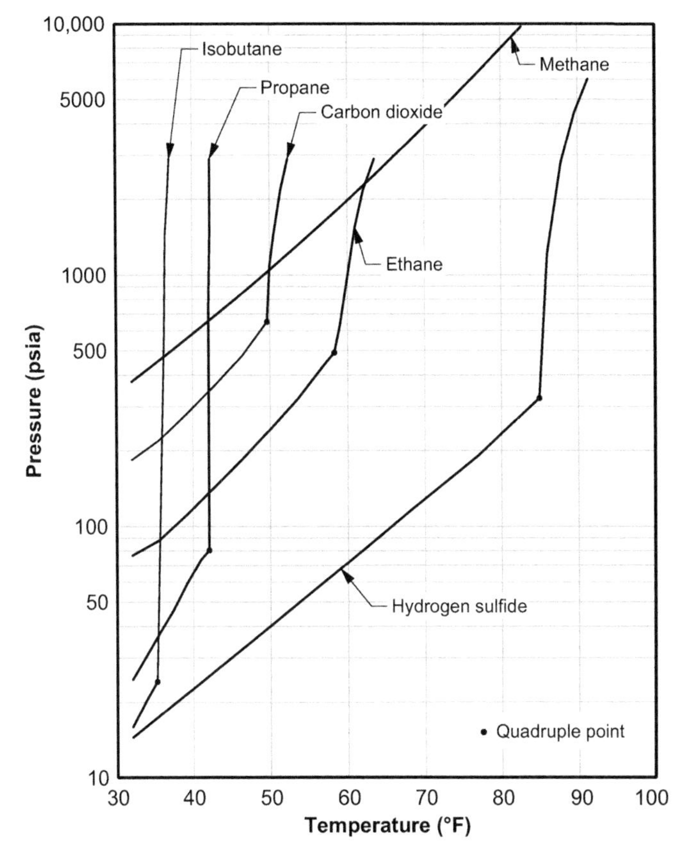

Figure 2.4 ***The Hydrate Loci for Several Components Found in Natural Gas (American Engineering Units).***

methane is the most soluble of the hydrocarbons listed. The values listed show significantly larger concentrations in the aqueous phase. However, the solubility is a function of the temperature and the pressure and the values for methane are at higher pressures than the other hydrocarbons.

Details about how the composition affects whether or not a hydrate will form are discussed in Chapter 9, which is on phase diagrams.

2.12.3 Caution

The values in these tables are not experimental data and should not be substituted for experimental data. It is useful to compare these values to experimental values, but it would be unwise to build correlations based upon these values. These tables are for use in engineering calculations.

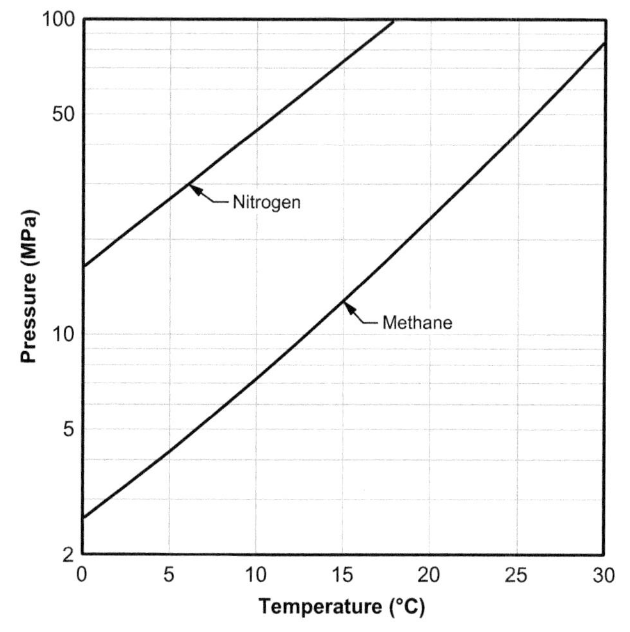

Figure 2.5 *Hydrate Loci for Methane and Nitrogen.*

2.12.4 Nitrogen

Figure 2.5 shows the hydrate locus for nitrogen. For comparison purposes, the hydrate locus of methane is also plotted. Note that significantly higher pressures are required to form a hydrate with nitrogen than with methane.

2.12.5 Ethylene

The hydrate locus for ethylene has been investigated several times. The hydrate locus for ethylene is plotted on Fig. 2.6. The shape of the hydrate locus is a little unusual, when compared with the hydrate loci of other common hydrate forming gases, such as methane, ethane, propane, carbon dioxide, and hydrogen sulfide. At first, it appears that perhaps there is a liquid-liquid-hydrate-vapor quadruple point at about 18 °C (64 °F). However, the system ethylene + water does not have such a point. The hydrate locus passes through the pressure-temperature plane beyond the critical point of ethylene (9 °C and 5060 kPa).

Another possible explanation for the shape of the curve is that possibly the hydrate changes crystal structure at about 18 °C. The somewhat unusual

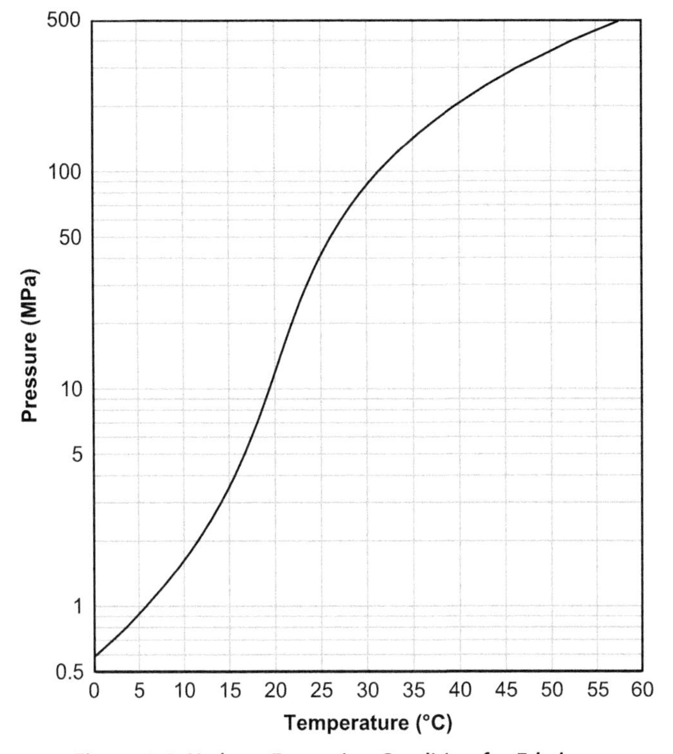

Figure 2.6 ***Hydrate Formation Condition for Ethylene.***

shape of the hydrate locus can simply be explained in terms of the fluid properties of ethylene.

2.12.6 Propylene

The hydrate locus of propylene is very short, only 1 °C (1.8 °F), in fact, many researchers had difficulty finding this locus. Accurate data were obtained by Clarke et al. (1964) and they established that the locus extended from −0.16 to +0.96 °C. The hydrate locus for propylene is shown in Fig. 2.7.

Above about 1 °C, the hydrate is formed with liquid propylene and the L_H–L_A–H locus is quite steep. There have been no studies of this locus.

2.13 V + L_A + H CORRELATIONS

The most commonly encountered hydrate forming conditions are those involving a gas, an aqueous liquid, and the hydrate.

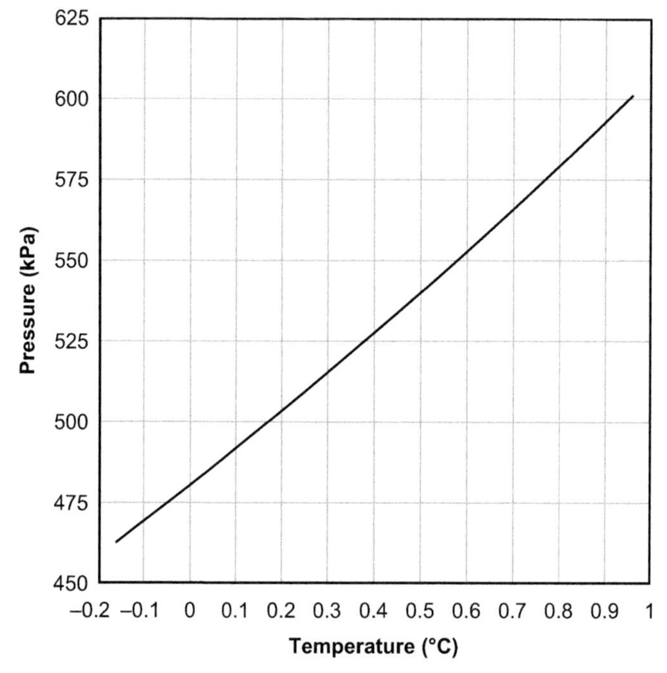

Figure 2.7 ***Hydrate Forming Conditions for Propylene.***

The experimental vapor-aqueous liquid-hydrate loci were correlated using the following equation:

$$\ln P = A + BT + C/T + D \ln T \tag{2.1}$$

where T is in Kelvin and P is in MPa. This is a semiempirical equation, which has its basis in the Clausius–Clapeyron equation. The coefficients, A, B, C, and D, for the various gases are summarized in Table 2.8. The correlation for hydrogen sulfide is taken from Carroll and Mather (1991), the correlation for propylene and acetylene are from Carroll (2006), and those for the other gases are from Carroll and Duan (2002).

2.13.1 Ethylene

An attempt was made to correlate the pressure and temperature using the usual approach of correlating the pressure (or the logarithm of the pressure) as a function of the temperature, such as those presented earlier for the components of natural gas. This approach was not very successful. It was found that a better correlation was obtained by correlating the temperature as a function of the pressure. The resulting expression is:

$$\frac{1}{T} = \sum_{i=0}^{4} a_i (\ln P)^i \tag{2.2}$$

where the temperature is in Kelvin and the pressure is in MPa and the polynomial coefficients are:

$a_0 = +3.585\ 2538 \times 10^{-3}$
$a_1 = -1.241\ 3537 \times 10^{-4}$
$a_2 = +3.090\ 7775 \times 10^{-5}$
$a_3 = -3.416\ 2547 \times 10^{-6}$
$a_4 = -1.121\ 0772 \times 10^{-7}$

This correlation is valid for the temperature range from 0 to 55 °C, or for pressures from about 0.5 to 500 MPa. This function is the curve plotted in Fig. 2.6.

As was mentioned earlier, the shape of the hydrate locus for ethylene is a little unusual. To demonstrate this further, Fig. 2.8 shows the hydrate loci for methane, ethane, and ethylene. At low temperature, the hydrate loci of ethane and ethylene are quite similar. However, at about 15 °C, the ethane locus has a quadruple point. This is approximately the point where the hydrate locus intersects the vapor pressure curve of pure ethane. At temperatures beyond this point, the hydrate locus rises very steeply. Ethylene has no such quadruple point.

At higher temperatures (greater than about 25 °C), the hydrate loci of methane and ethylene are similar (at least up to about 100 MPa).

Table 2.8 Summary of the Correlation Coefficients for Eqn 2.1

	A	B	C	D
Methane	−146.1094	+0.3165	+16,556.78	0
Ethane	−278.8474	+0.5626	+33,996.53	0
Propane	−259.5822	+0.5800	+27,150.70	0
Isobutane	+469.1248	−0.7523	−72,608.26	0
Propylene	+63.2863	0	−17,486.30	0
Acetylene	+34.0727	0	−9428.80	0
CO_2	−304.7103	+0.6138	+37,486.96	0
N_2	+26.1193	+0.0103	−7141.92	0
H_2S	−19.9874	+0.1514	+2788.88	−3.5786

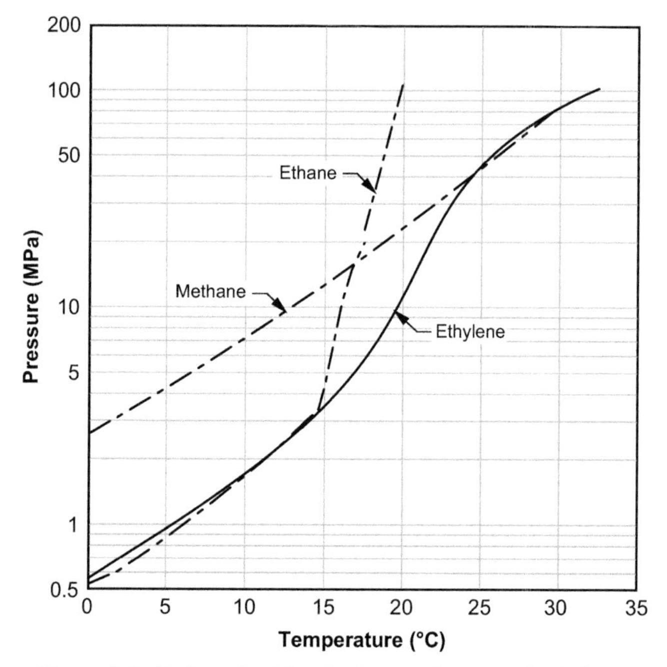

Figure 2.8 ***Hydrate Loci for Methane, Ethane, and Ethylene.***

More discussion of the somewhat unusual behavior of the ethylene hydrate locus is given in the study by Carroll (2006).

2.14 $L_A + L_N + H$ CORRELATIONS

It is somewhat unfortunate that it is common to refer to these solids as "gas hydrates" because it is possible for hydrates to form in the presence of liquids. Not only is it possible, it is common. Ethane, propane, and isobutane all form hydrates in the liquid form. Therefore, it is possible for liquefied petroleum gas (LPG) to form hydrates.

The hydrate locus for the liquid-liquid-hydrate region is much more difficult to correlate than the $V + L_A + H$ locus. The first problem is that the data are a little less reliable. Second, these loci tend to be very steep (dP/dT is very large). For these reasons, a simpler equation was used for this locus:

$$P = E + FT \tag{2.3}$$

where P is in MPa and T is in Kelvin. For the substances considered here, the coefficients are listed in Table 2.9. The values for hydrogen sulfide are

Table 2.9 Summary of the Correlation Coefficients for Eqns (2.2)–(2.3)

	E	F
Ethane	−1831.10	6.370
Isobutane	−9218.60	33.478
CO_2	−2604.71	9.226
H_2S	−3352.515	11.083

from Carroll and Mather (1991) and the others were determined as a part of this study.

Unfortunately, the data for propane are poor and a good correlation was not obtained.

2.15 QUADRUPLE POINTS

The intersection of four three-phase loci is a quadruple point. Table 2.10 summarizes the quadruple points for hydrate formers commonly encountered in natural gas. Most of these values were taken from Sloan (1998), with the exception of H_2S, which was taken from Carroll and Mather (1991), and propylene that is from Clarke et al. (1964).

The quadruple points are noted by the phases that are in equilibrium. The phase designations used are as follows: L_A = aqueous liquid, V = vapor, I = ice, H = hydrate, and L_H = nonaqueous liquid (this includes the L_S and L_C used earlier).

Table 2.10 Quadruple Points for Common Components in Natural Gas

	L_A–V–I–H		L_A–L_H–V–H	
	Temperature (°C)	**Pressure (MPa)**	**Temperature (°C)**	**Pressure (MPa)**
Methane	−0.3	2.563	No Q_2	
Ethane	−0.1	0.530	14.6	3.390
Propane	−0.1	0.172	5.6	0.556
Isobutane	−0.1	0.113	1.8	0.167
Propylene	−0.16	0.463	0.96	0.604
Carbon dioxide	−0.1	1.256	9.8	4.499
Hydrogen sulfide	−0.4	0.093	29.4	2.24
Nitrogen	−1.3	14.338	No Q_2	

Table 2.11 The Quadruple Points of Cyclopropane

Type	Temperature (°C)	Pressure (kPa)
H_I–H_{II}–I–V	−16.0	30
H_I–L_A–I–V	−0.5	63
H_I–H_{II}–L_A–V	1.5	86
H_I–L_H–L_A–V	16.2	566

Reprinted from Hafemann and Miller (1969).

Note that neither methane nor nitrogen has the second type of quadruple point (L_A–L_H–V–H). That is because neither of these gases liquefy at the conditions required to form such a quadruple point.

The compositions of the coexisting phases at the second quadruple point (L_A–L_H–V–H) are given in the tables presented earlier.

The first of these quadruple points is approximately where the hydrate locus intercepts the melting curve of pure water. Thus, they are all at approximately 0 °C. The exception to this is the nitrogen quadruple point, which is at an elevated pressure and a lower temperature.

The second quadruple point is approximately where the hydrate locus intercepts the vapor pressure curve of the pure hydrate former. As mentioned earlier, because the vapor pressure curves of nitrogen and methane do not extend to these temperatures, they do not exhibit the second quadruple point.

2.15.1 Cyclopropane

Because cyclopropane can form both type I and II hydrates, it has several quadruple points. Table 2.11 lists four quadruple points for cyclopropane.

2.16 OTHER HYDRATE FORMERS

Although the main focus of this work is the hydrates of natural gas, it is worth noting that many other compounds form hydrates.

2.16.1 Freons®

Freons, organic compounds of chlorine and fluorine, were once commonly used as refrigerants. Because of environmental concerns, their use has been curtailed. However, many of the Freons are hydrate formers, especially the smaller ones (Chinworth and Katz, 1947). It is likely that the newer, more environmentally friendly Freons are also hydrate formers. Therefore, hydrate formation may be a problem in a refrigeration loop if it is not dry.

The name Freon is a registered trademark of the E.I. du Pont de Nemours & Company (DuPont) of Wilmington, Delaware.

2.16.2 Halogens

The halogens are the elements in column 7A of the periodic table (see the appendix to Chapter 1). Of these elements, chlorine and bromine are known hydrate formers. It is likely that fluorine also forms a hydrate based on its size and chemical properties. Historically, chlorine was the first component definitely shown to form a hydrate.

Iodine, another halogen, can form a hydrate only in the presence of another hydrate former (similar to *n*-butane).

2.16.3 Noble Gases

The following noble gases (the rightmost group in the periodic table), also called inert gases, argon, krypton, xenon, and radon, all form hydrates. As was mentioned earlier, helium, another of the noble gases, does not form a hydrate. It is unlikely that neon, also a small gas, forms a hydrate.

This group of gases is remarkable for their chemical stability. Only under extreme conditions can they be made to react to form compounds. The fact that they form hydrates is a good indication that there is no chemical bonding between the host and guest molecules in a hydrate.

2.16.4 Air

Among the other important compounds that form a hydrate is oxygen. Because it is known that nitrogen also forms a hydrate, then air also forms a hydrate. Both oxygen and nitrogen form hydrates at very high pressures and, for this reason, it was once thought that they did not form hydrates.

A question that is frequently asked is that if air can form a hydrate, is any of the "ice" on the surface of the Earth composed of air hydrate? The answer is no. It requires very high pressure for the components of air to form hydrates (see Fig. 2.4 for the conditions required to form a nitrogen hydrate). Such pressures do not exist on the surface of the Earth.

2.16.5 Others

Sulfur dioxide also forms a hydrate. This is somewhat surprising because SO_2 is fairly soluble in water. This is probably the most soluble component that still forms a hydrate. As a rule of thumb, gases more soluble than SO_2 do not form hydrates.

Small mercaptans, such as methanethiol, ethanethiol, and possibly propanethiol, are also hydrate formers.

Another interesting compound that forms a hydrate is ethylene oxide. Ethylene oxide is an important industrial chemical, usually as a precursor to other chemicals.

Other hydrate formers include N_2O, H_2Se, SF_6, PH_3, AsH_3, SbH_3, and ClO_3F. Obviously, this list of compounds is of little interest to the natural gas industry. It is interesting to see the wide spectrum of components that do form hydrates.

2.17 HYDRATE FORMATION AT 0 °C

In an attempt to make a simple comparison between the various hydrate formers, the components in natural gas and other hydrocarbons, in particular, it was decided to form a common database. The question arose, what would be the best reference point for such a database. The quadruple point initially seems like a good reference, but methane and nitrogen do not have a L_A–L_H–V–H quadruple point, which immediately eliminates this point. The other quadruple point might be a good reference, but there is a slight variation in the temperature. Although not important by itself, why not eliminate the temperature variation simply by selecting the hydrate pressure at 0 °C as the reference.

Thus, a database was established for the hydrate formation at 0 °C. The hydrate pressure for several components at this temperature are given in Table 2.12 along with several physical properties of the various hydrate formers. The densities given in Table 2.12 were calculated with the Peng–Robinson equation of state. In addition, carbon dioxide does not have a normal boiling point, at atmospheric pressure it sublimes. The boiling point in Table 2.12 for CO_2 was estimated by extrapolating the vapor pressure.

Also given in Table 2.12 is the solubility of the various gases at 0 °C and a gas partial pressure of 1 atm. The values were taken from the review paper of Wilhelm et al. (1977). From the values given, it can be seen that there is no correlation between the solubility and the hydrate formation temperature.

2.18 MIXTURES

Although the behavior of pure formers is interesting, usually, in industrial practice, we must deal with mixtures. What type of hydrate is formed in a mixture? What is the effect when a non-former is in the mixture?

Table 2.12 Physical Properties and Hydrate Formation of Some Common Natural Gas Components

	Hydrate Structure	Molar Mass (g/mol)	Hydrate Pressure at 0 °C (MPa)	Normal Boiling Point (K)	Density (kg/m^3)	Solubility (mol frac × 10^4)
Methane	Type I	16.043	2.603	111.6	19.62	0.46
Ethane	Type I	30.070	0.491	184.6	6.85	0.80
Propane	Type II	44.094	0.173	231.1	3.49	0.74
Isobutane	Type II	58.124	0.113	261.4	3.01	0.31
Acetylene	Type I	26.038	0.557	188.4	6.70	14.1
Ethylene	Type I	28.054	0.551	169.3	7.11	1.68
Propylene	Type II	42.081	0.480	225.5	9.86	3.52
c-Propane	Type II	42.081	0.0626	240.3	1.175	2.81
CO_2	Type I	44.010	1.208	194.7†	25.56	13.8
N_2	Type II	28.013	16.220	77.4	196.6	0.19
H_2S	Type I	34.080	0.099	213.5	1.50	38.1

†Carbon dioxide does not have a normal boiling point. This value is estimated from the vapor pressure.

We have already encountered one interesting situation that arises with a mixture and that was the behavior of *n*-butane.

2.18.1 Mixtures of the Same Type

As a rule of thumb, if the gas mixture contains hydrate formers of only one type, then the hydrate formed will be of that type. Therefore, for example, a mixture of methane, hydrogen sulfide, and carbon dioxide, all type I formers, will form a type I hydrate. However, the behavior of hydrates is both complex and surprising. One day someone will probably discover a pair of type I formers that form a type II mixture.

One of the interesting things about hydrates is their unusual behavior. The behavior of mixtures is another example. Recently, Subramanian et al. (2000) showed that mixtures of methane and ethane, each that form type I as a pure former, forms type II in the range 72–99.3% methane.

2.18.2 Type I Plus Type II

However, if the mixture contains a type I former and a type II former, what is the nature of the hydrate formed? The thermodynamically correct response to this is that the hydrate formed is the one that results in a minimization of the free energy of the system. That is, the type formed by the mixture is the one that is thermodynamically stable. This is useful from a computational point of view, but fairly useless from a practical point of view. However, there are no hard and fast rules; each case must be examined on its own.

Holder and Hand (1982) studied the hydrate formation conditions in mixtures of ethane, a type I former, and propane, a type II former. They developed a map, reproduced here as Fig. 2.9, showing which regions each type of hydrate would form. As an approximation of their results, if the mixture is greater than 80% ethane, then the hydrate is type I, otherwise it is type II. When the hydrate is type I, the propane does not enter the crystal lattice, only the ethane does.

On the other hand, mixtures of methane, a type I former, and propane, a type II former, almost always form a type II. Only mixtures very rich in methane (99+%) will form a type I hydrate. This conclusion is based largely on the advanced hydrate models that will be discussed in Chapter 4.

2.18.3 Azeotropy

Another interesting phenomenon is the possibility of an "azeotropic" hydrate, which is a hydrate that forms at either a pressure greater than the

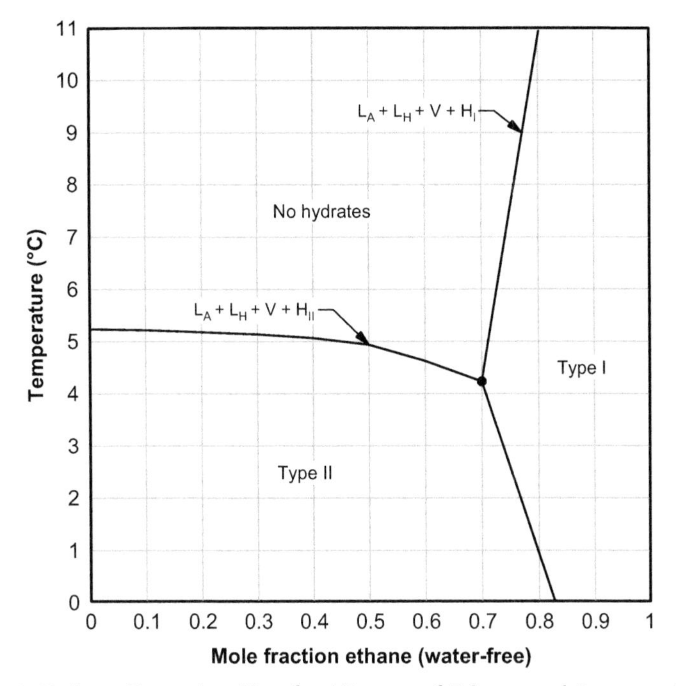

Figure 2.9 ***Hydrate Formation Map for Mixtures of Ethane and Propane.*** *Based on Holder and Hands (1982).*

pressures at which the pure components form hydrates or at a lower pressure than for the pure components.

For example, the hydrate in a mixture of hydrogen sulfide + propane forms at lower pressure than in either pure hydrogen sulfide or pure propane (Platteeuw and van der Waals, 1959). Using the computer software CSMHYD,[2] the hydrate pressure for pure H_2S at 3 °C is estimated to be 145 kPa and for pure propane it is 318 kPa. However, an equimolar mixture of the two is estimated to form a hydrate at only 64 kPa, which is significantly lower than those for the pure components.

It is worth noting that the system propane–hydrogen sulfide exhibits azeotropy in the traditional vapor–liquid sense as well (Carroll and Mather, 1992). It is probable that this is the reason that the hydrate also exhibits azeotropy.

[2] CSMHYD is copyright E. Dendy Sloan, Colorado School of Mines, Golden, CO.

2.18.4 Mixtures with Non-formers

Another important type of mixture is created when one of the components is a non-former. This is the more commonly encountered situation in industrial practice. We usually are not considering only mixtures containing hydrate formers.

It is difficult to generalize about the effect of non-formers. One problem with non-formers is that they tend to be heavy hydrocarbons, and thus will tend to liquefy at the conditions where a hydrate might be encountered. The hydrate formation conditions in a mixture of methane and pentane, a non-former, are governed by the potential of the mixture to liquefy as they are by the conditions at which pure methane forms a hydrate. Not that a liquid does not form a hydrate, but it is formed under different conditions than a gas.

Furthermore, there may be an azeotropic effect between the former and the non-former. Therefore, the hydrate formation pressure may not follow any simple rules of thumb.

Examples

Example 2.1

Will a hydrate form for methane at a temperature of 15 °C and 30 MPa?

Answer: From Table 2.2, we see that a hydrate of methane forms at a pressure of 12.79 MPa at 15 °C. Because 30 MPa is greater than 12.79 MPa, then we conclude that a hydrate will form.

Example 2.2

Will a hydrate form for ethane at a temperature of 10 °C and 0.5 MPa?

Answer: From Table 2.3, we see that a hydrate of ethane forms at a pressure of 1.68 MPa at 10 °C. Because 0.5 MPa is less than 1.68 MPa, then we conclude that a hydrate will not form.

Example 2.3

Will a hydrate form for propane at a temperature of 20 °C and 10 MPa?

Answer: From Table 2.4, we can safely assume that a propane hydrate will not form at 20 °C regardless of the pressure (within reason). The temperature is too high for a hydrate to form (also see Fig. 2.3.)

Example 2.4

At what temperature will a hydrate form for pure methane at a pressure of 1000 psia? Also, what are the compositions of the coexisting phases at this pressure? Express the water content of the gas in pounds per million standard cubic feet (lb/MMCF).

Answer: Convert 1000 psia to MPa:

$$1000 \times 0.101325/14.696 = 6.895 \text{ MPa}$$

From Table 2.2, the hydrate temperature lies between 7.5 and 10.0 °C. Linearly interpolating yields a temperature of 9.5 °C, which converts to 49 °F. Interpolating the compositions gives:

Aqueous	0.18 mol% CH_4
Vapor	0.025 mol% H_2O
Hydrate	14.4 mol% CH_4

Converting the water content of the gas to lb/MMCF, assuming standard conditions are 60 °F and 14.696 psi, then the volume of 1 lb mol of gas is:

$$V = nRT/P = (1)(10.73)(460 + 60)/14.696 = 379.7 \text{ ft}^3$$

Then converting from mole fraction to lb/MMCF gives:

$$\begin{aligned}&(0.025/100)\ \text{lb mol water}/379.7\ \text{ft}^3 \times 18.015\ \text{lb}/\text{lb mol}\\&= 0.0000119\ \text{lb}/\text{ft}^3\\&= 12\ \text{lb}/\text{MMCF}\end{aligned}$$

Therefore, the gas in equilibrium with the hydrate contains about 12 lb of water per million standard cubic feet of gas.

Example 2.5

Condensate and produced water are to be transported from a well site to a battery in a single pipeline. The design engineer postulates that as long as the condensate remains in the liquid phase, a hydrate will not form. So when they design the pipeline, the pressure is specified to be greater than the bubble point of the condensate (therefore, a gas will not form). Are they correct in believing that a hydrate will not form?

Answer: No! The three criteria for hydrate formation were given in Chapter 1: (1) a hydrate former must be present, (2) the correct combination of temperature and pressure, and (3) a sufficient amount of water.

Because this is a condensate, we can assume that methane, ethane, propane, and isobutane (and *n*-butane) are in the mixture—all hydrate formers. Because the condensate is being transported with water, it is a safe assumption that there will be sufficient water. Now the design engineer must be concerned with the temperature and pressure in the line. The right combination of temperature and pressure will result in hydrate formation.

Example 2.6

Return to Example 1.1, does this mixture contain any hydrate formers?

Answer: If you check the lists of hydrate formers given in this chapter, you will see that there are several hydrate formers in the mixture:

2. *Hydrate formers*: as was noted earlier, the gas mixture contains a number of hydrate formers: nitrogen, CO_2, H_2S, methane, ethane, propane, isobutane, and even *n*-butane, which is a special case.

Example 2.7

Hydrates of methane have been found on the ocean floor in various locations throughout the world. In this case, the pressure for hydrate formation is due to the hydrostatic head of the water.

Assuming that hydrates form at the same pressure and temperature in seawater as they do in pure water, estimate the depth at which you would have to go before encountering methane hydrates. Assume the seawater is at 2 °C (35.6 °F) and has a density of 1040 kg/m^3 (64.9 lb/ft^3). Also, assume that there is sufficient methane present to form a hydrate.

Answer: Linearly interpolating Table 2.2, we can obtain the hydrate forming pressure for methane at 2 °C. This yields 3.17 MPa, which the readers should verify for themselves. Atmospheric pressure at sea level is 101.325 kPa.

Now we use the hydrostatic equation to estimate the pressure as a function of the depth, as follows:

$$P - P_{atm} = \rho g h$$

where P is the pressure, P_{atm} is the atmospheric pressure, ρ is the density of the fluid, g is the acceleration because of gravity, and h is the depth of the fluid. Rearranging the above equation yields:

$$\begin{aligned} h &= (P - P_{atm})/\rho g \\ &= (3.17 - 0.101)/(1040 \times 9.81)\ (\mathrm{MPa})/\left(\mathrm{kg/m^3} \times \mathrm{m/s^2}\right) \\ &= 3.01 \times 10^{-4}\ \mathrm{Mm} \\ &= 301\ \mathrm{m} \end{aligned}$$

Therefore, at depths below 300 m (or about 1000 ft), we can expect to find methane hydrates, provided there is sufficient methane present.

APPENDIX 2A WATER CONTENT OF THE FLUID IN EQUILIBRIUM WITH HYDRATE FOR PURE COMPONENTS

The tables presented in this section are similar to those presented in the main portion of the text. The numbering system for these tables corresponds to the numbers in the main text, which is why they do not begin at one.

In the main text, the water content is presented in mol%. Here, three other units are given: parts per million (ppm), mg of water per standard cubic meter (mg/Sm^3), and pounds of water per million standard cubic feet (lb/MMCF). If the hydrate former is in the liquid state, then the volume in Sm^3 or MMCF is the gas equivalent.

The following notation is used in these tables to denote the various phases: H = hydrate, V = vapor, L_H = hydrocarbon-rich liquid, L_S = hydrogen sulfide-rich liquid, L_C = carbon dioxide rich liquid, and L_A = aqueous liquid. The composition of the hydrates and the aqueous phase are given in the main text.

There is some round-off error and, thus, some of the numbers may not appear to be the proper conversion. The round-off error includes the rounding of the hydrate pressure at a given temperature.

In addition, usually only two significant figures are used, but occasionally three. This also results in the numbers appearing somewhat unusual.

The caution stated in the text of the paper is worth repeating here. The values in these tables are not experimental data and should not be substituted for experimental data. It is useful to compare these values with experimental values, but it would be unwise to build correlations based upon these values. The tables are for use in engineering calculations.

The compositions for the aqueous phase, the vapor, and, if it exists, the nonaqueous liquid were calculated using AQUAlibrium 2 software.

Table 2.13 Hydrate Forming Conditions for Methane

Temperature (°C)	Pressure (MPa)	Phases	Vapor Composition		
			ppm	mg/Sm^3	lb/MMCF
0.0	2.60	L_A–H–V	270	20.6	12.6
2.5	3.31	L_A–H–V	260	20.0	12.3
5.0	4.26	L_A–H–V	260	19.4	12.0
7.5	5.53	L_A–H–V	250	18.9	11.6
10.0	7.25	L_A–H–V	240	18.6	11.4
12.5	9.59	L_A–H–V	240	18.5	11.4
15.0	12.79	L_A–H–V	250	18.7	11.5
17.5	17.22	L_A–H–V	250	19.4	11.9
20.0	23.4	L_A–H–V	270	20.2	12.4
22.5	32.0	L_A–H–V	280	21.1	13.0
25.0	44.1	L_A–H–V	290	21.7	13.4
27.5	61.3	L_A–H–V	290	22.0	13.5
30.0	85.9	L_A–H–V	290	21.7	13.3

Table 2.14 Hydrate Forming Conditions for Ethane

Temperature (°C)	Pressure (MPa)	Phases	Nonaqueous Phase Composition		
			ppm	mg/Sm^3	lb/MMCF
0.0	0.53	L_A–H–V	1160	881	54
2.0	0.61	L_A–H–V	1170	889	55
4.0	0.77	L_A–H–V	1070	813	50
6.0	0.99	L_A–H–V	960	734	45
8.0	1.28	L_A–H–V	860	652	40
10.0	1.68	L_A–H–V	750	571	35
12.0	2.23	L_A–H–V	650	492	30
14.0	3.10	L_A–H–V	520	396	24
14.6	3.39	L_A–L_H–V–H	490-V	369	23
			250-L_H	192	12
15.0	4.35	L_A–L_H–H	250	189	12
16.0	10.7	L_A–L_H–H	230	173	11
16.7	15.0	L_A–L_H–H	220	168	10
17.5	20.0	L_A–L_H–H	220	165	10

Table 2.15 Hydrate Forming Conditions for Propane

Temperature (°C)	Pressure (MPa)	Phases	Nonaqueous Phase Composition		
			ppm	mg/Sm^3	lb/MMCF
0.0	0.17	L_A–H–V	3600	274	169
1.0	0.21	L_A–H–V	3100	236	145
2.0	0.26	L_A–H–V	2700	204	125
3.0	0.32	L_A–H–V	2300	174	107
4.0	0.41	L_A–H–V	1900	148	91
5.0	0.51	L_A–H–V	1700	126	77
5.6	0.55	L_A–H–V–L_H	1580-V	123	75
			94-L_H	7.1	4.4
5.6	1.0	L_A–L_H–H	93	7.1	4.4
5.6	5.0	L_A–L_H–H	88	6.7	4.1
5.7	10.0	L_A–L_H–H	83	6.3	3.9
5.7	15.0	L_A–L_H–H	79	6.0	3.7
5.7	20.0	L_A–L_H–H	75	5.7	3.5

It is interesting to note that with ethane, propane, and isobutane (and in particular the last two), there is a dramatic change in the water content of the fluid when the nonaqueous phase changes from a gas to a liquid. In the gaseous state, the pressure is quite low and, thus, the fluid can hold significant water. In the liquid state, the water content is very low. This is because the solubility of water in a liquid hydrocarbon is quite small.

Table 2.16 Hydrate Forming Conditions for Isobutane

Temperature (°C)	Pressure (MPa)	Phases	Nonaqueous Phase Composition		
			ppm	mg/Sm³	lb/MMCF
0.0	0.11	L_A–H–V	5500	4160	256
0.5	0.12	L_A–H–V	5200	3950	243
1.0	0.14	L_A–H–V	4600	3500	215
1.5	0.15	L_A–H–V	4400	3380	208
1.8	0.17	L_A–L_H–V–H	4000-V	3040	187
			67-L_H	5.1	3.1
2.0	0.5	L_A–L_H–H	67	5.1	3.1
2.3	1.0	L_A–L_H–H	68	5.2	3.2
2.3	2.0	L_A–L_H–H	67	5.1	3.1
2.4	4.0	L_A–L_H–H	66	5.0	3.1
2.4	6.0	L_A–L_H–H	64	4.9	3.0
2.5	8.0	L_A–L_H–H	63	4.8	3.0
2.5	10.0	L_A–L_H–H	63	4.7	2.9
2.7	15.0	L_A–L_H–H	60	4.5	2.8
2.8	20.0	L_A–L_H–H	57	4.3	2.7

Table 2.17 Hydrate Forming Conditions for Hydrogen Sulfide

Temperature (°C)	Pressure (MPa)	Phases	Nonaqueous Phase Composition		
			ppm	g/Sm³[1]	lb/MMCF
0.0	0.10	L_A–H–V	6200	4.70	289
5.0	0.17	L_A–H–V	5400	4.12	253
10.0	0.28	L_A–H–V	4600	3.50	215
15.0	0.47	L_A–H–V	3800	2.93	180
20.0	0.80	L_A–H–V	3200	2.45	151
25.0	1.33	L_A–H–V	2800	2.07	127
27.5	1.79	L_A–H–V	2500	1.92	118
29.4	2.24	L_A–L_S–V–H	2400-V	1.79	110
			16,200-L_S	12.4	761
30.0	8.41	L_A–L_S–H	16,900	12.8	789
31.0	19.49	L_A–L_S–H	17,700	13.5	828
32.0	30.57	L_A–L_S–H	18,300	13.9	857
33.0	41.65	L_A–L_S–H	18,800	14.3	880

[1]Note the change of units. These are in g/Sm³ not mg/Sm³ as in the other tables.

In practice, this means that it requires significantly less water to form a hydrate in an LPG than it does in a gaseous mixture of hydrocarbons.

On the other hand, the acid gases (hydrogen sulfide and carbon dioxide) behave in an opposite fashion to the hydrocarbons. With acid gases, the

Table 2.18 Hydrate Forming Conditions for Carbon Dioxide

Temperature (°C)	Pressure (MPa)	Phases	Nonaqueous Phase Composition		
			ppm	mg/Sm³	lb/MMCF
0.0	1.27	L_A–H–V	560	426	26
2.0	1.52	L_A–H–V	560	423	26
4.0	1.94	L_A–H–V	530	402	25
6.0	2.51	L_A–H–V	510	385	24
8.0	3.30	L_A–H–V	490	376	23
9.8	4.50	L_A–L_C–V–H	510-V	385	24
			2100-L_C	1570	97
10.0	7.5	L_A–L_C–H	2200	1710	105
10.3	10.0	L_A–L_C–H	2400	1790	110
10.8	15.0	L_A–L_C–H	2500	1900	117
11.3	20.0	L_A–L_C–H	2700	2020	124

water content is larger in the liquid phase than it is in the vapor (Tables 2.13–2.18).

REFERENCES

Carroll, J.J., April 23–27, 2006. Hydrate Formation in Ethylene, Acetylene, and Propylene. AIChE Spring National Meeting, Orlando, FL.

(in Chinese) Carroll, J.J., Duan, J., 2002. Relational expression of the conditions forming hydrates of various components in natural gas. Nat. Gas. Ind. 22 (2), 66–71.

Carroll, J.J., Mather, A.E., 1991. Phase equilibrium in the system water-hydrogen sulphide: hydrate-forming conditions. Can. J. Chem. Eng. 69, 1206–1212.

Carroll, J.J., Mather, A.E., 1992. An examination of the vapor-liquid equilibrium in the system propane-hydrogen sulfide. Fluid Phase Equilib. 81, 187–204.

Chinworth, H.E., Katz, D.L., 1947. Refrigerant hydrates. J. ASRE, 1–4.

Clarke, E.C., Ford, R.W., Glew, D.N., 1964. Propylene gas hydrate stability. Can. J. Chem. 42, 2027–2029.

Hafemann, D.R., Miller, S.L., 1969. The clathrate hydrates of cyclopropane. J. Phys. Chem. 73, 1392–1397.

Holder, G.D., Hand, J.H., 1982. Multi-phase equilibria in hydrates from methane, ethane, propane, and water mixtures. AIChE J. 28, 440–447.

Holder, G.D., Kamath, V.A., 1984. Hydrates of (methane + *cis*-2-butene) and (methane + *trans*-2-butene). J. Chem. Thermodyn. 16, 399–400.

Majid, Y.A., Garg, S.K., Davidson, D.W., 1969. Dielectric and nuclear magnetic resonance properties of a clathrate hydrate of cyclopropane. Can. J. Chem. 47, 4697–4699.

Mao, W.L., Mao, H., 2004. Hydrogen storage in molecular compounds. Proc. Nat. Acad. Sci. 101, 708–710.

Ng, H.-J., Robinson, D.B., 1976. The role of n-butane in hydrate formation. AIChE J. 22, 656–661.

Platteeuw, J.C., van der Waals, J.H., 1959. Thermodynamic properties of gas hydrates II : phase equilibria in the system H_2S-C_3H_8-H_2O at −3°C. Rec. Trav. Chim. Pays-Bas 78, 126–133.

Sloan, E.D., 1998. Clathrate Hydrates of Natural Gases, second ed. Marcel Dekker, New York, NY.
Subramanian, S., Kini, R.A., Dec, S.F., Sloan, E.D., 2000. Evidence of structure II hydrate formation from methane + ethane mixtures. Chem. Eng. Sci. 55, 1981–1999.
von Stackelberg, M., 1949. Feste gashydrate. Naturwissenschaften 36, 327–333.
Wilhelm, E., Battino, R., Wilcock, R.J., 1977. Low-pressure solubility of gases in liquid water. Chem. Rev. 77, 219–262.

CHAPTER 3

Hand Calculation Methods

The first problem when designing processes involving hydrates is to predict the conditions of pressure and temperature at which hydrates will form. To begin the discussion of this topic, there are a series of methods that can be used without a computer. These are the so-called "hand calculation methods" because they can be performed with a pencil, paper, and a calculator.

Hand calculation methods are useful for rapid estimation of the hydrate formation conditions. Unfortunately, the drawback to these methods is that they are not highly accurate and, in general, the less information required as input, the less accurate the results of the calculation. In spite of this, these methods remain quite popular.

There are two commonly used methods for rapidly estimating the conditions at which hydrates will form. Both are attributed to Katz and co-workers. This leads to some confusion because both methods are often referred to as "the Katz method" or "Katz charts," although both methods actually involve charts. Here, the two methods will be distinguished by the names "gas gravity" and "*K*-factor." Both of these methods will be presented here in some detail.

A third chart method proposed by Baillie and Wichert (1987) will also be discussed in this chapter. This is basically a gas gravity approach but it includes a correction for the presence of hydrogen sulfide and, therefore, is more useful for sour gas mixtures.

Finally, there is a discussion of local models. Simple models that are developed for specific gas mixtures over limited ranges of temperature and pressures.

3.1 THE GAS GRAVITY METHOD

The gas gravity method was developed by Professor Katz and co-workers in the 1940s. The beauty of this method is its simplicity—involving only a single chart.

Wilcox et al. (1941) measured the hydrate loci for three gas mixtures, which they designated gas B, gas C, and gas D. These mixtures were composed of hydrocarbons methane through pentane with one mixture

Natural Gas Hydrates
ISBN 978-0-12-800074-8
http://dx.doi.org/10.1016/B978-0-12-800074-8.00003-X

having 0.64% nitrogen (gas B) and another with 0.43% nitrogen and 0.51% carbon dioxide (gas D). The gravities of these mixtures were gas B: 0.6685, gas C: 0.598, and gas D: 0.6469.

Figure 3.1 shows the hydrate loci for these three mixtures and for pure methane. The pure methane curve is the same as presented earlier. The curves for the three mixtures are simply smoothing correlations of the raw data. This figure is the basis for the charts that follow and are frequently repeated, in one form or another, throughout the hydrate literature.

The gas gravity chart is simply a plot of pressure and temperature with the specific gravity of the gas as a third parameter. Two such charts, one in SI units and the other in engineering units, are given here in Figs 3.2 and 3.3.

The first curve on these plots, that is the one at the highest pressure, is for pure methane. This is the same pressure-temperature locus presented in Chapter 2.

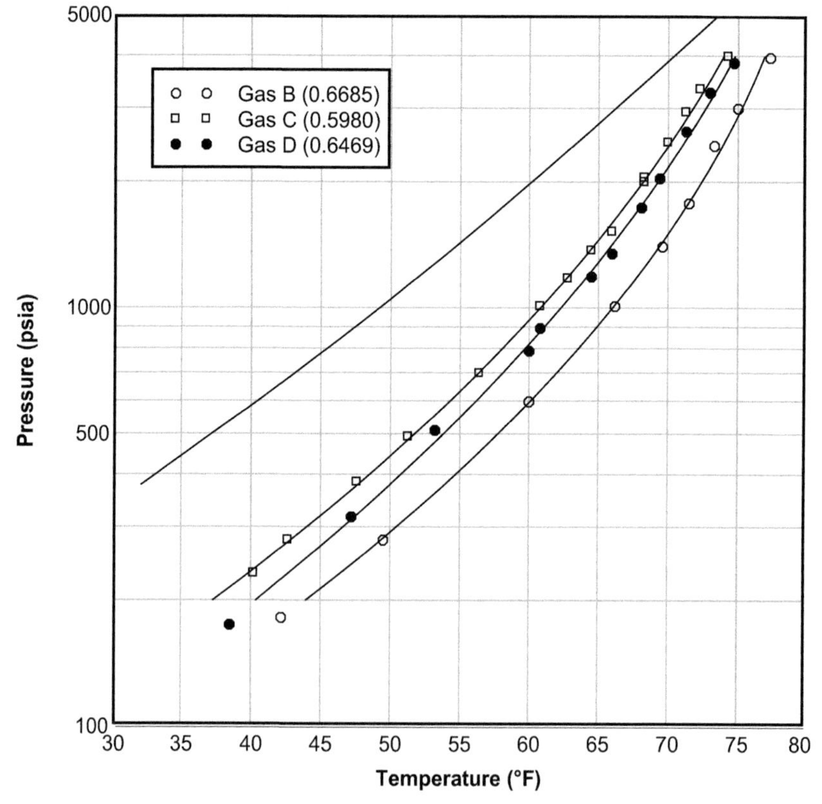

Figure 3.1 ***Hydrate Curves for Three of the Mixtures Studied by Wilcox et al. (1941).***

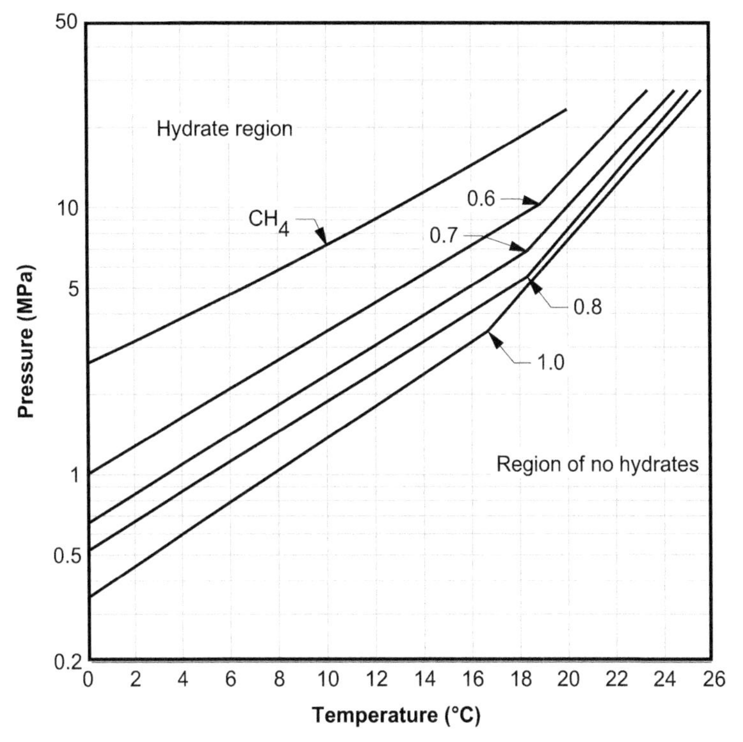

Figure 3.2 ***Hydrate Locus for Sweet Natural Gas Using the Gas Gravity Method—SI Units.***

The chart is very simple to use. First, you must know the specific gravity of the gas, which is also called the relative density. Given the molar mass (molecular weight) of the gas, M, the gas gravity, γ, can be calculated as follows:

$$\gamma = \frac{M}{28.966} \tag{3.1}$$

The 28.966 is the standard molar mass of air. For example, the gravity of pure methane is 16.043/28.966 or 0.5539.

It is a very simple procedure to use this chart. For example, if you know the pressure, temperature, and gas gravity and you want to know if you are in a region where a hydrate will form, first you locate the pressure-temperature point on the chart. If this point is to the left and above the appropriate gravity curve, then you are in the hydrate-forming region. If you are to the right and below, then you are in the region where a hydrate will not form. Remember that hydrate formation is favored by high pressure and low temperature.

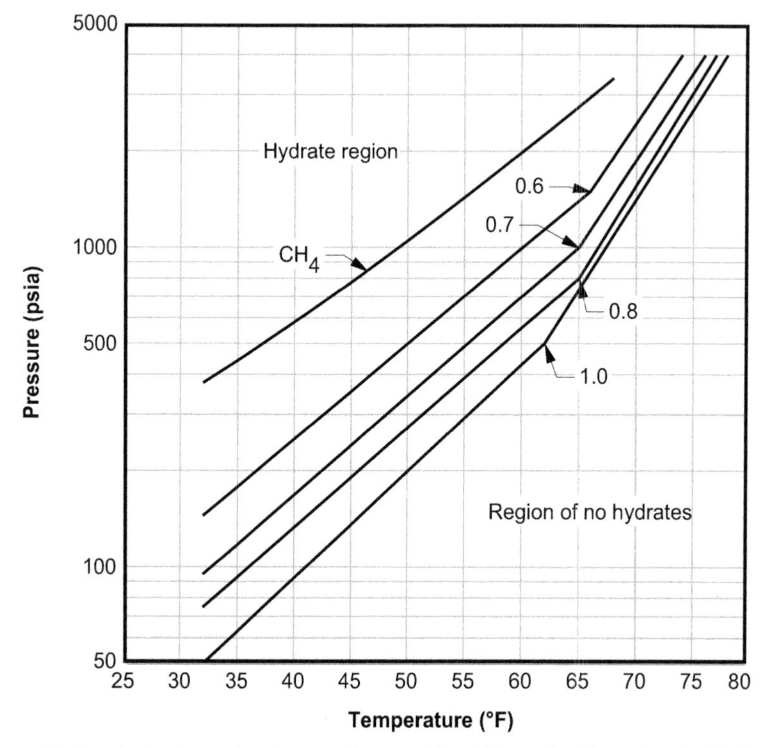

Figure 3.3 ***Hydrate Locus for Sweet Natural Gas Using the Gas Gravity Method—American Engineering Units.***

On the other hand, if you want to know what pressure a hydrate forms, you enter the chart on the abscissa (x-axis) at the specified temperature. You go up until you reach the appropriate gas gravity curve. This may require some interpolation on the part of the user. Then you go to the left and read the pressure on the ordinate (y-axis).

Finally, if you want to know the temperature at which a hydrate will form, you enter the ordinate and reverse the process. Ultimately, you read the temperature on the abscissa.

This method does not indicate the composition or type of the hydrate. However, usually all we are interested in is the condition at which a hydrate will form, and this chart rapidly provides this information.

It is surprising how many people believe that this chart is the definitive method of estimating hydrate formation conditions. They point to this chart and declare, "Hydrates will not form!" This chart is an approximation and should only be used as such.

3.1.1 Verifying the Approach

Carroll and Duan (2002) showed that for paraffin hydrocarbons, there was a strong correlation between the pressure at which a hydrate formed at 0 °C and the molar mass of the hydrate former. Furthermore, hydrogen sulfide, carbon dioxide, and nitrogen deviated significantly from this trend. Here, we will look at this in a little more detail and examine the use of other simple properties in place of the gravity.

As a measure of the tendency of a component to form a hydrate, we will use the hydrate-forming conditions at 0 °C and the data given in Table 2.12.

3.1.1.1 Molar Mass

Figure 3.4 shows the hydrate formation temperature at 0 °C as a function of the molar mass. For the hydrocarbon components, there is a good

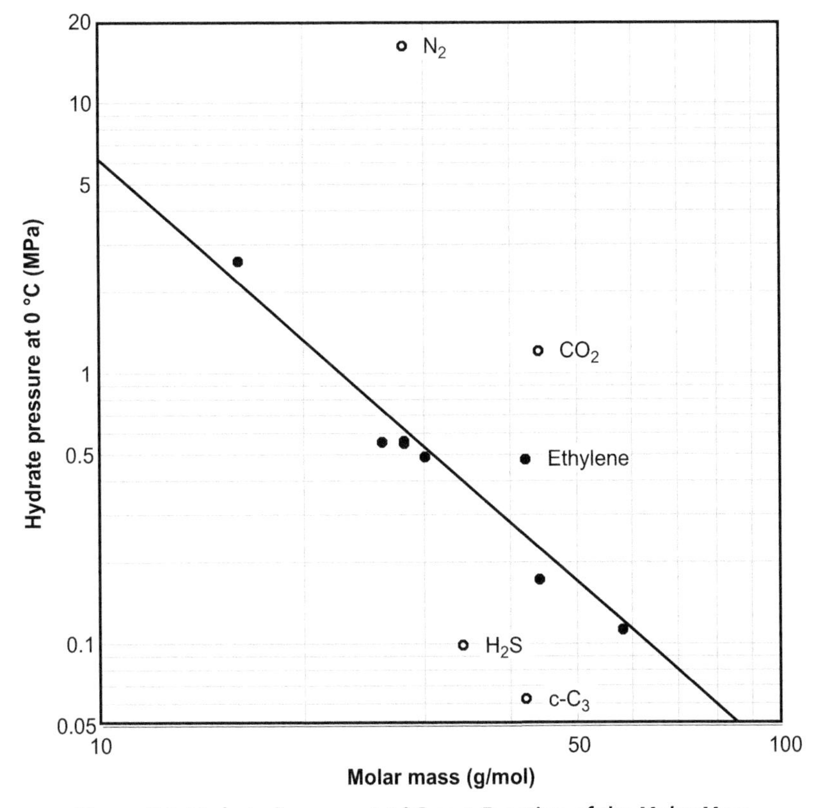

Figure 3.4 *Hydrate Pressure at 0 °C as a Function of the Molar Mass.*

correlation between these two quantities. A simple correlation between the molar mass and the hydrate pressure is:

$$\log P = 3.03470 - 2.23793 \log M \tag{3.2}$$

where P is in MPa and M is in kg/kmol (or equivalently g/mol).

The only hydrocarbon component that deviates significantly from this trend is propylene, but ethylene exhibits some deviation as well.

The three non-hydrocarbon components deviate significantly from this trend. Equation (3.2) dramatically overpredicts the hydrate pressure of hydrogen sulfide and dramatically underpredicts the hydrate pressure for both nitrogen and carbon dioxide. Therefore, it should come as no surprise that the simple gas gravity method for predicting hydrates requires that nitrogen, carbon dioxide, and hydrogen sulfide be used as special components.

3.1.1.2 Boiling Point

Next, consider the hydrate formation as a function of the boiling point. The boiling point represents a measure of the volatility of the hydrate former.

Figure 3.5 shows the hydrate formation conditions as a function of the normal boiling points.

3.1.1.3 Density

The next approach was to use the density of the hydrate former at the given conditions, rather than the gravity. Figure 3.6 shows a plot of the hydrate conditions at 0 °C as a function of the density.

All of the data are well correlated with the following equation:

$$\log P = -1.27810 + 1.09714 \log \rho \tag{3.3}$$

where P is in MPa and ρ is in kg/m^3. This includes nitrogen, hydrogen sulfide, and carbon dioxide, which did not follow previous trends.

3.1.1.4 Discussion

This section demonstrates that there are some basic relations between the molar mass and the hydrate formation for hydrocarbons only. Because the gas gravity is directly related to the molar mass, it is safe to extrapolate this conclusion to the gravity. This reinforces earlier comments that this approach is not applicable to gas mixtures that contain significant amounts of hydrogen sulfide and carbon dioxide.

The use of the boiling point did not improve the situation.

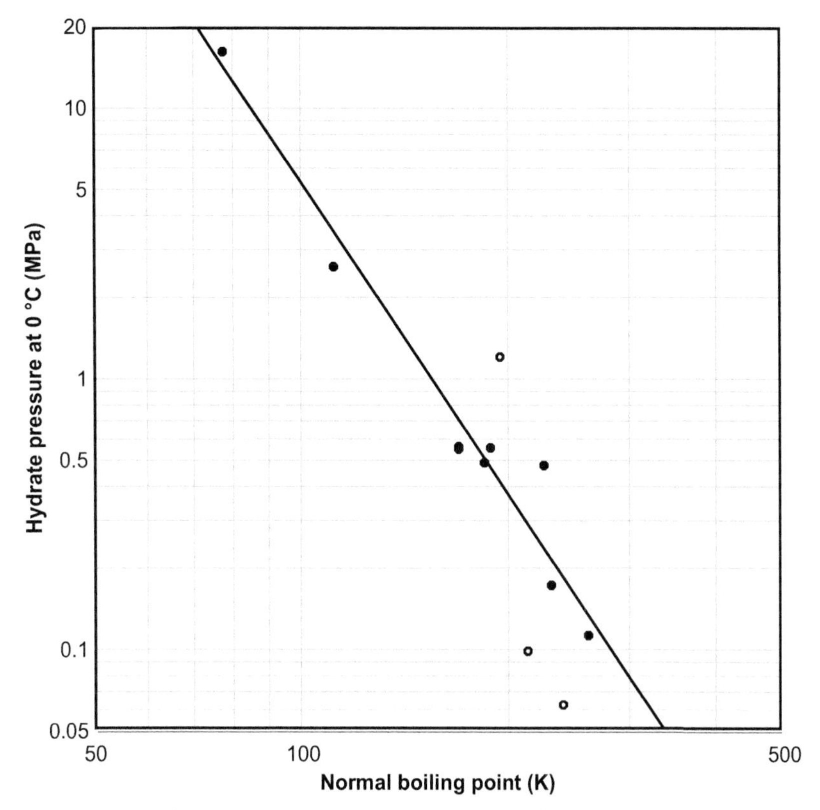

Figure 3.5 *Hydrate Pressure at 0 °C as a Function of the Normal Boiling Point.*

On the other hand, the use of density instead of gravity improves the situation. For the limited database, there is a correlation between the density of the gas and the hydrate pressure—at least at 0 °C.

If we could find a better correlation of the hydrate formation conditions as a function of some simple properties of the gas mixture and applicable to all gas mixtures (sweet, sour, paraffin, olefin, etc.) and type of hydrate, this would be very useful to process engineers and operating personnel in the natural gas business. Often, engineers and operators require a quick estimate of the hydrate formation conditions in order to deal with immediate problems, but such estimates still need to be sufficiently accurate. There is no point in making a correlation that is no more accurate than the existing ones. At this point, such a correlation has not been achieved, but we are inching closer.

It has been demonstrated that there appears to be some correlation between the density and hydrate formation that is applicable to pure

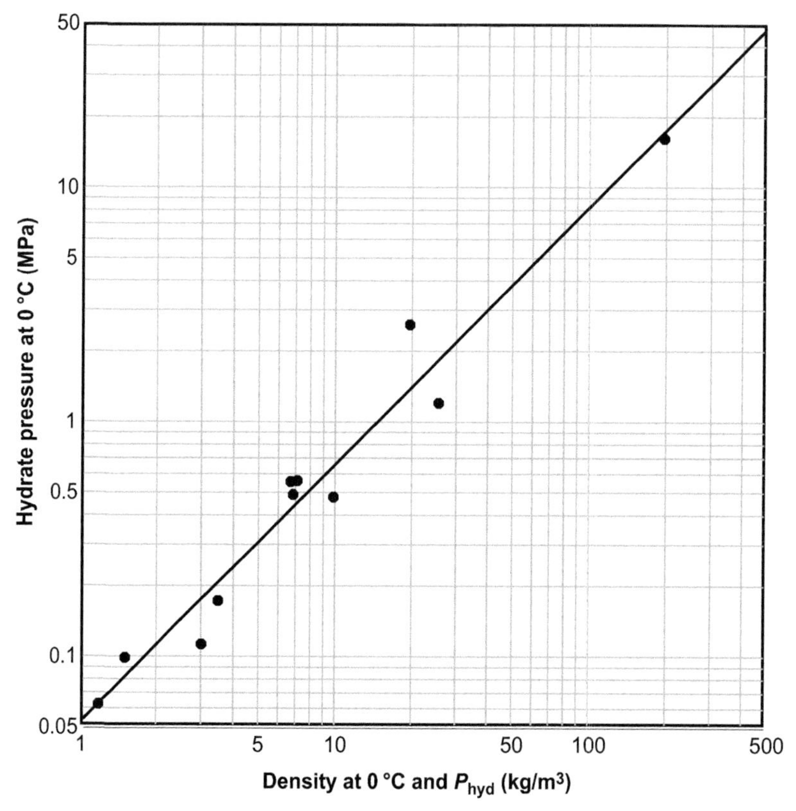

Figure 3.6 ***Hydrate Pressure at 0 °C as a Function of the Density.***

components, but which is applicable to all components. Other methods, including the gas gravity, seem to be useful only for a subset of the data. The next stage is to extend the range of temperature and pressure and, if this is successful, to test this with mixture data.

3.2 THE *K*-FACTOR METHOD

The second method that lends itself to hand calculations is the *K*-factor method. This method originated with Carson and Katz (1942) (also see Wilcox et al., 1941), although there have been additional data and charts reported since then. One of the ironies of this method is that the original charts of Carson and Katz (1942) have been reproduced over the years even though they were originally marked as "tentative" by the authors.

The *K*-factor is defined as the distribution of the component between the hydrate and the gas:

$$K_i = \frac{y_i}{s_i} \tag{3.4}$$

where y_i and s_i are the mole fractions of component i in the vapor and hydrate, respectively. These mole fractions are on a water-free basis and water is not included in the calculations. It is assumed that sufficient water is present to form a hydrate.

There is a chart available for each of the components commonly encountered in natural gas that is a hydrate former: methane, ethane, propane, isobutane, *n*-butane, hydrogen sulfide, and carbon dioxide. Versions of these charts, one set in SI units and another in American engineering units, are appended to this chapter.

All nonformers are simply assigned a value of infinity, because, by definition, $s_i = 0$ for nonformers, there is no nonformer in the hydrate phase. This is true for both light nonformers, such as hydrogen, and heavy ones, such as *n*-pentane and *n*-hexane.

At first glance, this seems a little awkward, but it comes out almost naturally when performing the calculations.

3.2.1 Calculation Algorithms

The K-charts are usually used in three methods: (1) given the temperature and pressure calculate the composition of the coexisting phases, (2) given the temperature calculate the pressure at which the hydrate forms and the composition of the hydrate, and (3) given the pressure calculate the temperature at which the hydrate forms and the composition of the hydrate.

3.2.1.1 Flash

The first type of calculation is a flash. In this type of calculation, the objective is to calculate the amount of the phases present in an equilibrium mixture and to determine the composition of the coexisting phases. The temperature, pressure, and compositions are the input parameters.

The objective function to be solved, in the Rachford–Rice form, is:

$$f(V) = \sum \frac{z_i\,(1 - K_i)}{1 + V(K_i - 1)} \tag{3.5}$$

where z_i is the composition of the feed on a water-free basis. An iterative procedure is used to solve the vapor phase fraction, V, such that the function equals zero. This equation is applicable to all components, but may cause numerical problems. The following equation helps to alleviate such problems:

$$f(V) = \sum_{\text{former}} \frac{z_i(1 - K_i)}{1 + V(K_i - 1)} + \sum_{\text{nonformer}} \frac{z_i}{V} \tag{3.6}$$

Once you have calculated the phase fraction, the vapor phase compositions can be calculated as follows:

$$y_i = \frac{z_i K_i}{1 + V(K_i - 1)} \tag{3.7}$$

for formers and

$$y_i = \frac{z_i}{V} \tag{3.8}$$

for nonformers. From the vapor phase, the composition of the solid is calculated from:

$$s_i = \frac{y_i}{K_i} \tag{3.9}$$

This may be a little difficult to understand, but the s_i is not really meant to be the composition of the hydrate phase. These are merely an intermediate value in the process of calculating the hydrate pressure as a function of the temperature. The objective is to calculate the hydrate locus and not to estimate the hydrate composition.

Only rarely would this type of calculation be used. The calculations in the next section are much more common.

3.2.1.2 Incipient Solid Formation

The other two methods are incipient solid formation points and are equivalent to a dew point. This is the standard hydrate calculation. The purpose of this calculation it to answer the question "Given the temperature and the composition of the gas, at what pressure will a hydrate form?" A similar calculation is to estimate the temperature at which a hydrate will form given the pressure and the composition. The execution of these calculations is very similar.

The objective functions to be solved are:

$$f_1(T) = 1 - \sum y_i/K_i \tag{3.10}$$

$$f_2(P) = 1 - \sum y_i/K_i \tag{3.11}$$

Depending upon whether you want to calculate the pressure or the temperature, the appropriate function, either Eqn (3.6) or (3.7) is selected. Iterations are performed on the unknown variable until the summation is equal to unity. So to use the first equation (Eqn (3.6)), the pressure is known and iterations are performed on the temperature.

Figure 3.7 shows a simplified pseudo code description of the algorithm for performing a hydrate formation pressure calculation using the K-factor method.

3.2.2 Liquid Hydrocarbons

The K-factor method is designed for calculations involving a gas and a hydrate. In order to extend this method to liquid hydrocarbons, the vapor–liquid K-factor should be incorporated. For the purposes of this book, these K-factors will be denoted K_V in order to distinguish them from the K-factor defined earlier. Therefore, we present the following:

$$K_{Vi} = \frac{y_i}{x_i} \tag{3.12}$$

where x_i is the mole fraction of component i in the nonaqueous liquid.

If the equilibria involve a gas, a nonaqueous liquid and a hydrate, then the following equations should be solved to find the phase fractions, L and V:

$$f_1(V, L) = \sum \frac{z_i(1 - K_{Vi})}{L(1 - V) + (1 - V)(1 - L)(K_{Vi}/K_i) + VK_{Vi}} \tag{3.13}$$

$$f_2(V, L) = \sum \frac{z_i(1 - K_{Vi}/K_i)}{L(1 - V) + (1 - V)(1 - L)(K_{Vi}/K_i) + VK_{Vi}} \tag{3.14}$$

This is a nontrivial problem that requires the solution of two nonlinear equations in two unknowns, L and V, the phase fractions of the liquid and vapor.

```
1.  Input the temperature, T

2.  Input the vapor composition, yi

3.  Assume a value for the pressure, P

4.  Set the K-factors for all non-formers to infinity

5.  Given P and T, obtain K-factors from the Katz charts (or
    from correlations) for the hydrate forming components in the
    mixture.

6.  Calculate the summation:
```

$$\sum y_i/K_i$$

```
    Note for non-formers the expression yi/Ki is zero.

7.  Does the summation equal unity?

     That is, does Σ yi/Ki = 1?

    7a. Yes - Goto Step 10
    7b. No - Goto Step 8

8.  Update the pressure estimate

    8a. If the sum is greater than 1, reduce the pressure
    8b. If the sum is less than 1, increase the pressure
    8c. Use caution if the sum is significantly different from 1

9.  Goto Step 4

10. Convergence! Current P is the hydrate pressure.

11. Stop
```

Figure 3.7 ***Pseudo Code for Performing a Hydrate Pressure Estimate Using the Katz K-factor Method.***

On the other hand, for equilibria involving a hydrate and a nonaqueous liquid, the K-factors are as follows:

$$K_{Li} = \frac{K_{Vi}}{K_i} = \frac{s_i}{x_i} \tag{3.15}$$

To determine the incipient solid formation point, the following function must be satisfied:

$$\sum \frac{K_{Vi}\, x_i}{K_i} = 1 \tag{3.16}$$

and, as with the vapor–solid calculation, you can either iterate on the temperature for a given pressure or iterate on the pressure for a given temperature.

The vapor–liquid *K*-factors can be obtained from the *K*-factor charts in the *GPSA Engineering Data Book* or one of the other simple or complex methods available in the literature.

Typically, the problem of calculating the hydrate formation conditions in the presence of liquid hydrocarbon is too difficult for hand calculations. In such cases, it is wise to use one of the commercially available software packages.

3.2.3 Computerization

Ironically, although we have classified these as "hand" calculation methods, the charts have been converted to correlations in temperature and pressure. Sloan (1998) presents a series of correlations based on the charts that are suitable for computer calculations. The correlations are quite complex and will not be repeated here. Because of their complexity, these correlations are not suitable for hand calculations.

The accompanying Web site contains several versions of the *K*-factor method for computers. They are a series of stand-alone disk operating system programs for performing various calculations, including a hydrate locus prediction. There is one set that uses SI units and the other uses American engineering units.

3.2.4 Comments on the Accuracy of the *K*-Factor Method

Until about 1975, the *K*-factor method represented the state of the art for estimating hydrate formation conditions. This method was only supplanted by the more rigorous methods, which are discussed in the next chapter, with the emergence of inexpensive computing power.

The *K*-factor method, as given on the accompanying Web site, is surprisingly accurate for predicting the hydrate locus of pure methane, ethane, carbon dioxide, and hydrogen sulfide. Figure 3.8 shows a comparison between the hydrate locus based on a correlation of the experimental data (the same as shown in Figure 2.3) and the prediction from the *K*-factor method for methane and ethane. Figure 3.9 is a similar plot for hydrogen sulfide and carbon dioxide.

These calculations are performed in spite of the warning in the *GPSA Engineering Data Book* (1998) that these methods should not be used for pure components. It is demonstrated here that the method is surprisingly accurate for methane, ethane, hydrogen sulfide, and carbon dioxide. Although not

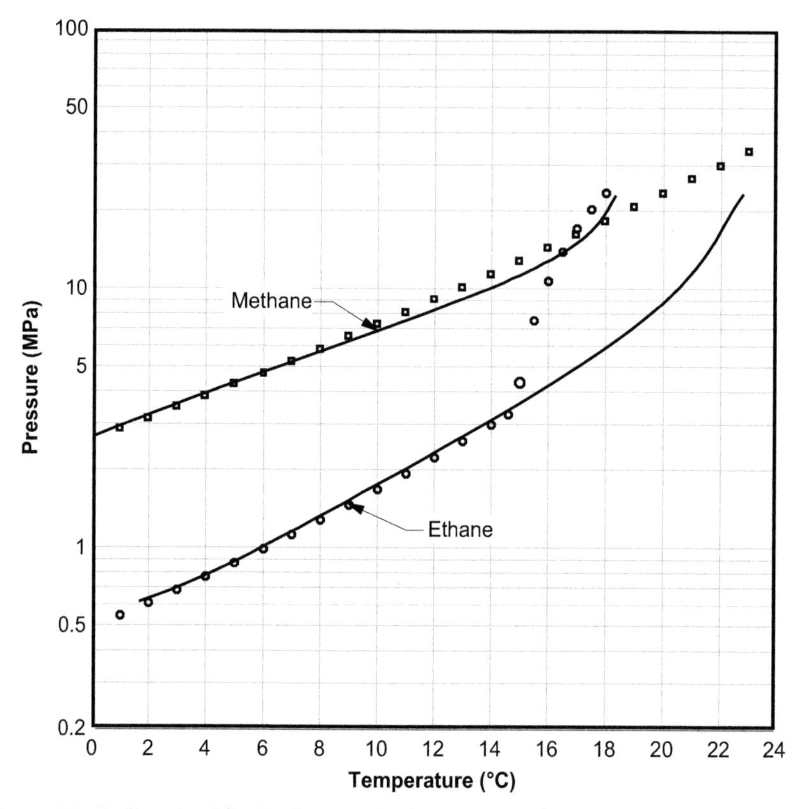

Figure 3.8 ***Hydrate Loci for Methane and Ethane.*** (points from correlation, curves from *K*-factor.)

shown, the method does not work for pure propane, isobutane, and *n*-butane.

The *K*-factor method is poor at low pressure and, thus, it does not predict the hydrate loci for either propane or isobutane in the pure state.

In addition, in order to get a prediction for hydrogen sulfide, the temperature must be at least 10 °C (50 °F), which corresponds to 0.3 MPa (45 psia). Similarly, for ethane, the temperature must be 1.7 °C (35 °F), which is a hydrate pressure of 0.62 MPa (90 psia). In general, it is recommended that the minimum pressure for the use of the correlation is 0.7 MPa (100 psia).

Although the *K*-factor method works quite well for pure hydrogen sulfide, it should be used with caution (and probably not at all) for sour gas mixtures. H_2S forms a hydrate quite readily and it exerts a large influence on the hydrate forming of mixtures that contains it as a component.

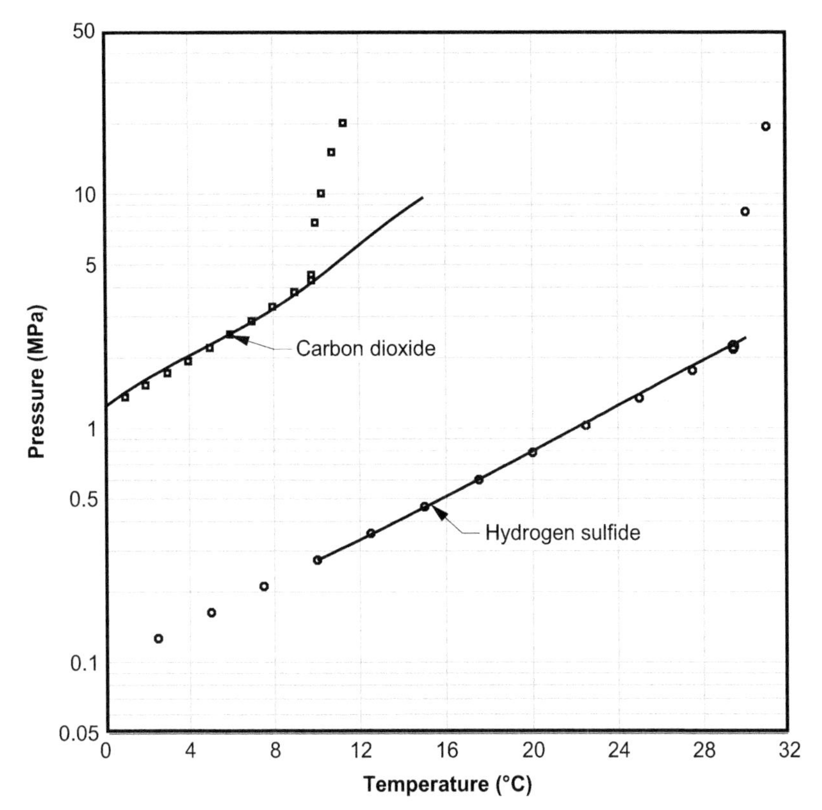

Figure 3.9 ***Hydrate Loci for Hydrogen Sulfide and Carbon Dioxide.*** (Points from correlation, curves from *K*-factor.)

In addition, the method cannot predict the hydrate for liquids (as noted in the text). At the experimental quadruple point, the *K*-factor method tends to continue to extrapolate as if the fluid were a vapor.

On the other hand, the method is not good for high pressure. For pure methane, the *K*-factor method does not give results at pressures greater than about 20 MPa (3000 psia), which is about 18 °C (64.5 °F). Fortunately, this range of pressure and temperature is sufficient for most applications. Having said that, it is probably wise to limit the application of this method to pressures less than 7 MPa (1000 psia).

In summary, the recommended ranges for the application of the *K*-factor method are:

$0 < t < 20$ °C	$32 < t < 68$ °F
$0.7 < P < 7$ MPa	$100 < P < 1000$ psia

It is probably safe to extrapolate this pressure and temperature range to mixtures. However, the method tends to be less accurate for mixtures.

3.2.4.1 Ethylene

In his studies of hydrate formation in methane-ethylene mixture, Otto (1959) attempted to generate a *K*-factor chart for ethylene. Although he successfully generated a chart, it was not very successful for predicting the hydrate conditions in ethylene mixtures. Errors as large as 1000–1500 kPa (150–220 psia) were observed when predicting the hydrate pressures for mixtures containing ethylene.

It was demonstrated in Chapter 2 that among hydrate formers, ethylene is unique. This further demonstrates the unusual nature of this component.

3.2.5 Mann et al.

It is possible to include the hydrate type in the *K*-factor method. For example, Mann et al. (1989) presented two sets of *K*-factors for the hydrate formers, one for each crystal structure. However, this method has not gained acceptance in the gas processing industry.

The Hydrate + software package from FlowPhase incorporated the Mann et al. method as one of its optional calculation packages.

3.3 BAILLIE–WICHERT METHOD

Another chart method for hydrate prediction was developed by Baillie and Wichert (1987). The basis for this chart is the gas gravity, but the chart is significantly more complex than the Katz gravity method. The chart is for gases with gravity between 0.6 and 1.0.

In addition to the gravity, this method accounts for the presence of hydrogen sulfide (up to 50 mol%) and propane (up to 10%). The effect of propane comes in the form of a temperature correction, which is a function of the pressure and the H_2S concentration.

Of the simple methods presented in this chapter, only the Baillie–Wichert method is designed for use with sour gas. This is a significant advantage over both the gas gravity and *K*-factor methods.

Figure 3.10 shows the chart for this method in SI units and Fig. 3.11 is in American engineering units.

The chart was designed to predict the hydrate temperature of a sour gas of known composition at a given pressure. The H_2S content of the sour gas for the application of the chart can be from 1% to 50%, with the H_2S to CO_2

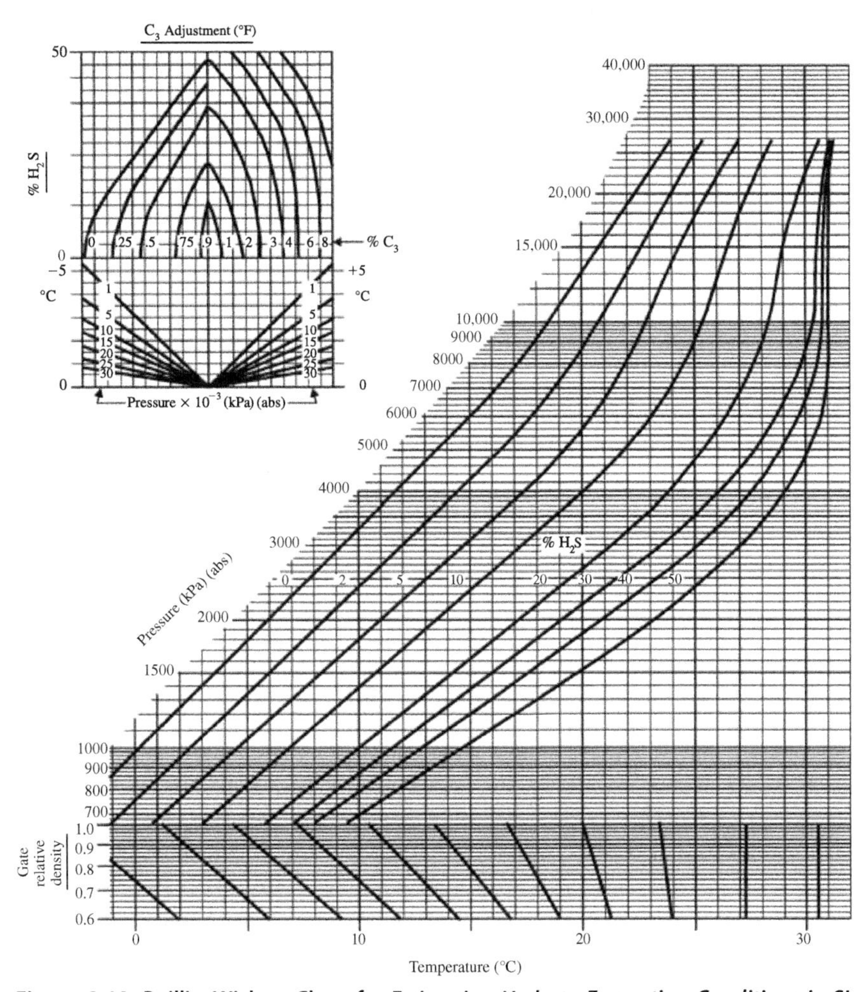

Figure 3.10 ***Baillie–Wichert Chart for Estimating Hydrate Formation Conditions in SI Units.*** *Reprinted from the* GPSA Engineering Data Book, *11th ed.—reproduced with permission.*

ratio between 10:1 and 1:3. Under these conditions, the chart method usually predicts a hydrate temperature of ±2 °F for 75% of the cases (Wichert, 2004). Baillie and Wichert (1987) state that, for a given pressure, their chart estimates the hydrate temperature to within 1.7 °C (3 °F) for 90% of their tests.

Furthermore, from the chart itself, it can be seen that the pressure is limited to 4000 psia (27.5 MPa), the gas gravity must be between 0.6 and 1.0, and the propane composition must be less than 10 mol%.

In a study of the hydrate formation in sour gas mixtures, and bearing in mind the limit on the ratio of H_2S to CO_2 given above, Carroll (2004)

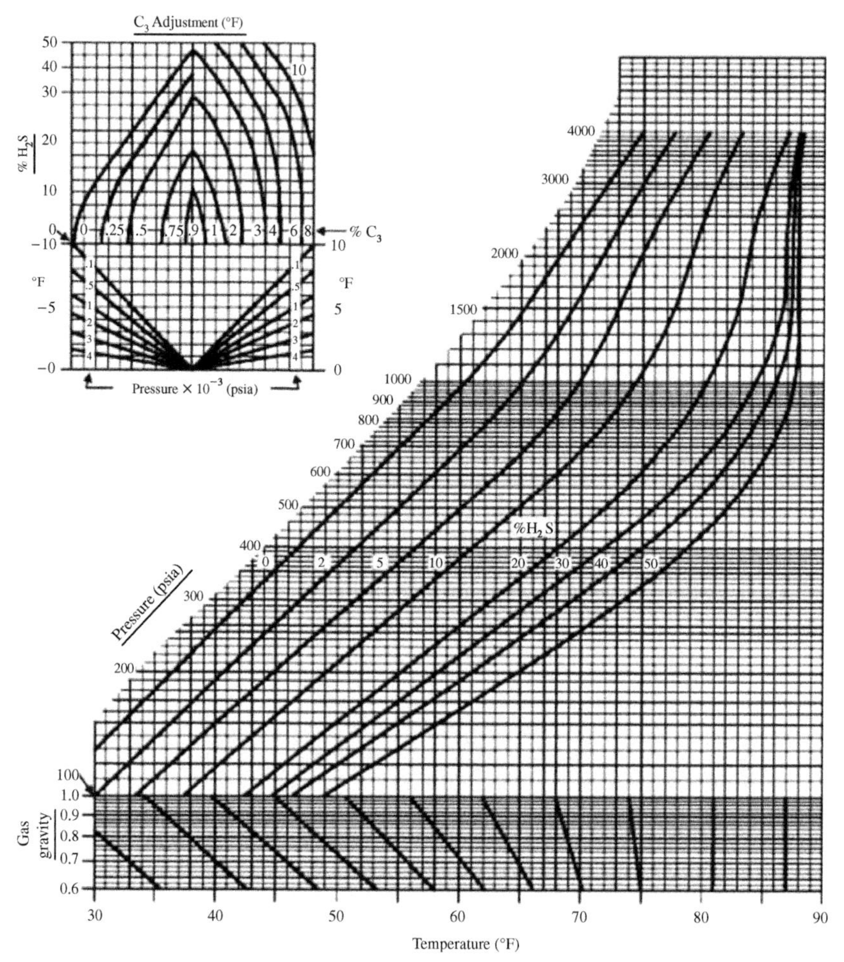

Figure 3.11 ***Baillie–Wichert Chart for Estimating Hydrate Formation Conditions in American Engineering Units.*** *Reprinted from the* GPSA Engineering Data Book, *11th ed.—reproduced with permission.*

found that the Baillie–Wichert method has an average error of 2.0 °F (1.1 °C). This method predicts the experimental hydrate temperature to within 3 °F about 80% of the time. This compares well with more rigorous methods, which will be discussed in the next chapter.

The study presented by Carroll (2004) only included data values in the composition ranges stated by Wichert (2004). If the composition is outside the range given, the errors increase significantly. More discussion of the study of Carroll (2004) is presented later in this chapter.

The use of these charts is neither simple nor intuitive. Figure 3.12 gives the pseudo code for the procedure for estimating the hydrate temperature. This procedure is a once through method resulting directly in the hydrate temperature. Figure 3.13 gives the pseudo code for

```
1.  Input gas composition gravity and pressure

2.  Calculate the gas gravity

3.  Enter the main graph at the given pressure (the sloping
    axis)

4.  Move to the right to the appropriate H2S concentration,
    interpolating as necessary

5.  From that point go straight down to the appropriate gas
    gravity.

6.  In the gravity region, follow the sloping lines down to read
    the temperature off the axis at the bottom of the graph.
    This is the base temperature.

7.  Go to the propane correction portion of the chart (small
    chart in the upper left)

8.  Enter the correction portion of the chart on the upper left
    with the H2S concentration.

9.  Move straight to the left until reach the appropriate
    propane concentration.

10. Got straight down to the appropriate pressure curve.

11. The pressure section is in two parts.
      11a. Did you enter the left section? Goto Step 12
      11b. Did you enter the right section? Goto Step 14

12. Read the temperature correction from the left axis. In this
    case the temperature correction is a negative value.

13. Goto Step 15

14. Read the temperature correction from the right axis. In this
    case the temperature correction is a positive value.

15. This is the temperature correction.

16. Obtain the hydrate formation temperature by adding the
    temperature correction to the base temperature, keeping in
    mind the sign of the temperature correction.
```

Figure 3.12 ***Pseudo Code for Estimating Hydrate Pressure Using the Baillie–Wichert Method.***

```
1.  Input gas composition gravity and pressure

2.  Calculate the gas gravity

3.  Assume a value for the hydrate pressure

4.  Go to the propane correction portion of the chart (small
    chart in the upper left).

5.  Enter the correction portion of the chart on the upper left
    with the H2S concentration.

6.  Move straight to the left until you reach the appropriate
    propane concentration.

7.  Got straight down to the appropriate pressure curve.

8.  The pressure section is in two parts.
      8a. Did you enter the left section? Goto Step 9
      8b. Did you enter the right section? Goto Step 11

9.  Read the temperature correction from the left axis. In this
    case the temperature correction is a negative value.

10. Goto Step 12

11. Read the temperature correction from the right axis. In this
    case the temperature correction is a positive value.

12. Subtract the temperature correction from the input
    temperature to obtain the base temperature

13. Use the base temperature to enter the gas gravity section of
    the main chart

14. Follow parallel to the sloping lines up to the appropriate
    gas gravity point

15. From that point go straight up to the appropriate H2S
    concentration

16. From the H2S concentration read the pressure off the sloping
    axis

17. Does this pressure equal the assumed value for the pressure?
      17a. Yes - Goto Step 20
      17b. No - Goto Step 18

18. Set the pressure obtain in Step 16 to the assumed pressure

19. Goto Step 4

20. Solution reached! The pressure obtained in Step 16 is the
    hydrate formation pressure.
```

Figure 3.13 *Pseudo Code for Estimating Hydrate Temperature Using the Baillie-Wichert Method.*

estimating the hydrate pressure. To use the Baillie–Wichert method to estimate the hydrate pressure requires an iterative procedure. You must start with an estimate of the pressure and iterate until you reach a solution. You can use the Katz gas gravity method to obtain a value for the starting pressure. From this starting point, the method converges in only a few iterations.

3.4 OTHER CORRELATIONS

Although the gas gravity method is not highly accurate, it has a high level of appearance because of its simplicity. Therefore, many authors have attempted to build correlations to describe the relation between the gas gravity and the hydrate pressure. Some of these are described in this section.

These correlations are useful for spreadsheet calculations in addition to hand calculations, but the user is advised to be cautious. These correlations are no more accurate than the original charts. In addition, they are not applicable to sour gas mixtures.

When using these equations, the reader is cautioned to be careful regarding common logarithms (log), which are base 10, and natural logarithms (ln), which are base e.

3.4.1 Makogon

Makogon (1981) provided a simple correlation for the hydrate formation pressure as a function of temperature and gas gravity for paraffin hydrocarbons. His correlation is:

$$\log P = \beta + 0.0497\left(t + kt^2\right) - 1 \tag{3.17}$$

where P is in MPa and t is in Celsius. Makogon provided graphic correlations for β and k, but Elgibaly and Elkamel (1998) give the following simple correlations:

$$\beta = 2.681 - 3.811\gamma + 1.679\gamma^2 \tag{3.18}$$

$$k = -0.006 + 0.011\gamma + 0.011\gamma^2 \tag{3.19}$$

Note the complete formulation of this correlation, as given in Elgibaly and Elkamel (1998), has errors but their correlation of β and k are correct.

3.4.2 Kobayashi et al.

Kobayashi et al. (1987) proposed the following, rather complicated, equation for estimating hydrate formation conditions as a function of the gas gravity:

$$\begin{aligned}\frac{1}{T} = {} & 2.7707715 \times 10^{-3} - 2.782238 \times 10^{-3} \ln P - 5.649288 \times 10^{-4} \ln \gamma \\ & - 1.298593 \times 10^{-3} \ln P^2 + 1.407119 \times 10^{-3} \ln(P)\ln(\gamma) \\ & + 1.785744 \times 10^{-4} \ln(\gamma)^2 + 1.130284 \times 10^{-3} \ln(P)^3 \\ & + 5.9728235 \times 10^{-4} \ln(P)^2 \ln(\gamma) - 2.3279181 \times 10^{-4} \ln(P)\ln(\gamma)^2 \\ & - 2.6840758 \times 10^{-5} \ln(\gamma)^3 + 4.6610555 \times 10^{-3} \ln(P)^4 \\ & + 5.5542412 \times 10^{-4} \ln(P)^3 \ln(\gamma) - 1.4727765 \times 10^{-5} \ln(P)^2 \ln(\gamma)^2 \\ & + 1.393808 \times 10^{-5} \ln(P)\ln(\gamma)^3 + 1.4885010 \times 10^{-5} \ln(\gamma)^4 \end{aligned} \tag{3.20}$$

With this set of coefficients, the temperature, T, is in Rankine, the pressure, P, is in psia, and γ is the gas gravity, dimensionless.

Unfortunately, there appears to be something wrong with this equation or with the coefficients given. No matter what value is entered for the pressure, the resulting temperature is always approximately 0 R (−460 °F) (see the example section of this chapter). Much effort was expended trying to find the error (including using natural instead of common logarithms), but the problem could not be completely isolated.

3.4.3 Motiee

Motiee (1991) presented the following correlation for the hydrate temperature as a function of the pressure and the gas gravity:

$$\begin{aligned} T = {} & -283.24469 + 78.99667 \log(P) - 5.352544 \log(P)^2 \\ & + 349.473877\gamma - 150.854675\gamma^2 - 27.604065 \log(P)\gamma \end{aligned} \tag{3.21}$$

where T is the hydrate temperature in K, P is the pressure in kPa, and γ is the gas gravity.

3.4.4 Østergaard et al.

Another correlation was proposed by Østergaard et al. (2000). These authors began with a relatively simple function of the hydrate formation conditions using the gas gravity method, which is applicable to sweet gas.

Table 3.1 Parameters for the Østergaard et al. Correlation for Hydrate Formation

c_1	4.5134×10^{-3}	c_6	3.6625×10^{-4}
c_2	0.46852	c_7	−0.485054
c_3	2.18636×10^{-2}	c_8	−5.44376
c_4	-8.417×10^{-4}	c_9	3.89×10^{-3}
c_5	0.129622	c_{10}	−29.9351

$$\ln(P) = \left(c_1(\gamma + c_2)^{-3} + c_3 F_m + c_4 F_m^2 + c_5\right)T + c_6(\gamma + c_7)^{-3} + c_8 F_m + c_9 F_m^2 + c_{10} \tag{3.22}$$

where P is the pressure in kPa, γ is the gas gravity, T is the temperature in K, and F_m is the mole ratio between nonformers and formers in the mixture. The constants for this equation are given in Table 3.1.

They then provide a correction for nitrogen and carbon dioxide. Those interested in these corrections should consult with the original paper. In addition, this method is not applicable to pure methane or to pure ethane, both of these gases have gravities outside the range of their correlation.

It is interesting to note that Østergaard et al. (2000) state that they attempted to include H_2S in their correlation, but they were not successful.

A spreadsheet is available from Østergaard et al. to perform calculations using their correlation.

3.4.5 Towler and Mokhatab

Towler and Mokhatab (2005) proposed a relatively simple equation for estimating hydrate temperatures as a function of the pressure and the gas gravity:

$$T = 13.47\ln(P) + 34.27\ln(\gamma) - 1.675\ln(P)\ln(\gamma) - 20.35 \tag{3.23}$$

Note that in their original paper, they use log, but it is clear when you use this equation that these are natural logarithms.

3.5 COMMENTS ON ALL OF THESE METHODS

In spite of their relative simplicity, these methods are surprisingly accurate. For sweet natural gas mixtures, the gas gravity method is accurate to within 20% or better for estimating hydrate pressures. However, as will be demonstrated, larger errors can be encountered.

The *K*-factor charts are probably accurate to about 10–15% for similar mixtures. Another inaccuracy is the user's ability to read the chart—they are frustratingly difficult to read! This can also contribute significantly to the error.

The Baillie–Wichert method is better than the gas gravity method when applied to sweet gas. The reason for this is the inclusion of a correction factor for propane. The real advantage of this method is that it is applicable to sour gas mixtures. Of the three methods presented in this chapter, the Baillie–Wichert chart is the method of choice for sour gas mixtures.

However, the question arises "How can these methods be even approximately correct when they do not account for the two different types of hydrates?" The short answer is that they cannot. It is thermodynamically inconsistent for the models not distinguish between the two types.

The long answer to this question is that, for natural gas mixtures, the type II hydrate predominates. Whenever there is only a small amount of type II former present, the resultant hydrate is type II. This change from type I to type II can have a significant effect on the hydrate-forming pressure. For example, pure methane forms a type I hydrate at 15 °C and 12.8 MPa (see Table 2.2). The presence of only 1% propane, a type II former, results in a mixture that forms a type II hydrate. This mixture is estimated to form a hydrate at 15 °C and 7.7 MPa. The mixture calculation is performed using CSMHYD.[1] CSMHYD is one of the computer programs discussed in the next chapter.

Before the widespread use of computers and the availability of software (pre-1970s), the *K*-factor method of Katz and co-workers was the state of the art, and it remains very popular in spite of its drawbacks.

Doing calculations when liquid hydrocarbons are present are very difficult via methods designed for hand calculations, and the results are usually not highly accurate. Therefore, it is usually not worth the time to do these types of calculations by hand.

3.5.1 Water

All of the chart methods assume that there is plenty of water present in the system. Thus, these methods predict the worst-case for the hydrate-forming conditions. These methods should not be used to estimate the hydrate-forming conditions in a dehydrated gas.

[1] CSMHYD copyright is held by E. Dendy Sloan, Colorado School of Mines, Golden, CO.

3.5.2 Nonformers

It is interesting to contrast the effect of nonformers on the predicted formation conditions from these methods.

A light nonformer, such as hydrogen, will reduce the gravity of the gas. In the gas gravity method, the presence of such a gas decreases the gravity of the gas and, thus, it is predicted to increase the pressure at which a hydrate will form for a given temperature. On the other hand, the presence of a heavy nonformer, such as *n*-pentane, would increase the gravity of the gas and, thus, it is predicted to decrease the hydrate pressure. The gravity effect of the nonformers is approximately the same for the Baillie–Wichert method, except for the two components for which there are correction factors.

On the other hand, the *K*-factor method handles all nonformers the same way. All nonformers are assigned a *K*-factor of infinity. Therefore, the presence of any nonformer, be it heavy or light, is predicted to increase the pressure at which a hydrate will form for a given temperature.

Note there is a contrast for the prediction of the effect of a heavy nonformer on the hydrate pressure. The gas gravity method predicts that the heavy component will decrease the hydrate pressure, whereas the *K*-factor method predicts an increase in the pressure. What do the experimental data say?

Ng and Robinson (1976) measured the hydrate locus for a mixture of methane (98.64 mol%) and *n*-pentane (1.36%). They found that the locus for the mixture was at higher pressure than the locus for pure methane. This is consistent with the *K*-factor approach and is contrary to the behavior predicted by the gas gravity method.

On the other hand, both the gas gravity and *K*-factor methods predict that a light nonformer would increase the hydrate pressure. Again, it is interesting to examine the experimental evidence.

Zhang et al. (2000) measured the hydrate-forming conditions for several hydrocarbon + hydrogen mixtures. In every case, the presence of hydrogen increased the hydrate formation pressure. This is consistent with the predictions from both the gas gravity and *K*-factor methods.

One word of caution about this simplified discussion. The addition of a heavy nonformer may lead to the formation of a hydrocarbon liquid phase. The effect of the formation of this phase should not be overlooked.

3.5.3 Isobutane vs *n*-Butane

As was discussed in Chapter 2, isobutane is a true hydrate former inasmuch as it will form a hydrate without other hydrate formers present. On the other hand, *n*-butane will only enter the hydrate lattice in the presence of another hydrate former. In some sense, both of these components are hydrate formers and, therefore, it is interesting to compare how these simple methods for estimating hydrate formation handle these two components.

In both the gas gravity and Baillie–Wichert methods, it does not matter whether the butane is in the iso form, the normal form, or a mixture of the two. A mixture containing 1 mol% isobutane will have the same molar mass as a mixture containing 1 mol% *n*-butane. Because the molar masses are the same, then the gas gravities are the same. Also, because the gas gravities are the same, then the predicted hydrate formation conditions are the same for these two mixtures. The same is true for the Baillie–Wichert method. Their method distinguishes propane, but does not include the effect of the butanes, except through their effect on the gravity.

On the other hand, the *K*-factor method treats these components as separate and distinct components. There are separate charts for the *K*-factors of isobutane and *n*-butane, and from these charts one obtains unique *K*-factors for these two components.

Consider the simple example of two gas mixtures: (1) 96.8 mol% methane and 3.2% isobutane and (2) 96.8 mol% methane and 3.2% *n*-butane. Both of these mixtures have a gravity of 0.600.

From the gas gravity method, both of these mixtures would have the same hydrate formation conditions. For example, at 2 MPa, the hydrate temperature is estimated to be about 5.7 °C. The Baillie–Wichert method does not distinguish between these two mixtures either. Using their chart, the hydrate temperature is also estimated to be about 5.7 °C.

On the other hand, the *K*-factor method produces significantly different results. Using the programs on the accompanying Web site, the hydrate temperature for the first mixture is estimated to be 10.6 °C and 2.1 °C for the second mixture.

The reader should attempt to reproduce these calculations for themselves and, therefore, to verify these observations.

It is a little difficult at this point to answer the question "Which one is correct?" However, in a subsequent section of this chapter, the accuracy of the *K*-factor method for mixtures of methane and *n*-butane is discussed. At this point, it is sufficient to say that the *K*-factor method predicts the experimental data for these mixtures to better than 2 °C.

In the next chapter, we will examine some more advanced methods for calculating hydrate formation, including some commercial software packages. One of these software packages is *EQUI-PHASE Hydrate*.[2] This program estimates that the hydrate temperature for the first mixture is 9.6 °C and for the second mixture it is 1.3 °C, which are in reasonable agreement with the *K*-factor method. Therefore, it is fair to conclude that the *K*-factor better reflects the actual behavior than those methods based on gas gravity.

3.5.4 Quick Comparison

The next several examples will be compared with the predictions largely from the *K*-factor method. The mixtures examined are as follows: (1) a synthetic natural gas consisting only of light hydrocarbons, (2) two mixtures rich in carbon dioxide, and (3) mixtures of methane and *n*-butane.

3.5.4.1 Mei et al. (1998)

Mei et al. (1998) obtained the hydrate locus for a synthetic natural gas mixture. The mixture is largely methane (97.25 mol%), but includes ethane (1.42%), propane (1.08%), and isobutane (0.25%). This gas has a gravity of 0.575. These data are so new that they could not have been used in the development of the models and thus provide a good test of said models.

Figure 3.14 shows the experimental data from Mei et al. (1998), the locus for pure methane (from Chapter 2), and the predicted hydrate locus from the gas gravity and *K*-factor methods.

The *K*-factor method is surprisingly good. For a given pressure, the *K*-factor method predicts the hydrate temperature to within 1 °C; or looking at it from the other direction, for a given temperature the method predicts the pressure to within 10%.

On the other hand, the gravity method exhibits a significant error. For a given temperature, the gravity method predicts a hydrate pressure that is approximately double the measured value. This is a simple yet realistic composition for a natural gas, and this example clearly demonstrates the potential errors from the simple gas gravity method. It should now be clear that this chart does not represent the definitive map of the hydrate formation regions.

[2] EQUI-PHASE Hydrate copyright is held by D.B. Robinson & Associates, Edmonton, Alberta, Canada.

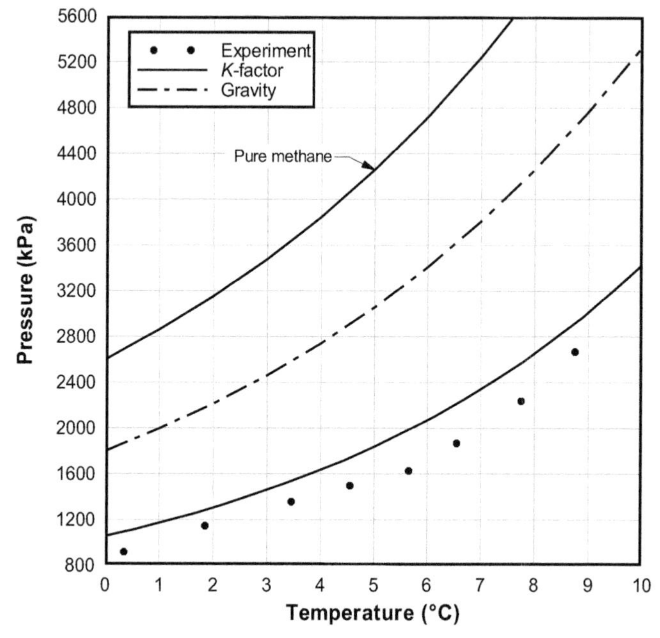

Figure 3.14 ***Hydrate Locus for a Synthetic Natural Gas Mixture.*** (SG = 0.575) (CH_4 97.25 mol%, C_2H_6 1.42%, C_3H_8 1.08, *i*-C_4H_{10} 0.25%).

The gravity of this mixture is lighter than the minimum used with the Baillie–Wichert chart, so no comparison was made. However, it is worth noting that the Baillie–Wichert method does include a correction for propane.

The hydrate locus for pure methane is included for comparison purposes. It is interesting to note the effect of a seemingly small amount of "other" components can have on the hydrate locus. The experimental mixture contains 97.25% methane, which for many other purposes would be safe to assume that this mixture has the properties of pure methane. Yet, the hydrate for this mixture forms at pressures almost 2/5 that of pure methane. For example, the experimental hydrate pressure for this mixture at 2 °C is about 1200 kPa, whereas for pure methane it is about 3100 kPa—a ratio of 0.39.

3.5.4.2 Fan and Guo (1999)

Fan and Guo (1999) measured the hydrate loci for two mixtures rich in carbon dioxide. Again, none of these data was used to generate the *K*-factor charts. These mixtures should prove to be a difficult test because it is commonly believed that the *K*-factor method is not highly accurate for CO_2-rich fluids.

These mixtures are too rich in CO_2 to expect either the gas gravity or Baillie–Wichert methods to be applicable. In addition, the gravity of these mixtures is much greater than the maximum for use with either the gas gravity or Baillie–Wichert methods, so no comparison is made.

Figure 3.15 shows the hydrate locus for a binary mixture of CO_2 (96.52 mol%) and CH_4 (3.48%). Both the experimental values and the prediction using the *K*-factor method are plotted.

At low temperature, the *K*-factor method tends to overpredict the hydrate pressure. On the other hand, at high temperature, the *K*-factor method tends to underpredict the hydrate pressure. The transition temperature is at about 7 °C (45 °F), although the somewhat sparse nature of the data make the exact point difficult to determine.

Overall, the average absolute error in the estimated hydrate pressure is 0.25 MPa (36 psia).

The second mixture is composed of CO_2 (88.53 mol%), CH_4 (6.83%), C_2H_6 (0.38%), and N_2 (4.26%). Figure 3.16 shows the experimental data

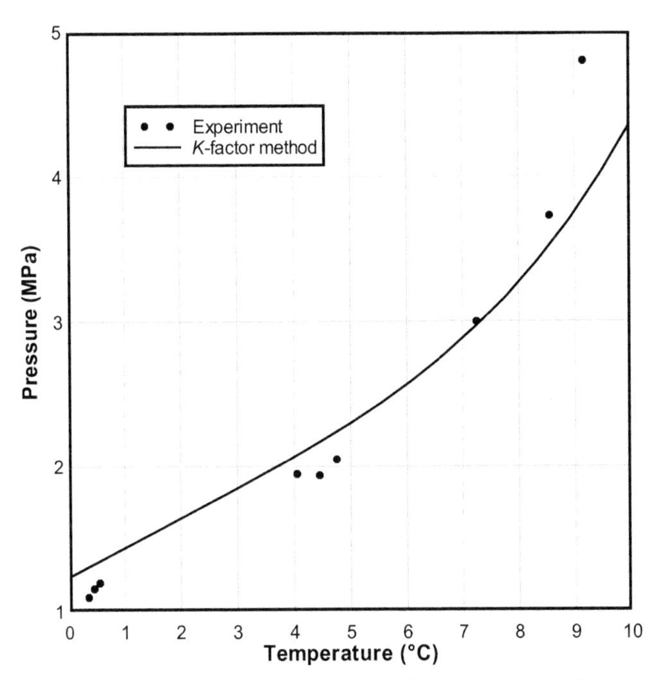

Figure 3.15 ***Hydrate Locus for a Mixture of Carbon Dioxide.*** (96.52 mol%) and methane (3.48 mol%).

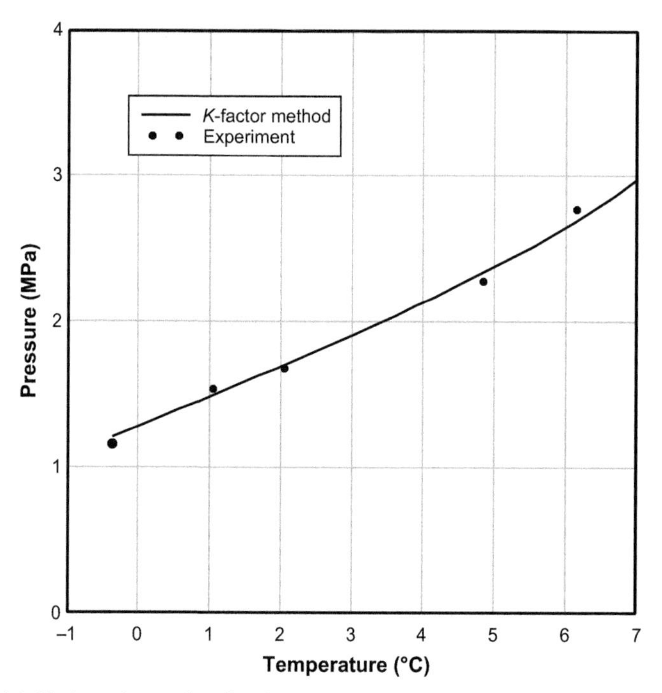

Figure 3.16 ***Hydrate Locus for the Quaternary Mixture.*** (88.53 mol% CO_2, 6.83% CH_4, 0.38% C_2H_6, 4.26% N_2).

and the *K*-factor prediction for this mixture. Over this range of temperature, the *K*-factor method is an excellent prediction of the experimental data. The average absolute error is only 0.05 MPa (7 psia).

3.5.4.3 Ng and Robinson (1976)

An important study is that of Ng and Robinson (1976). This study was discussed earlier because of its importance in determining the true role of *n*-butane in the formation of hydrates. Contrary to the other experimental data examined here, it is likely that these data were used to develop the *K*-factor chart for *n*-butane. However, they are an interesting set of data nonetheless.

They measured the hydrate loci for four binary mixtures of methane and *n*-butane. Three of these hydrate loci are shown in Fig. 3.17. The fourth locus was omitted for clarity.

It is clear from these plots that something unusual occurs with the *K*-factor prediction at about 11 °C (52 °F). The curve for the mixture leanest in *n*-butane has a noticeable hump at approximately this temperature. The two mixtures richest in *n*-butane show a more dramatic transition.

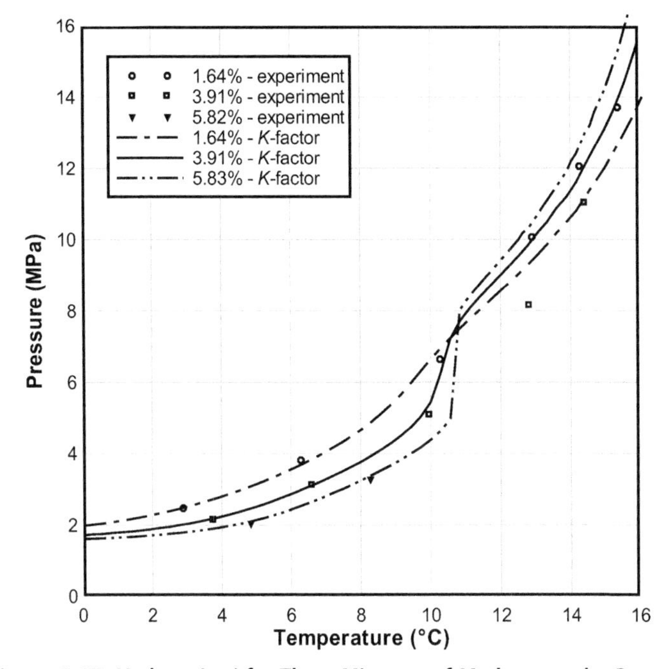

Figure 3.17 ***Hydrate Loci for Three Mixtures of Methane and n-Butane.***

Furthermore, above 11 °C, there is an inversion in the predicted behavior. At low temperature, the hydrate pressure decreases with increasing *n*-butane content. This is the behavior observed from the experimental data, regardless of the temperature. The prediction inverts for temperatures greater than 11 °C. For high temperatures, the hydrate pressure increases with increasing *n*-butane content.

The reason for this strange prediction can at least be partially explained by examining the *K*-factor plot for *n*-butane. At this temperature, all of the isobars converge to a single curve that extends to a higher temperature.

3.5.5 Sour Natural Gas

Carroll (2004) performed a study to determine the accuracy of hydrate prediction methods for sour gas mixtures. A database of experimental points was obtained from the literature. It included measurements from three different laboratories and a total of 125 points. The temperatures ranged from approximately 3 to 27 °C (37°–80 °F). All of the mixtures contained some hydrogen sulfide (and hence were sour) and a few also contained carbon dioxide.

Carroll (2004) found that the *K*-factor method performed poorly. The simple *K*-factor method, also designed for hand calculations, had an average error of 2.7 °F (1.5 °C). The method predicted the experimental hydrate temperature to within 3 °F 60% of the time.

On the other hand, the modified *K*-factor method of Mann et al. (1989) was as accurate as the more rigorous computer models. The average error for the method of Mann et al. (1989) was 1.5 °F (0.8 °C) and it predicted the hydrate temperature to within 3 °F (1.7 °C) about 90% of the time. These errors are comparable to those from the more rigorous models presented in the next chapter.

The study of Carroll (2004) also included predictions from the Baillie–Wichert chart. These results are discussed in the section on the Baillie–Wichert method.

3.6 LOCAL MODELS

There are several rigorous models for calculating hydrate conditions; many will be discussed in the next chapter. Local models are simplified models used over a small range of temperature and pressure. They are used when the calculation device has limited computing and data storage capabilities, such as a control device (a programmed logic controller) or a spreadsheet, or perhaps, in a worst-case scenario, a hand-held calculator. We can use some of the information provided in this chapter and the preceding chapter to develop these models.

For a fixed composition, the general form of the function for the hydrate locus is:

$$P_{\text{hyd}} = f(T, x) \tag{3.24}$$

There is some theoretical basis to an equation of the form:

$$\ln P_{\text{hyd}} = a + b/T \tag{3.25}$$

which is similar to the Clapeyron correlation for vapor pressures. The temperature range (and hence the pressure range) depends upon the system under consideration. The model developer should verify the applicability of such an equation for their given application. As will be shown, failure to do so can lead to significant errors.

In addition, this approach works quite well for a fixed composition. Adjusting it for variable composition is discussed in a subsequent section of this paper.

Assuming that a is a linear function of the temperature gives the following equation:

$$\ln P_{hyd} = a + \frac{b}{T} + cT \quad (3.26)$$

Assuming that a is a constant and b is a function of the temperature results in an equation with a form exactly the same as Eqn (3.24).

$$\ln P_{hyd} = a + \frac{b}{T + c} \quad (3.27)$$

In the natural gas business, it is common to require the hydrate formation temperature given the pressure. Thus, Eqn (3.25) can be rearranged to obtain:

$$T_{hyd} = \frac{b}{\ln P - a} \quad (3.28)$$

In the selection of a local model, it is convenient for the equation to be explicit in both temperature and pressure. It would be inconvenient to have a model that required an iterative solution. Thus, an equation like Eqn (3.28) is useful for calculating hydrate pressure; it cannot be rearranged to get a form explicit in the temperature.

Equation (3.28) could be rearranged into a quadratic, but this is still too complex for our purposes. The solution to such an equation is too cumbersome.

3.6.1 Wilcox et al. (1941)

Consider the hydrate data taken by Wilcox et al. (1941) for what they labelled gas B. This is a relatively light, sweet gas mixture. In total, there are nine points for this composition ranging in pressure from 182 to about 4000 psia.

If one simply uses statistical software, using Eqn (3.25) to fit the all of the data appears to give a good correlation ($r^2 = 0.98624$). The resulting correlation is:

$$\ln P_{hyd} = 51.66789 - \frac{23,42433}{T} \quad (3.29)$$

where P_{hyd} is in psia and T is in R. However, a plot of the data shows systematic deviations between the correlation and the experimental data. This is shown in Fig. 3.18. At high and low pressure, the correlation underpredicts the hydrate pressure, whereas at intermediate pressure it overpredicts. The average absolute error (AAE) in the predicted hydrate temperature is 1.8 °F.

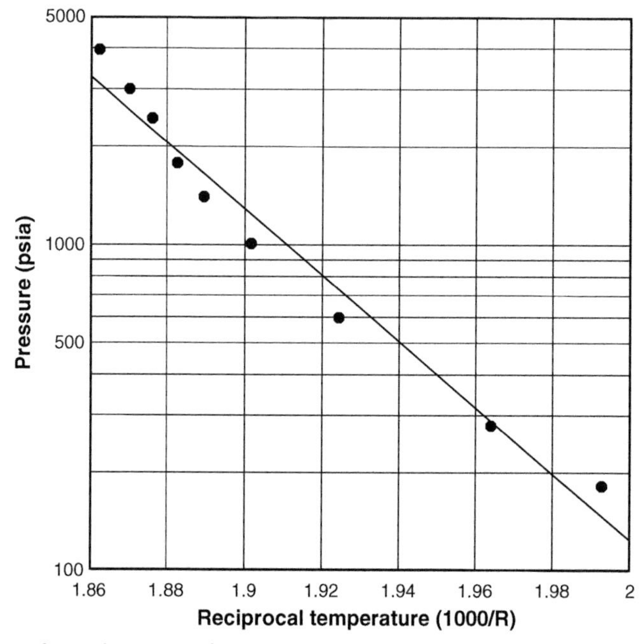

Figure 3.18 ***The Hydrate Data for the Wilcox et al. (1941) Gas B and the Fit of the Data Given by Eqn (3.29).***

Next, the data were localized—examined over a narrower range of pressure (and hence temperature). The data were separated into three pressure regions: low (180–1000 psia), intermediate (600–1750 psia), and high (1400–4000 psia). Data from the three regions were then fit to an equation of the form given by Eqn (3.25). The results of the fitting are summarized in Table 3.2 and are shown graphically in Fig. 3.19.

Again, merely looking at the statistical results from the data, fitting may not reveal exactly how good or bad the correlation is. However, from the graph, it is clear that the fit is very good if one only considers the range of pressure used to develop the correlation. Extrapolating the curve results in very large errors. However, this is the definition of a local model.

Table 3.2 Summary of Fitting Parameters for the Local Models for the Wilcox et al. (1941) Gas B

	Pressure Range (psia)	a	b	r^2
Low Pressure	180 – 1000	42.694 907	−18.841 542	0.99690
Intermediate	600 – 1750	55.855 342	−25.714 757	0.99698
High Pressure	1400 – 4000	80.912 574	−38.987 619	1.00000
All	180 – 4000	51.667 820	−23.424 368	0.98624

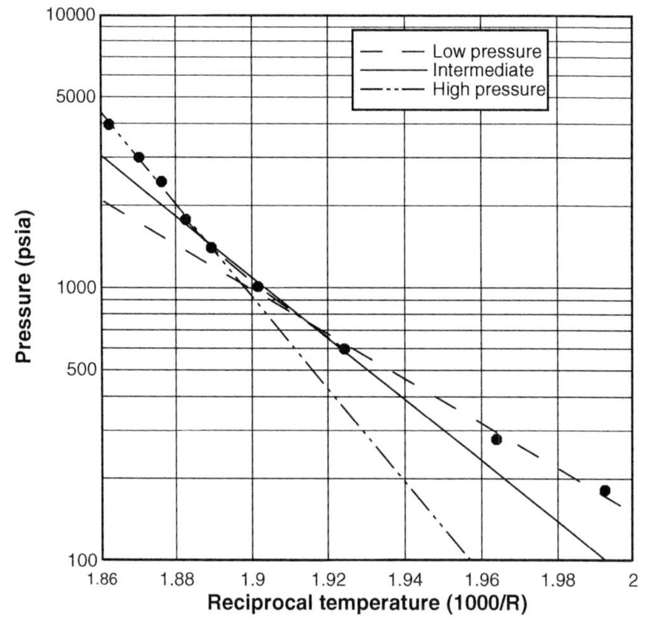

Figure 3.19 ***Local Modelling of the Hydrate Data from Wilcox et al. (1941) for Gas B.***

3.6.2 Composition

The common parameter for characterizing the composition of a natural gas mixture is the gas gravity, γ, or equivalently the molecular weight. Therefore, for small variations in the composition, we can propose a local model of the form:

$$P_{hyd} = f(T, x, \gamma) \tag{3.30}$$

Based on Eqn (3.25), perhaps we could use something of the form:

$$\ln P_{hyd} = a + b\gamma + \frac{c + d\gamma}{T} \tag{3.31}$$

Basically, this equation is constructed by assuming that the parameters in Eqn (3.25) are functions of the gas gravity. However, if we are considering a relatively narrow range of temperature (and remember T is the absolute temperature), then this equation becomes:

$$\ln P_{hyd} = a + b\gamma + \frac{c}{T} \tag{3.32}$$

Alternatively, we could develop and equation of the form:

$$\frac{1000}{T} = a + bP + c \ln P + d\gamma \tag{3.33}$$

Other functions could be used for the functional relationship between the temperature, pressure and the composition.

3.6.2.1 Sun et al.

Sun et al. (2003) took a set of measurements for sour gas mixtures; remember sour gas being a mixture containing H_2S. These data are from 1 to 26.5 °C and 0.58 to 8.68 MPa. The specific gravity of these mixtures range from 0.656 to 0.787. The data set is approximately 60 points in total. It was noted earlier that the simple gas gravity method is not applicable to sour gas mixtures, thus, this set of data provide a severe test for our simplified local models.

Using least-squares regression to fit the set of data, one obtains the following correlation:

$$\frac{1000}{T} = 4.343295 + 1.07340 \times 10^{-3} P - 9.19840 \times 10^{-2} \ln P - 1.071989\gamma \tag{3.34}$$

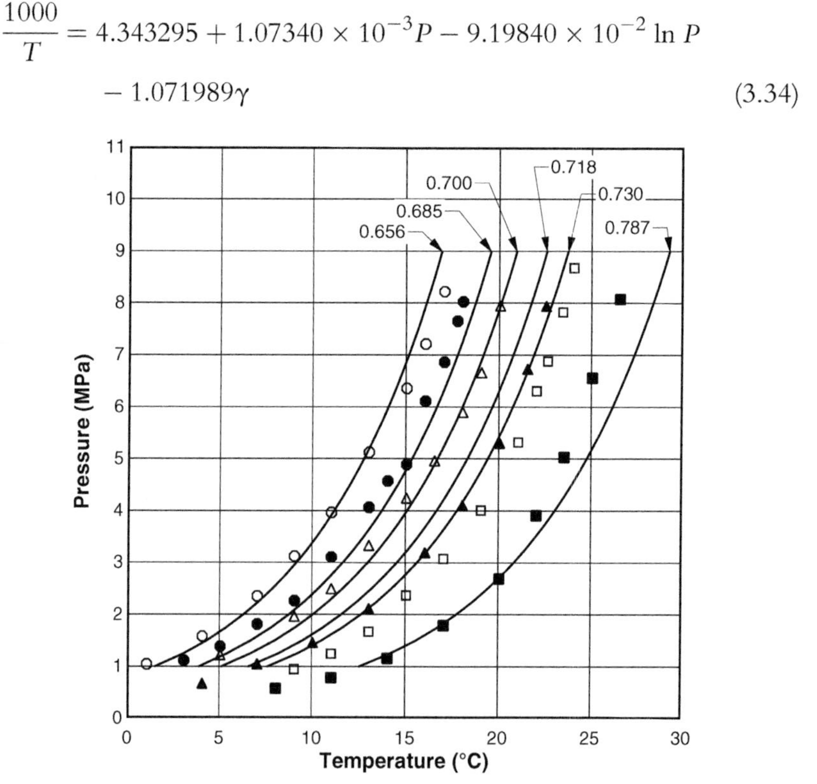

Figure 3.20 ***Local Modelling for the Sour Gas Data from Sun et al. (2003) Showing the Fit of Eqn (3.34).***

The average error for this equation is 0.01 °C, the AAE is 1.16 and the maximum error is 2.8 °C. One would expect an average error near zero for this type of correlation and that is what was obtained.

Figure 3.20 shows the data of Sun et al. (2003) and the calculation from the local model. The first observation we can make from this set of data is that as the gas gravity increases so does the hydrate temperature.

We will revisit the data of Sun et al. (2003) in the next chapter.

Examples

Example 3.1

Use the gas gravity method to calculate the hydrate formation pressure of ethane at 10 °C. The value in Table 2.3 is about 1.68 MPa.

Answer: The molar mass of ethane is 30.070 and thus:

$$\gamma = 30.070/28.966 = 1.038$$

Extrapolating the gravity chart, one reads about 1.25 MPa. This is an error of about 25%, which seems unreasonably large.

Example 3.2

Use the gravity method to calculate the pressure at which a hydrate will form at 14.2 °C for the following mixture:

CH_4	0.820
CO_2	0.126
H_2S	0.054

The experimental value is 4.56 MPa.

Answer: Calculate the molar mass of the gas mixture:

$$0.820 \times 16.043 + 0.126 \times 44.011 + 0.054 \times 34.082 = 20.541$$

$$\gamma = 20.541/28.966 = 0.709$$

From Fig. 3.2, we read slightly less than 4.0 MPa—very good agreement, especially considering that this is a sour gas.

Example 3.3

Repeat the above calculation using the *K*-factor method.

Answer: As a first guess, assume that the pressure is 4 MPa, selected because of ease of reading the chart. However, the values for CO_2 must be extrapolated at this pressure. A second iteration is performed at 5 MPa.

The iterations are summarized in the table below. The charts are not easily read to a greater accuracy than 1 MPa or so, so there is no point in repeating the iteration beyond this seemingly crude level.

	4 MPa		5 MPa	
	K_i	y_i/K_i	K_i	y_i/K_i
CH_4	1.5	0.547	1.35	0.607
CO_2	~3	0.042	2	0.063
H_2S	0.21	0.257	0.18	0.300
		Sum = 0.846		Sum = 0.970

Linearly extrapolating from the results in the table yields 5.24 MPa. The result is somewhat greater than the experimental value. A significant portion of this error lies in one's ability to read the charts. This includes the need to extrapolate some of the curves.

Example 3.4

Use the correlations from Sloan (1998) to redo Example 3.2. An Excel spreadsheet is available for these calculations; however, it requires engineering units. Converting 14.2 °C gives 57.6 °F.

Answer: Based on the previous calculations, assume a pressure of 700 psi and iterate from there.

700 psi	Sum = 0.925
750 psi	Sum = 0.945
800 psi	Sum = 0.983
825 psi	Sum = 1.001

Therefore, the answer is approximately 825 psi or 5.69 MPa, which represents an error of about 25%.

Clearly, using the correlations of Sloan (1998) and some computing power makes these calculations significantly easier.

Example 3.5

For the mixture given in Example 3.2, calculate the composition of the hydrate at 15 °C and 6.5 MPa. Use the correlations of Sloan (1998) to obtain the K-factors.

Answer: Again, convert to engineering units: 59 °F and 943 psi. At these conditions, the K-factors are:

CH_4	1.228
CO_2	2.289
H_2S	0.181

Iterate on Eqn (3.2) until the sum is zero. The Excel spreadsheet is set up to facilitate this iteration.

$V = 0.5$	Sum = −0.1915
$V = 0.75$	Sum = −0.1274
$V = 0.9$	Sum = −0.0620
$V = 0.95$	Sum = −0.0274
$V = 0.99$	Sum = +0.0098 finally spanned the answer!
$V = 0.9795$	Sum = −0.0011
$V = 0.9806$	Sum = +0.0000 solution reached!

As an initial starting point, I usually select a phase fraction of 0.5. From there, I iterate toward the answer in a somewhat arbitrary manner. Once I get close to the answer (preferably spanning the answer), I switch to a linear interpolation method. I use the two iterates that span the answer and linearly interpolate for my next estimates.

Therefore, the mixture is 98 mol% gas. The compositions of the two phases, which are also calculated in the spreadsheet, are as follows:

	Feed	Vapor	Solid
CH_4	0.820	0.8230	0.6704
CO_2	0.126	0.1274	0.0556
H_2S	0.054	0.0496	0.2740

Again, it is important to note that all of these compositions are on a water-free basis.

Example 3.6

Estimate the hydrate temperature for the following mixture at 5 MPa using the Baillie–Wichert chart.

Methane	86.25 mol%
Ethane	6.06%
Propane	2.97%
Isobutane	0.31%
n-Butane	0.63%
Pentanes	0.20%
Hexanes	0.02%
CO_2	1.56%
H_2S	2.00%

Answer: First, estimate the molar mass and gravity of the gas. This is done by multiplying the mole fraction of a component times that molar mass of the component and then summing all of these quantities. Therefore:

Methane	$0.8625 \times 16.043 = 13.887$
Ethane	$0.0606 \times 30.070 = 1.822$
Propane	$0.0297 \times 44.097 = 1.310$
Isobutane	$0.0031 \times 58.125 = 0.180$
n-Butane	$0.0063 \times 58.125 = 0.366$
Pentanes	$0.0020 \times 72.150 = 0.144$
Hexanes	$0.0002 \times 86.177 = 0.017$
CO_2	$0.0156 \times 44.010 = 0.687$
H_2S	$0.0200 \times 34.080 = 0.682$
	$M = 19.045$ g/mol

$$\gamma = M/28.966 = 19.0145/28.966 = 0.6575$$

Enter the main chart along the sloping axis at 5 MPa and go across to the 2 mol% H_2S curve (the second of the family of curves). From that point, you go straight down until you reach the 0.6575 gravity point. Then you follow the sloping lines down to the temperature axis. This value is the base temperature. In this case, the base temperature is about 16.7 °C.

Now, go to the temperature correction chart in the upper left. Enter the chart at 2 mol% H_2S and move across to the 2.97 mol% propane (essentially 3 mol%). From that point, you go down to the 5 MPa curve. This is on the right side of the chart so you go to the right axis. At that point, you read that the temperature correction is +1.5 °C—a positive value because it comes from the right half of the correction chart.

The hydrate temperature is the sum of these two terms:

$$T_{\text{hyd}} = 16.7 + 1.5 = 18.2\ °\text{C}$$

Example 3.7

Estimate the hydrate formation temperature at 1000 psia for a gas with a 0.6 gravity using the Kobayashi et al. equation. Assume that there is CO_2, H_2S, and N_2 in the gas. From Fig. 3.3, the temperature is 60 °F.

Answer: Substituting 1000 psia and 0.6 gravity onto the Kobayiashi et al. correlation yields:

$$\begin{aligned}
\frac{1}{T} = {} & 2.7707715 \times 10^{-3} - 2.782238 \times 10^{-3}\ln(1000) - 5.649288 \times 10^{-4}\ln(0.6) \\
& - 1.298593 \times 10^{-3}\ln(1000)^2 + 1.407119 \times 10^{-3}\ln(1000)\ln(0.6) \\
& + 1.785744 \times 10^{-4}\ln(0.6)^2 + 1.130284 \times 10^{-3}\ln(1000)^3 \\
& + 5.9728235 \times 10^{-4}\ln(1000)^2\ln(0.6) - 2.3279181 \times 10^{-4}\ln(1000)\ln(0.6)^2 \\
& - 2.6840758 \times 10^{-5}\ln(0.6)^3 + 4.6610555 \times 10^{-3}\ln(1000)^4 \\
& + 5.5542412 \times 10^{-4}\ln(1000)^3\ln(0.6) - 1.4727765 \times 10^{-5}\ln(1000)^2\ln(0.6)^2 \\
& + 1.393808 \times 10^{-5}\ln(1000)\ln(0.6)^3 + 1.4885010 \times 10^{-6}\ln(0.6)^4
\end{aligned}$$

If we examine this term-by-term, the following table results:

	Coefficient	C_{ij} log(P)^i log(γ)^j	
C1	2.7707715E-03	2.7707715E-03	
C2	−2.7822380E-03	−1.9219019E-02	
C3	−5.6492880E-04	2.8858011E-04	
C4	−1.2985930E-03	−6.1965070E-02	
C5	1.4071190E-03	−4.9652423E-03	
C6	1.7857440E-04	4.6597707E-05	
C7	1.1302840E-03	3.7256187E-01	←
C8	5.9728235E-04	−1.4558822E-02	
C9	−2.3279181E-04	−4.1961402E-04	
C10	−2.6840758E-05	3.5777731E-06	
C11	4.6610555E-03	1.0612851E+01	←
C12	5.5542412E-04	−9.3520806E-02	
C13	−1.4727765E-05	−1.8338174E-04	
C14	1.3938080E-05	−1.2833878E-05	
C15	1.4885010E-06	1.0135375E-07	
		Sum = 1.0793677E+01	

Finally, from the sum, one can calculate the temperature: $T = 1/10.793677 = 0.0926$ R, which is clearly in error. There appears to be two terms that are problematic: the C7 term and, in particular, the C11 term.

APPENDIX 3A KATZ *K*-FACTOR CHARTS

The *K*-factor charts for performing hydrate calculations are collected in this appendix. There are two charts for each of the components commonly found in natural gas. One chart is in SI units and the other is in American engineering units. These figures are taken from the *GPSA Engineering Data Book* and are reproduced here with permission (Figs 3.1a–3.14a).

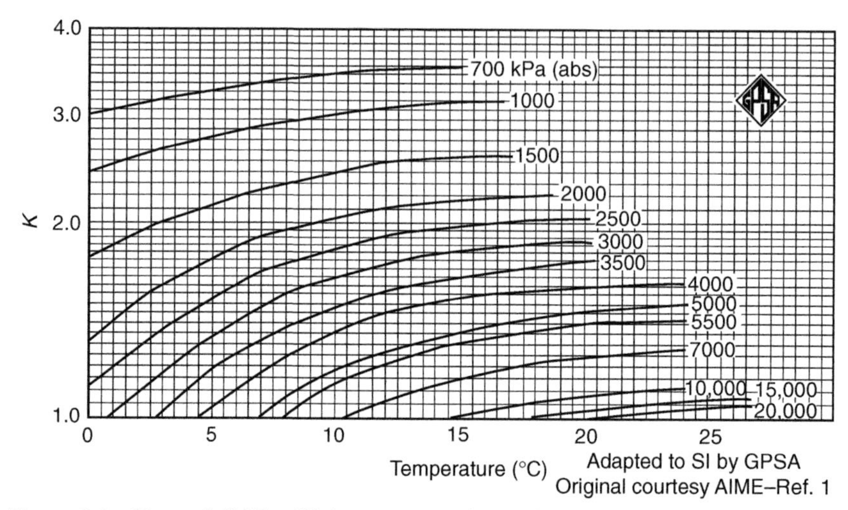

Figure 3.1a ***Vapor–Solid Equilibrium K-Factor for Methane in SI Units.*** *Reprinted from the* GPSA Engineering Data Book, *11th ed.—reproduced with permission.*

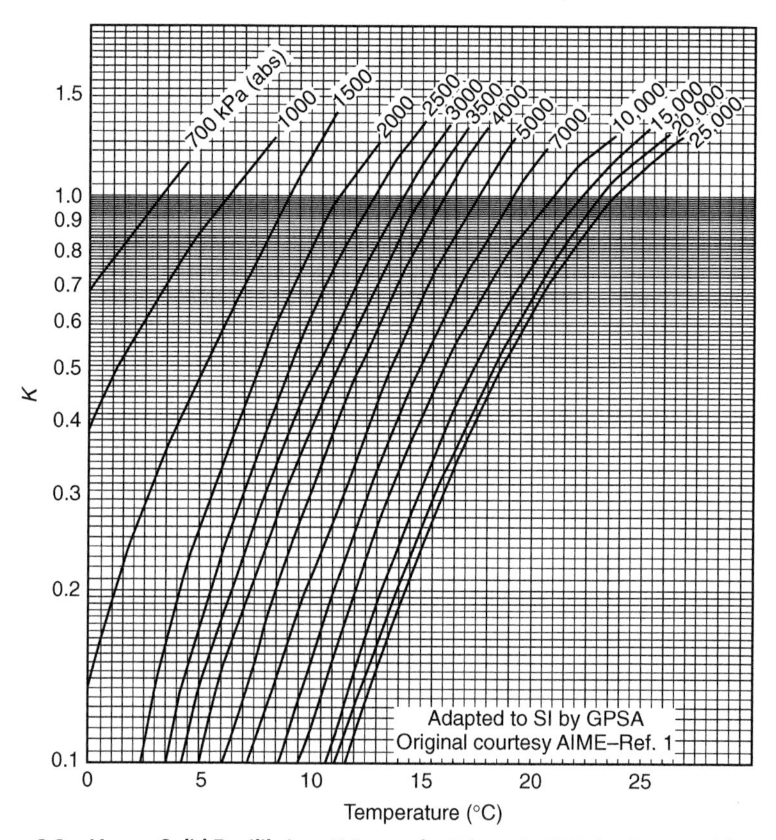

Figure 3.2a ***Vapor–Solid Equilibrium K-Factor for Ethane in SI Units.*** *Reprinted from the* GPSA Engineering Data Book, *11th ed.—reproduced with permission).*

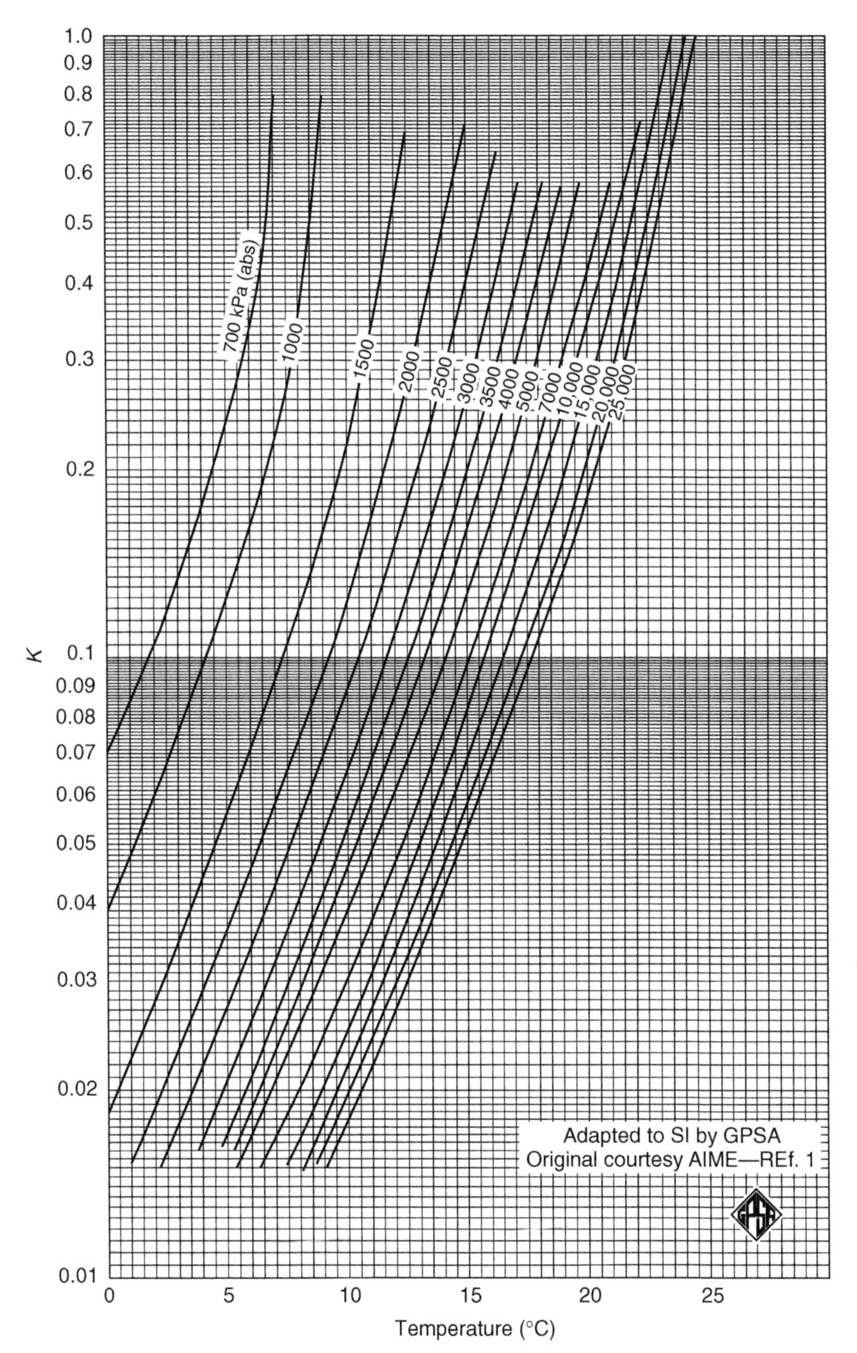

Figure 3.3a ***Vapor–Solid Equilibrium K-Factor for Propane in SI Units.*** *Reprinted from the* GPSA Engineering Data Book, *11th ed.—reproduced with permission.*

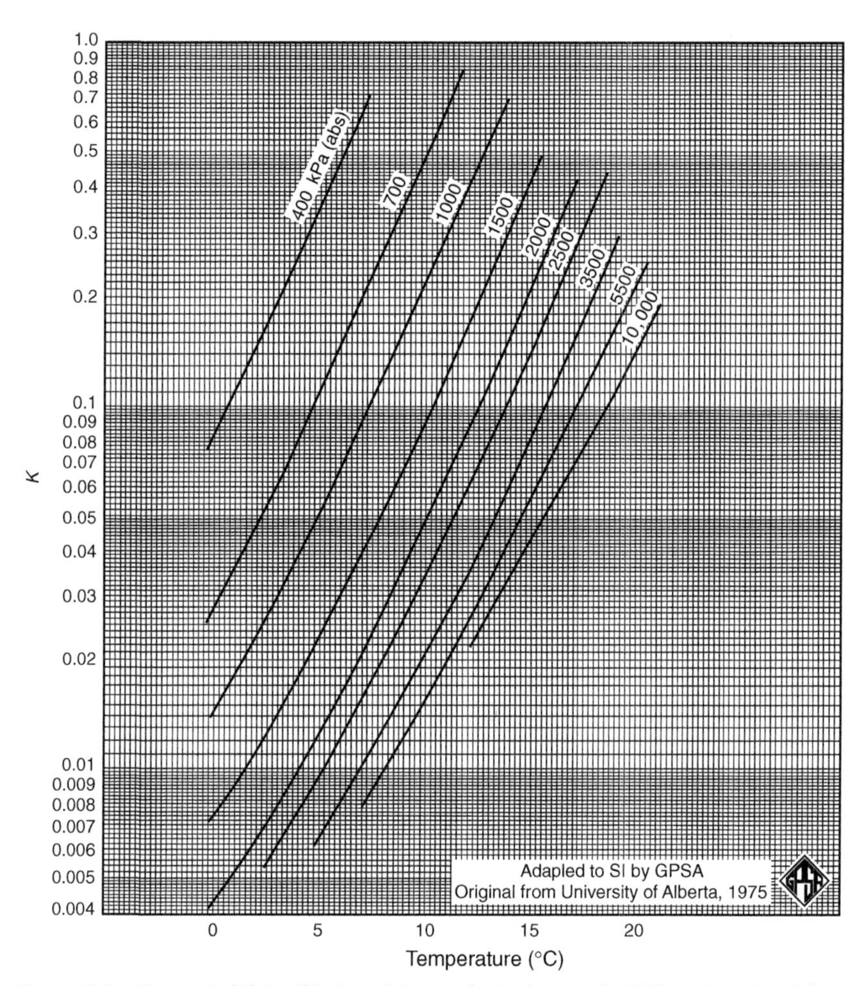

Figure 3.4a ***Vapor–Solid Equilibrium K-Factor for Isobutane in SI Units.*** *Reprinted from the* GPSA Engineering Data Book, *11th ed.—reproduced with permission.*

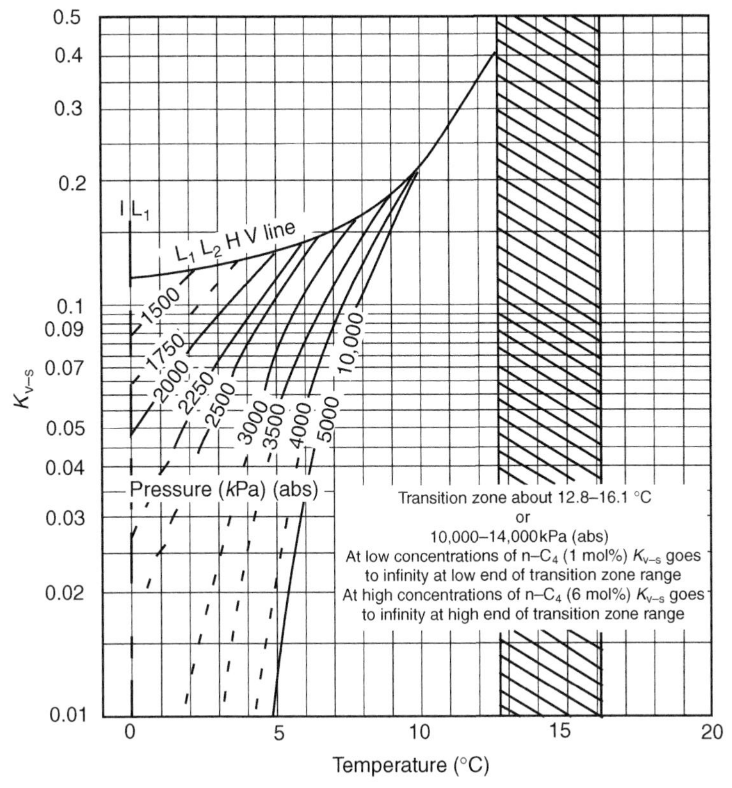

Figure 3.5a ***Vapor–Solid Equilibrium K-Factor for n-Butane in SI Units.*** *Reprinted from the* GPSA Engineering Data Book, *11th ed.—reproduced with permission.*

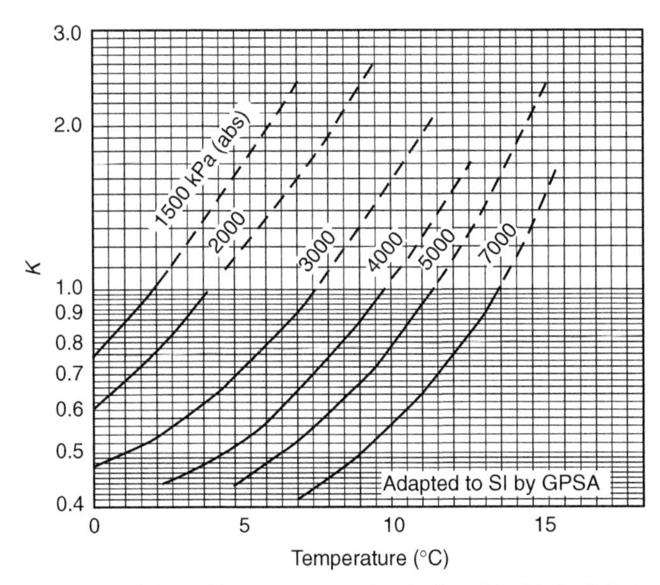

Figure 3.6a ***Vapor–Solid Equilibrium K-Factor for Carbon Dioxide in SI Units.*** *Reprinted from the* GPSA Engineering Data Book, *11th ed.—reproduced with permission.*

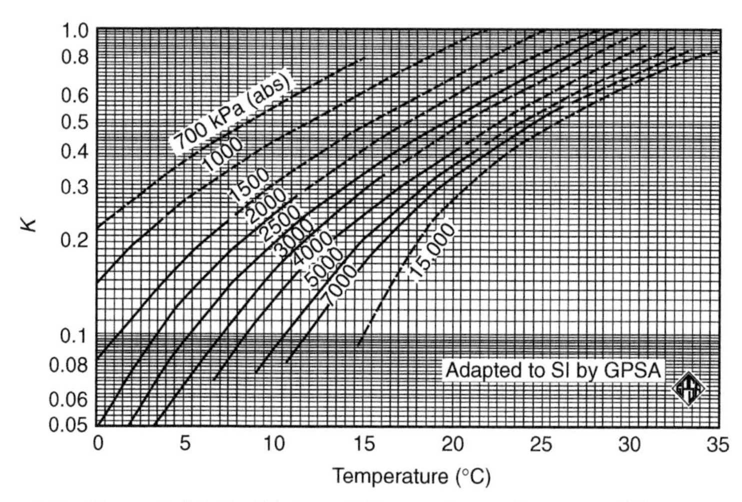

Figure 3.7a ***Vapor–Solid Equilibrium K-Factor for Hydrogen Sulfide in SI Units.*** *Reprinted from the* GPSA Engineering Data Book, *11th ed.—reproduced with permission.*

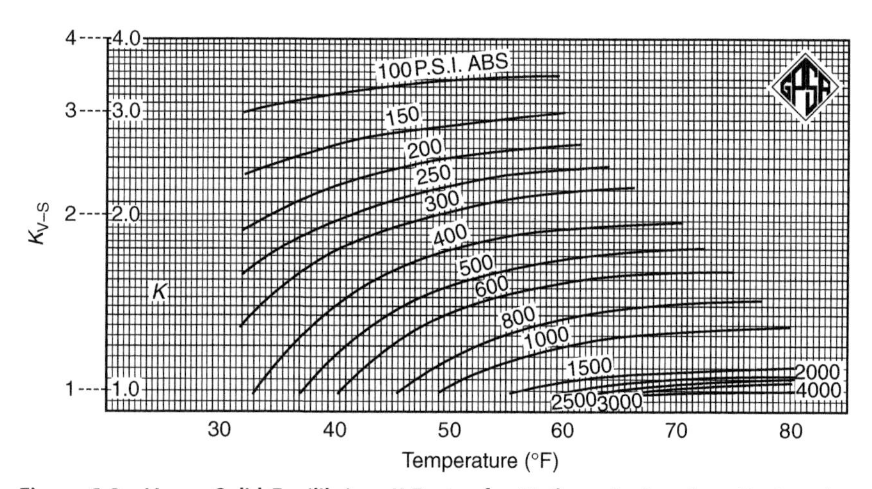

Figure 3.8a ***Vapor–Solid Equilibrium K-Factor for Methane in American Engineering Units.*** *Reprinted from the* GPSA Engineering Data Book, *11th ed.—reproduced with permission.*

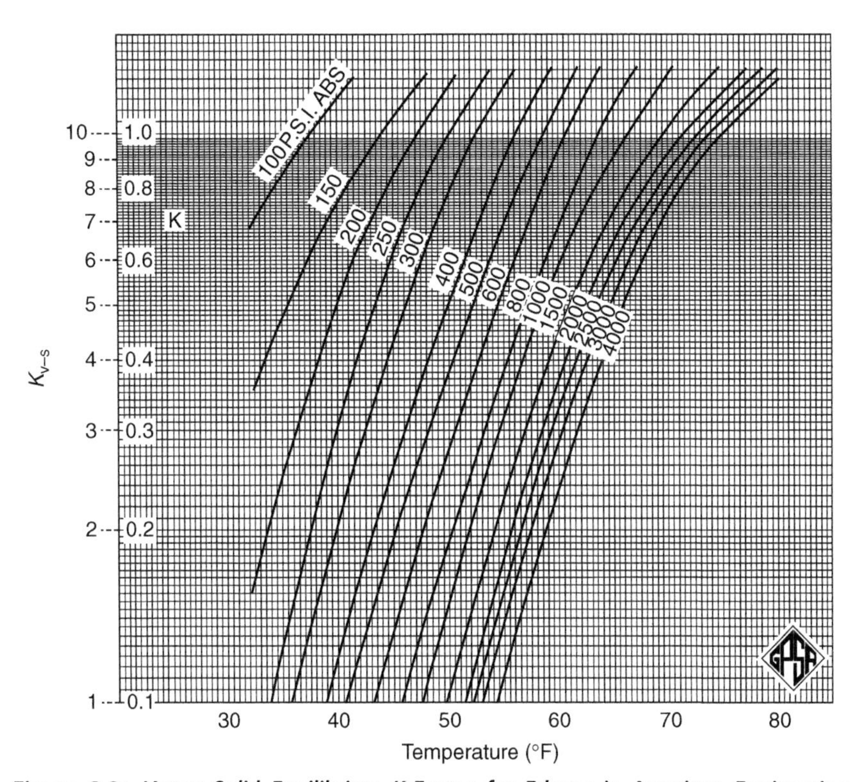

Figure 3.9a ***Vapor–Solid Equilibrium K-Factor for Ethane in American Engineering Units.*** *Reprinted from the* GPSA Engineering Data Book, *11th ed.—reproduced with permission.*

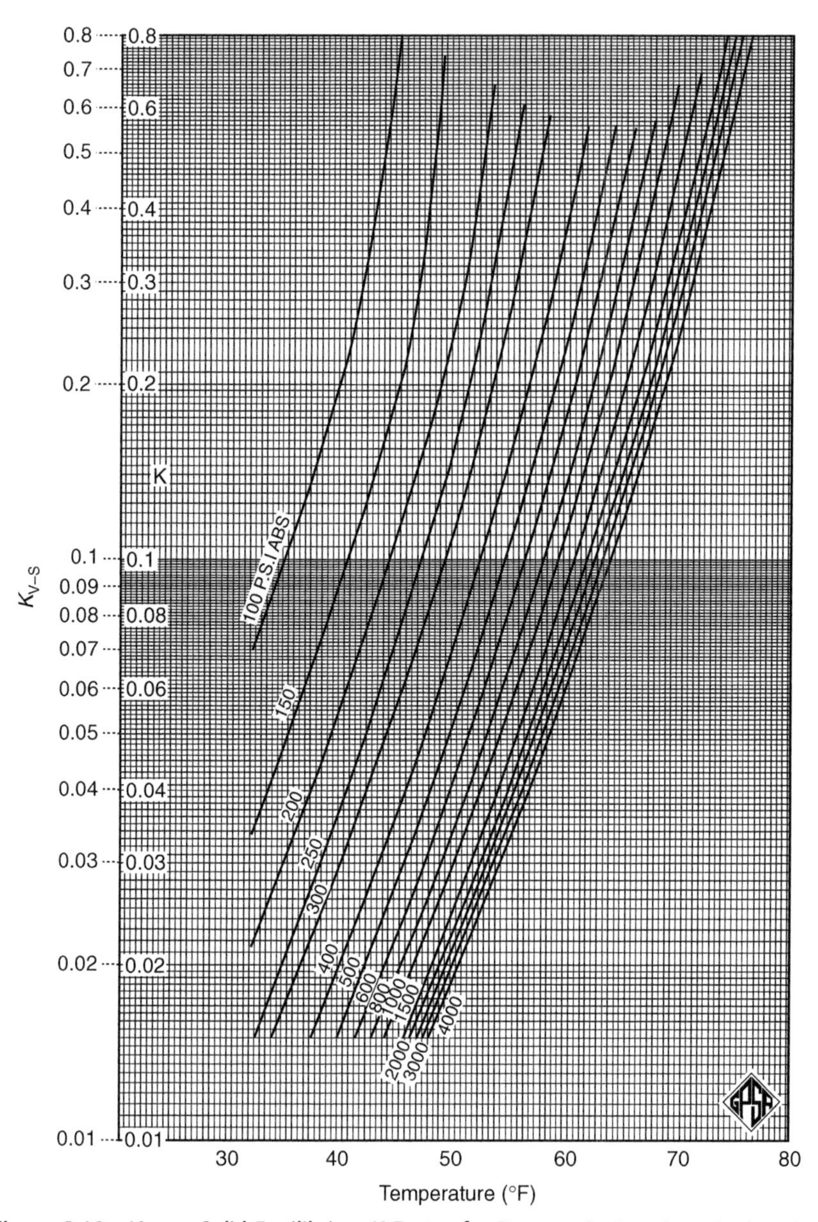

Figure 3.10a ***Vapor–Solid Equilibrium K-Factor for Propane in American Engineering Units.*** *Reprinted from the* GPSA Engineering Data Book, *11th ed.—reproduced with permission.*

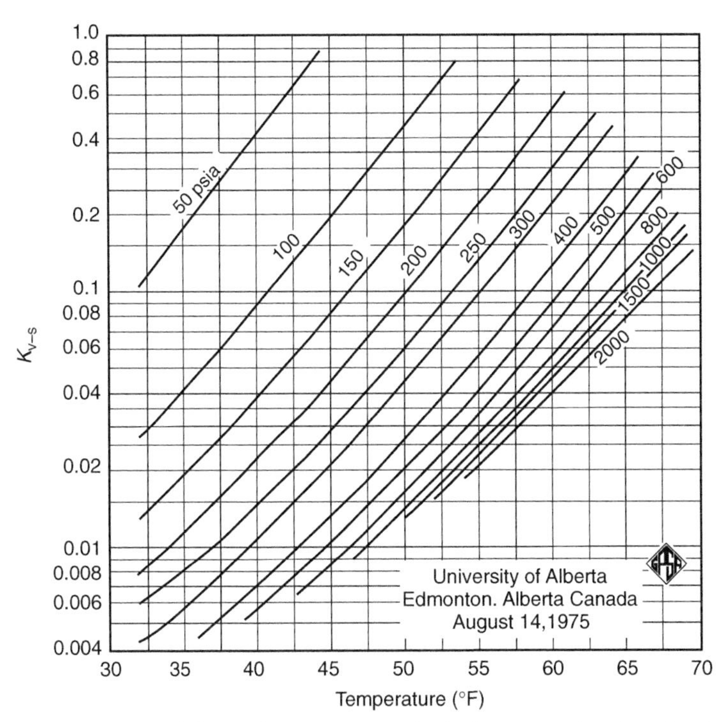

Figure 3.11a ***Vapor–Solid Equilibrium K-Factor for Isobutane in American Engineering Units.*** *Reprinted from the* GPSA Engineering Data Book, *11th ed.—reproduced with permission.*

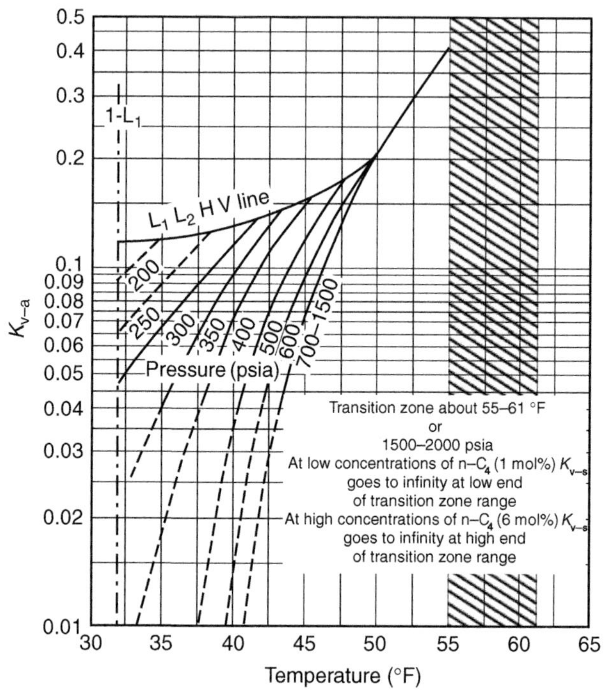

Figure 3.12a ***Vapor–Solid Equilibrium K-Factor for n-Butane in American Engineering Units.*** *Reprinted from the* GPSA Engineering Data Book, *11th ed.—reproduced with permission.*

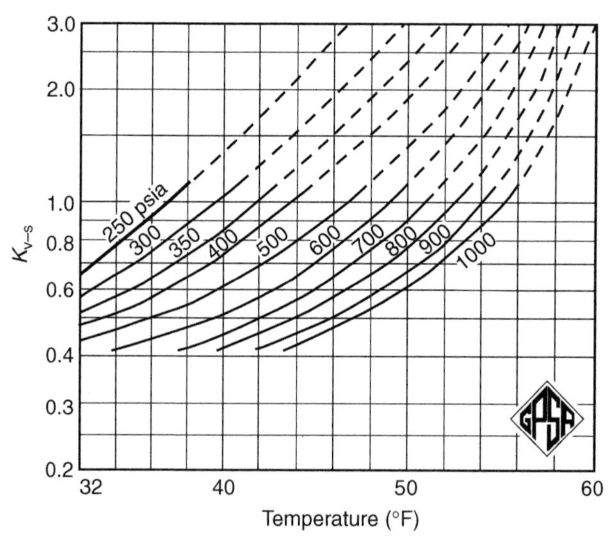

Figure 3.13a ***Vapor–Solid Equilibrium K-Factor for Carbon Dioxide in American Engineering Units.*** *Reprinted from the* GPSA Engineering Data Book, *11th ed.—reproduced with permission.*

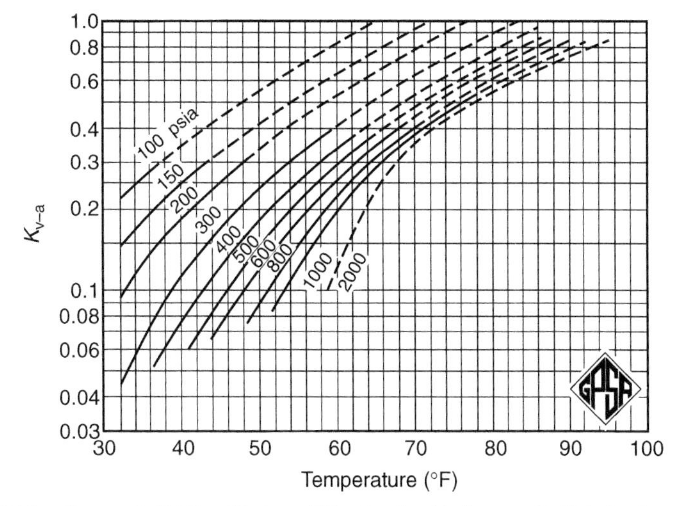

Figure 3.14a ***Vapor–Solid Equilibrium K-Factor for Hydrogen Sulfide in American Engineering Units.*** *Reprinted from the* GPSA Engineering Data Book, *11th ed.—reproduced with permission.*

REFERENCES

Baillie, C., Wichert, E., 1987. Chart gives hydrate formation temperature for natural gas. Oil Gas. J. 85 (4), 37–39.

Carroll, J.J., May 19–21, 2004. An Examination of the Prediction of Hydrate Formation Conditions in Sour Natural Gas. GPA Europe, Dublin, Ireland.

Carroll, J.J., Duan, J., 2002. Relational expression of the conditions forming hydrates of various components in natural Gas. Nat. Gas. Ind. 22 (2), 66–71 (in Chinese).

Carson, D.B., Katz, D.L., 1942. Natural gas hydrates. Trans. AIME 146, 150–158.

Elgibaly, A.A., Elkamel, A.M., 1998. A new correlation for predicting hydrate formation conditions for various Gas mixtures and inhibitors. Fluid Phase Equil. 152, 23–42.

Fan, S.-S., Guo, T.-M., 1998. GPSA Engineering Data Book, 11th ed. Gas Processors Suppliers Association, Tulsa, OK.

Fan, S.-S., Guo, T.-M., 1999. Hydrate formation of CO_2-rich binary and quaternary gas mixtures in aqueous sodium chloride solutions. J. Chem. Eng. Data 44, 829–832.

Kobayashi, R., Song, K.Y., Sloan, E.D., 1987. In: Bradley, H.B. (Ed.), Petroleum Engineers Handbook. SPE, Richardson, TX, pp. 25-1–25-28.

Makogon, Y.F., 1981. Hydrates of Hydrocarbons. PennWell, Tulsa, OK.

Mann, S.L., McClure, L.M., Poettmann, F.H., Sloan, E.D., March 13–14, 1989. Vapor–solid equilibrium ratios for structure I and structure II natural gas hydrates. In: Proc. 69th Annual Gas. Proc. Assoc. Conv., San Antonio, TX.

Mei, D.-H., Liao, J., Yang, J.-T., Guo, T.-M., 1998. Hydrate formation of a synthetic natural Gas mixture in aqueous solutions containing electrolyte, methanol, and (electrolyte + methanol). J. Chem. Eng. Data 43, 178–182.

Motiee, M., July 1991. Estimate possibility of hydrate. Hydro. Proc. 70 (7), 98–99.

Ng, H.-J., Robinson, D.B., 1976. The role of *n*-butane in hydrate formation. AIChE J. 22, 656–661.

Østergaard, K.K., Tohidi, B., Danesh, A., Todd, A.C., Burgass, R.W., 2000. A general correlation for predicting the hydrate-free zone of reservoir fluids. SPE Prod. Facil. 15, 228–233.

Otto, F.D., 1959. Hydrates of Methane–Propylene and Methane–Ethylene Mixtures (M.Sc. thesis). University of Alberta, Edmonton, AB.

Sloan, E.D., 1998. Clathrate Hydrates of Natural Gases, second ed. Marcel Dekker, New York, NY.

Sun, C.-Y., Chen, G.-J., Lin, W., Guo, T.-M., 2003. Hydrate formation conditions of sour natural gases. J. Chem. Eng. Data 48, 600–603.

Towler, B.F., Mokhatab, S., April 2005. Quickly estimate hydrate formation conditions in natural gases. Hydro. Proc., 61–62.

Wichert, E., 2004. Personal communication.

Wilcox, W.I., Carson, D.B., Katz, D.L., 1941. Natural gas hydrates. Ind. Eng. Chem. 33, 662–671.

Zhang, S.-X., Chen, G.-J., Ma, C.-F., Yang, L.-Y., Guo, T.-M., 2000. Hydrate formation of hydrogen + hydrocarbon gas mixtures. J. Chem. Eng. Data 45, 908–911.

CHAPTER 4

Computer Methods

The emergence of powerful desktop computers has made the design engineer's life significantly easier. No longer do engineers have to rely on the approximate hand calculation methods, such as those provided earlier for hydrate formation estimates. In addition, a wider spectrum of additional calculations is available to them. This is true for hydrate calculations in which a number of software packages are available.

However, engineers should not blindly trust these programs—it is the responsibility of the user to ensure the software selected is appropriate for the job and gives accurate results.

The bases of these computer programs are the rigorous thermodynamic models found in the literature. Three of the more popular ones will be briefly reviewed here. This is followed by a brief discussion of the use of some of the available software packages.

As was mentioned previously, one of the problems in the study of gas hydrates was the observation that they were nonstoichiometric. First, scientists had to come to terms with this as an observation. Any model for hydrate formation had to handle this somewhat unusual property. Next, the fact that there is more than one type of hydrate had to be addressed. Rigorous models would have to distinguish between the various types of hydrates.

4.1 PHASE EQUILIBRIUM

The criteria for phase equilibrium, established over 100 years ago by Gibbs, are that: (1) the temperature and pressure of the phases are equal, (2) the chemical potentials of each of the components in each of the phases are equal, and (3) the global Gibbs free energy is a minimum. These criteria apply to phase equilibrium involving hydrates and form the basis for the models for performing hydrate equilibrium calculations.

Most phase equilibrium calculations switch from chemical potentials to fugacities, but hydrate calculations are usually performed based on chemical potentials. In the calculation of hydrates, the free energy minimization is also important. The stable hydrate phase (type I, II, or even H) is the one that results in a minimum in the Gibbs free energy. Therefore, simply meeting the first two criteria is not sufficient to solve the hydrate problem.

Natural Gas Hydrates
ISBN 978-0-12-800074-8
http://dx.doi.org/10.1016/B978-0-12-800074-8.00004-1

From a thermodynamic point of view, the hydrate formation process can be modeled as taking place in two steps. The first step is from pure water to an empty hydrate cage. This first step is hypothetical, but is useful for calculation purposes. The second step is the filling of the hydrate lattice. The process is as follows:

Pure water(α) → empty hydrate lattice(β) → filled hydrate lattice(H)

The change in chemical potential for this process is given as:

$$\mu^{\mathrm{H}} - \mu^{\alpha} = \left(\mu^{\mathrm{H}} - \mu^{\beta}\right) + \left(\mu^{\beta} - \mu^{\alpha}\right) \quad (4.1)$$

where μ is the chemical potential and the superscripts refer to the various phases. The first term after the equal sign represents the stabilization of the hydrate lattice. It is the variation in the models used to estimate this term that separates the various models.

The second term represents a phase change for the water and can be calculated by regular thermodynamic means. This term is evaluated as follows:

$$\frac{\mu^{\beta} - \mu^{\alpha}}{RT} = \frac{\Delta\mu(T,P)}{RT} = \frac{\Delta\mu(T_{\mathrm{O}},P_{\mathrm{O}})}{RT_{\mathrm{O}}} - \int_{T_{\mathrm{O}}}^{T} \frac{\Delta H}{RT^2}\mathrm{d}T + \int_{P_{\mathrm{O}}}^{P} \frac{\Delta v}{R\overline{T}}\mathrm{d}P \quad (4.2)$$

where R is the universal gas constant, T is the absolute temperature, P is the pressure, H is the enthalpy, v is the molar volume, the subscript O represents a reference state, and the Δ terms represent the change from a pure water phase (either liquid or ice) to a hydrate phase (either type I or II). The bar over the temperature in the last term in Eqn (4.2) indicates that this is an average temperature. The various properties required for this calculation have been tabulated and are available in the literature (see Pedersen et al., 1989 for example).

This term is virtually the same regardless of the model used. Subtle changes are made to account for other changes made to the model, but, theoretically, the same equation and set of parameters should apply regardless of the remainder of the model.

4.2 VAN DER WAALS AND PLATTEEUW

The first model for calculating hydrate formation was that of van der Waals and Platteeuw (1959). They postulated a statistical model for hydrate formation. The concentration of the non-water species in the hydrate was treated in a manner similar to the adsorption of a gas into a solid. For a single guest molecule, this term is evaluated as follows:

$$\mu^{\mathrm{H}} - \mu^{\beta} = RT \sum_i \nu_i \ln (1 - Y_i) \tag{4.3}$$

where ν_i is the number of cavities of type I and Y is a probability function. The Y is the probability that a cavity of type I is occupied by a guest molecule and is given by:

$$Y_i = \frac{c_i P}{1 + c_i P} \tag{4.4}$$

The c_i in this equation is a function of the guest molecule and the cage occupied and P is the pressure. Although it is not obvious from this discussion, the c_is are also functions of the temperature.

A simple example—because ethane only occupies the small cages of a type I hydrate, then c_i for ethane for the small cages is zero. On the other hand, methane, which is also a type I former, occupies both the large and small cages. Therefore, both the c_is for this component are non-zero.

4.3 PARRISH AND PRAUSNITZ

The approach of the original van der Waals and Platteeuw (1959) method provided a good basis for performing hydrate calculations, but it was not sufficiently accurate for engineering calculations. One of the first models with the rigor required for engineering calculations was that of Parrish and Prausnitz (1972).

There are two major differences between the original van der Waals and Platteeuw (1959) model and that proposed by Parrish and Prausnitz (1972). First, they extended the model to mixtures of hydrate formers. This is done as follows:

$$\mu^{\mathrm{H}} - \mu^{\beta} = RT \sum_i \nu_i \ln \left(1 - \sum_K Y_{Ki} \right) \tag{4.5}$$

where the second sum is over all components. The probability function for a component becomes:

$$Y_{Ki} = \frac{c_i P_K}{1 + \sum_j c_{ij} P_j} \tag{4.6}$$

Here, the summation is also over the number of components and the P followed by a subscript is the partial pressure for a given component. The other components are included in this term because they are competing to occupy the same cages. Thus, the presence of another guest molecule reduces the probability that a given guest can enter the hydrate lattice.

Second, Parrish and Prausnitz (1972) replaced the partial pressure in Eqn (4.6) with the fugacity. There is no simple definition for the thermodynamic concept of fugacity. Usual definitions given in thermodynamics textbooks rely on the chemical potential, which is an equally abstract quantity. For our purposes, we can consider the fugacity as a "corrected" pressure, which accounts for nonidealities. Substituting the fugacity into Eqn (4.6) results in:

$$Y_{Ki} = \frac{c_i \widehat{f}_{\mathrm{I}}}{1 + \sum_j c_{ij} \widehat{f}_j} \tag{4.7}$$

where $\widehat{f}_{\mathrm{I}}$ is the fugacity of component I in the gaseous mixture. This allowed their model to account for nonidealities in the gas phase and thus to extend the model to higher pressures. In addition, some of the parameters in the model were adjusted to reflect the change from pressures to fugacities and to improve the overall fit of the model. That is, a different set of c_is is required for the fugacity model than for the pressure model.

It is interesting to note that at the time that Parrish and Prausnitz (1972) first presented their model, the role of *n*-butane in hydrate formation was not fully understood. Although they give parameters for many components (including most of the components important to the natural gas industry), they did not give parameters for *n*-butane. Later modifications of the Parrish and Prausnitz method correctly included *n*-butane.

4.4 NG AND ROBINSON

The next major advance was the model of Ng and Robinson (1977). Their model could be used to calculate the hydrate formation in equilibria with a hydrocarbon liquid.

First, this required an evaluation of the change in enthalpy and change in volume in Eqn (4.2), or at least an equivalent version of this equation.

In the model of Ng and Robinson (1977), the fugacities were calculated using the equation of state of Peng and Robinson (1976). This equation of state is applicable to both gases and the non-aqueous liquid. Again, small adjustments were made to the parameters in the model to reflect the switch to the Peng–Robinson equation.

Similarly, the Soave (1972) or any other equation of state applicable to both the gas and liquid could be used. However, the Soave and Peng–Robinson equations (or modifications of them) have become the workhorses of this industry.

It is important to note that later versions of the Parrish and Prausnitz method were also designed to be applicable to systems containing liquid formers.

4.5 CALCULATIONS

Now that one has these equations, how does the calculation proceed? For now, we will only consider the conditions for incipient solid formation. For example, given the temperature, at what pressure will a hydrate form for a given mixture?

First, you perform the calculations assuming the type of hydrate formed. Use the equations outlined above to calculate the free energy change for this process. This is an iterative procedure that continues until the following is satisfied:

$$\mu^{H} - \mu^{\alpha} = 0$$

Remember, at equilibrium, the chemical potentials of the two phases must be equal. For a pure component in which the type of the hydrate is known, this is when the calculation ends (provided you selected the correct hydrate to begin with).

Next, repeat the calculation for the other type of hydrate at the given temperature and the pressure calculated above. If the result of this calculation is:

$$\mu^{H} - \mu^{\alpha} > 0$$

then the type of hydrate assumed initially is the stable hydrate and the calculation ends. If the difference in the chemical potentials is less than zero, then the hydrate type assumed to begin the calculation is unstable. Thus, the iterative procedure is repeated assuming the other type of hydrate. Once this is solved, the calculation is complete.

This represents only one type of calculation, but it is obvious from even this simple example that a computer is required.

4.5.1 Compositions

From the development of the model, as discussed above, it appears as though the bases for obtaining the model parameters are experimentally measured compositions. However, accurate and direct measurements of the composition are rare. The compositions are usually approximated from pressure–temperature data. In reality, the model parameters are obtained by fitting the pressure–temperature loci and deducing the compositions.

However, using this type of model allows us to estimate the composition of hydrates. Again, this may seem obvious, but the parameters were obtained by fitting pressure–temperature data and not by fitting composition data, which would seem more logical.

Figure 4.1 shows the ratio of water to hydrogen sulfide in the hydrate at a temperature of 0 °C. The ratio is always greater than the theoretical limit 5¾ (see Chapter 2). Because this ratio is always greater than the theoretical limit, this means that there are more water molecules per H_2S molecules than the value that results if all of the cages are occupied. Physically, this translates to there being unoccupied cages.

At approximately 1 MPa, the hydrogen sulfide liquefies and the equilibrium changes from one between a liquid, a vapor, and a hydrate to two liquids and a hydrate.

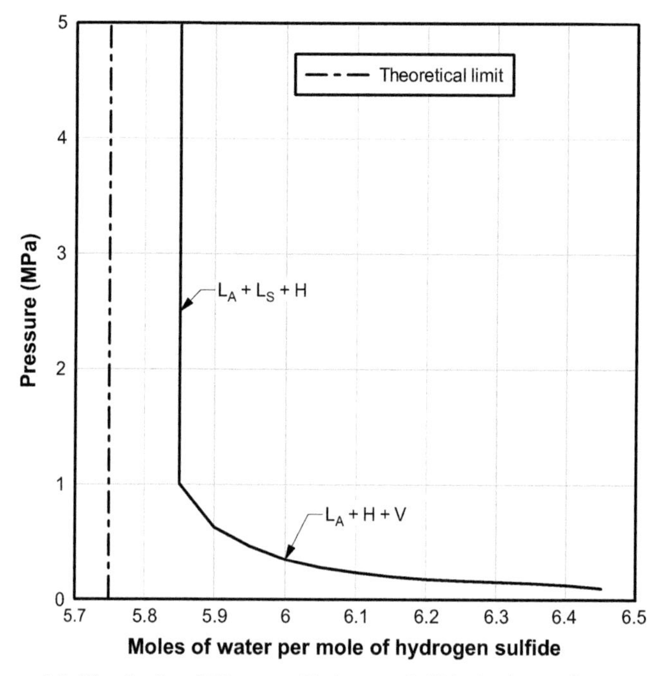

Figure 4.1 ***The Ratio of Water to Hydrogen Sulfide in the Hydrate at 0 °C.***

4.6 COMMERCIAL SOFTWARE PACKAGES

There are several software packages available that are dedicated to hydrate calculations. These include *EQUI-PHASE Hydrate* from Schlumberger in Canada, *PVTSim* from Calsep in Denmark, and *Multiflash* from Infochem (now KBC Advanced Technologies) in the United Kingdom. In addition, the packages *CSMHYD* and *CSMGEM* are available from the Colorado School of Mines in Golden, Colorado. Unlike most commercial hydrate software, a thorough description of *CSMGEM* is available in the literature (Ballard and Sloan, 2004).

Most of the popular, general-purpose process simulation programs include the capability to predict hydrate formation. Often, this includes warnings about streams where hydrate formation is possible. These include *Promax* (*Prosim* is a previous generation process simulator) from Bryan Research & Engineering (Bryan, TX), *Hysys* and *Aspen* from Aspen Technology (Cambridge, MA), *VMGSim* from Virtual Materials Group (Calgary, Canada), and others.

4.7 THE ACCURACY OF THESE PROGRAMS

Earlier, we compared the hand calculations against two models. We shall do the same here. The software programs examined will be: (1) CSMHYD (released August 5, 1996), (2) EQUI-PHASE Hydrate (v. 4.0), (3) Prosim (v. 98.2), and (4) Hysys (v. 3.2, Build 5029). Although these may not be the most recent versions of these software packages, the results obtained are still typical of what can be expected.

The comments presented in this section should not be interpreted as either an endorsement or as a criticism of the software. The predictions are presented and the potential ramifications are discussed. If the reader has access to a different software package, they are invited to make these comparisons for themselves. All of the software packages have their strengths and their weaknesses. General conclusions based on a small set of data (or worse, no data!) are usually not warranted.

4.7.1 Pure Components

Perhaps the simplest test of a correlation is its ability to predict the pure component properties. One would expect that the developers of these software packages used the same data to generate their parameters as were used to generate the correlations. Large deviations from the pure component loci would raise some questions about the accuracy of the predictions for mixtures.

The following pure components will be discussed and several plots will be presented. The points on these plots are the same as those on Figs 3.4 and 3.5 and are from the correlations presented in Chapter 2. They are not experimental data and should not be interpreted as such. The curves on the plots are generated using the above-mentioned software packages.

One thing that should be kept in mind when reviewing the plots in this section is that predicting the hydrate temperature less than the actual temperature could lead to problems. Potentially, it means that you predict that hydrates will not form, but, in reality, they do.

4.7.1.1 Methane

Figure 4.2 shows the hydrate locus for pure methane. Throughout the range of pressure shown on this plot, all three software packages are of acceptable error. Only at extreme pressures do the errors exceed 2 °C.

First, consider pressures below 10 MPa (1450 psia), which is a reasonable pressure limit for the transportation and processing of natural gas. However,

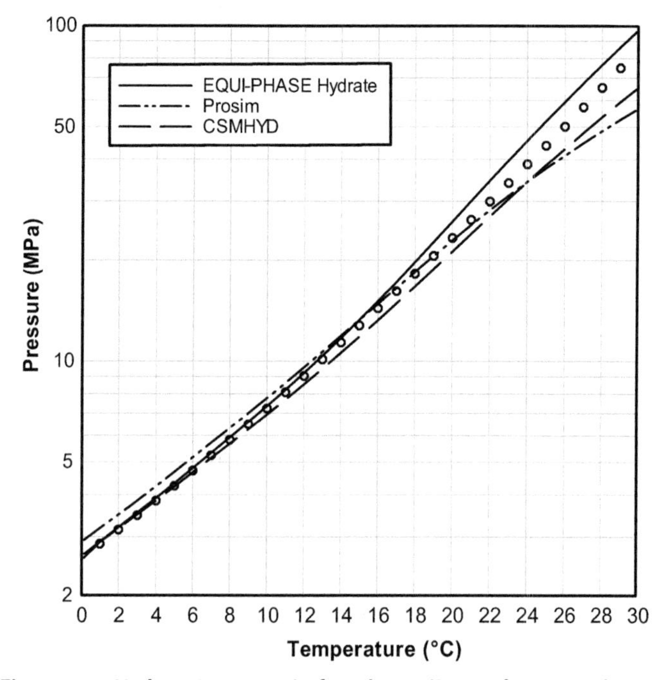

Figure 4.2 ***Hydrate Locus Loci of Methane.*** (Points from correlation.)

it is not sufficient for the production of gas. The high-pressure region will be discussed later in this section.

EQUI-PHASE Hydrate is an accurate prediction of the correlation in the low-pressure region, with errors much less than 1 °C. At the lowest pressures, CSMHYD is also a very good fit. However, as the pressure increases, the deviations become larger. Once the pressure reaches 10 MPa the CSMHYD predicts hydrate temperatures that are about 1 °C too high. Throughout this region, Prosim consistently underpredicts the hydrate temperature by about 1 °C.

At pressures greater than 10 MPa, none of the three software packages is highly accurate. EQUI-PHASE Hydrate predicts a hydrate temperature that is consistently less than the correlation. At extreme pressures, the error is as much as 1 °C. On the other hand, both Prosim and CSMHYD predict that the hydrate forms at higher temperatures than the correlation. At very high pressure, the errors from Prosim become quite large. For example, at 50 MPa (7250 psia), the difference is larger than 2 °C. With CSMHYD, for pressure up to 50 MPa, the errors are less than 2 °C. However, as the pressure continues to increase, so does the observed error.

4.7.1.2 Ethane

Figure 4.3 shows the hydrate locus for pure ethane. It is clear from Fig. 4.3 that this locus is different from that of methane. First, ethane tends to form a hydrate at a lower pressure than methane. More significantly, at approximately 3 MPa (435 psia) the curves show a transition from the $L_A + H + V$ region (L_A = aqueous liquid, H = hydrate, and V = vapor) to the $L_A + L_H + H$ (L_H = ethane-rich liquid). Therefore, we should examine the two regions separately. Experimental data for the ethane hydrate locus do not exist for pressure greater than about 20 MPa, and, thus, the discussion is limited to this pressure.

EQUI-PHASE Hydrate almost exactly reproduces the $L_A + H + V$ locus from the correlation. CSMHYD overpredicts the hydrate temperature, but only slightly. Errors are less than one-half of a Celsius degree. On the other hand, Prosim underpredicts the hydrate temperature, but, again, the error is very small, and also are less than one-half of a Celsius degree.

Errors are slightly larger for the region $L_A + L_H + H$. EQUI-PHASE Hydrate predicts that $L_A + L_H + H$ locus is very steep, steeper than the correlation indicates. Thus, as the pressure increases, so do the errors from EQUI-PHASE Hydrate. At 20 MPa, the error is about 3 °C. Both Prosim

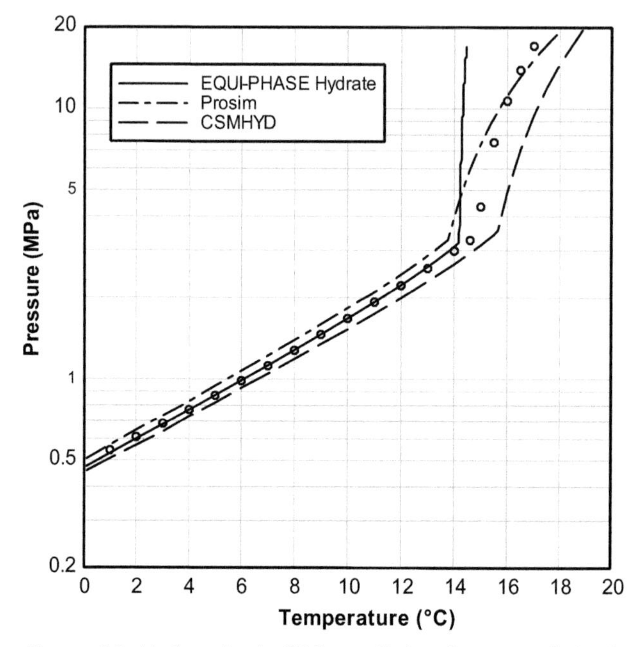

Figure 4.3 ***Hydrate Loci of Ethane.*** (Points from correlation.)

and CSMHYD better reflect the curvature of the $L_A + L_H + H$ shown by the correlation. The error from Prosim is less than 1 °C for this region. *CSMHYD* is slightly poorer with errors slightly larger than 1 °C.

4.7.1.3 Carbon Dioxide

Figure 4.4 shows the hydrate locus for carbon dioxide. The hydrate curve for CO_2 is similar to that for ethane. As with ethane, we will examine two regions: (1) the $L_A + H + V$, pressures less than about 4.2 MPa (610 psia), and (2) $L_A + L_C + H$ ($L_C = CO_2$-rich liquid) for pressures greater than 4.2 MPa.

For the $L_A + H + V$ region, both EQUI-PHASE Hydrate and *CSMHYD* accurately reproduce the correlation values. For these two packages, the errors are a small fraction of a Celsius degree. Prosim consistently underpredicts the hydrate temperature. The average deviation is about 1 °C.

In the $L_A + L_C + H$ region, all three packages are good predictions of the correlation data. However, in each case, the software packages over-predict the hydrate temperature. EQUI-PHASE Hydrate predicts the

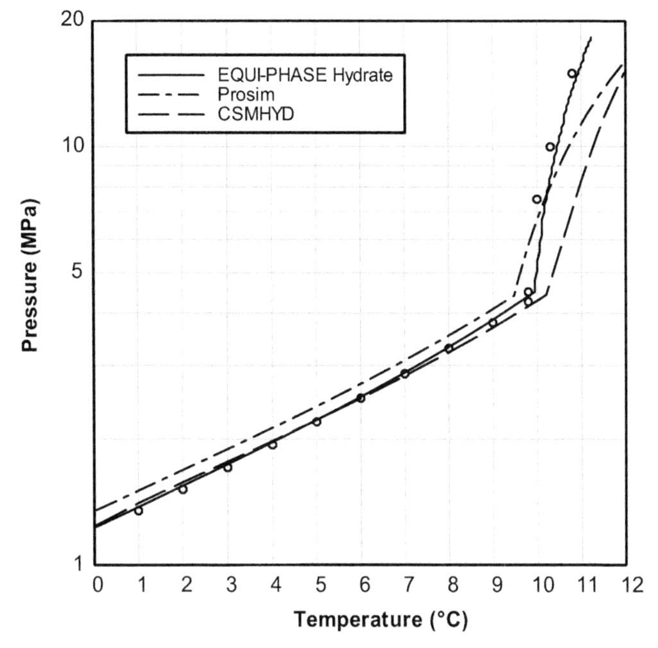

Figure 4.4 ***Hydrate Loci for Carbon Dioxide.*** (Points from correlation.)

correlation data to better than one half of a Celsius degree. The other two packages have a maximum error of about 1 °C.

4.7.1.4 Hydrogen Sulfide

The last of the pure components examined in this chapter will be hydrogen sulfide. Figure 4.5 presents the hydrate locus for H_2S. Again, this is similar to those for ethane and CO_2 inasmuch as they show the two regions: $L_A + H + V$ and $L_A + L_S + H$ (L_S is the hydrogen sulfide-rich liquid phase). As has been stated earlier, one of the important things about hydrogen sulfide is that it forms a hydrate at such low pressure and extends to high pressure.

In the $L_A + H + V$ region, both EQUI-PHASE Hydrate and CSMHYD accurately predict the hydrate formation region. The errors are less than 1 °C and, in the case of EQUI-PHASE Hydrate, they are much less. Prosim consistently predicts a lower hydrate temperature than the correlation data. The errors are slightly larger than 1 °C.

For the $L_A + L_S + H$ region, both EQUI-PHASE Hydrate and CSMHYD accurately reproduce the correlation data. Again, Prosim underpredicts the hydrate formation temperature, typically by about 2 °C.

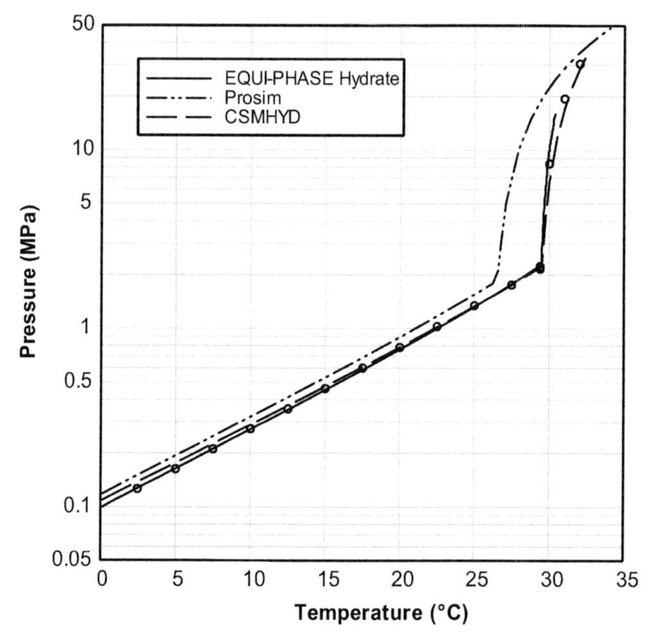

Figure 4.5 ***Hydrate Loci for Hydrogen Sulfide.*** (Points from correlation.)

4.7.2 Mixtures

In the previous chapter, several mixtures were examined. These same mixtures will be used here. The reader should cross-reference this section with the equivalent section in the previous chapter.

4.7.2.1 Data of Mei et al.

The data of Mei et al. (1998) were used in the previous chapter for comparison with the hand calculation methods. Figure 4.6 shows the data for Mei et al. (1998) and the prediction from the three software packages.

All three of the software packages predict that the hydrate formation temperature is less than the experimental data. The maximum error for the EQUI-PHASE Hydrate is slightly less than 1.8 °C. CSMHYD has a maximum error slightly larger than 2.1 °C and, finally, Prosim has a maximum error slightly larger than 3.0 °C.

Again, it is surprising that the errors from the models are as large as they are considering that the system is a very simple mixture. In fact, the K-factor method presented in Chapter 3, which is claimed to be simple and known to be not highly accurate, is as good as or better than the software packages.

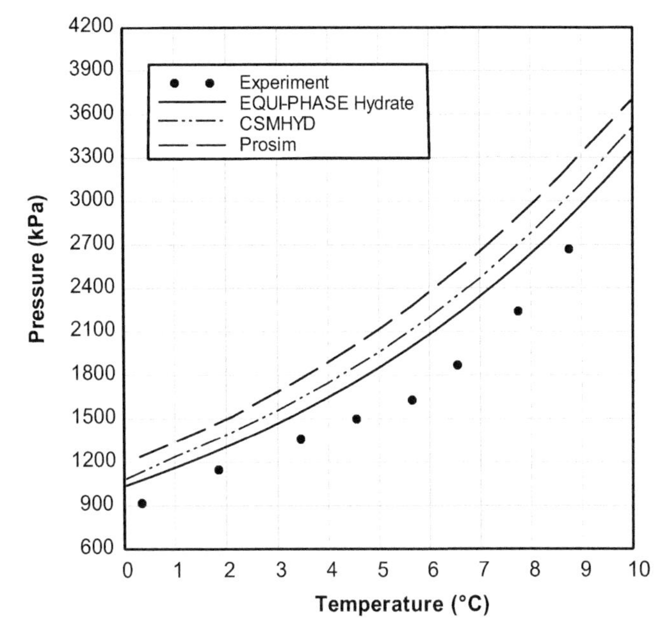

Figure 4.6 ***Hydrate Locus for a Synthetic Natural Gas Mixture (CH_4 97.25 mol%, C_2H_6 1.42%, C_3H_8 1.08%, i-C_4H_{10} 0.25%).***

4.7.2.2 Data of Fan and Guo

The data of Fan and Guo (1999) for two mixtures rich in carbon dioxide were introduced in the previous chapter.

Figure 4.7 shows the experimental data for the mixture of CO_2 (96.52 mol%) and methane (3.48%) and the predictions from the three software packages.

At low pressures (less than about 3000 kPa), all the models predict that the hydrate temperature is less that the experimental data. However, the difference is less than 1 °C for CSMHYD and EQUI-PHASE Hydrate and slightly more for Prosim. For the single point at higher pressure (approximately 5000 kPa), all three models predict a higher hydrate temperature than the experimental data.

Figure 4.8 shows the hydrate loci for the second mixture from Fan and Guo (1999). For the range of temperature shown, both CSMHYD and EQUI-PHASE Hydrate are excellent predictions of the data. However, EQUI-PHASE Hydrate had some trouble with this mixture. It was unable to calculate the hydrate locus over the entire range of temperature. Furthermore, it was unable to perform the point-by-point calculations for

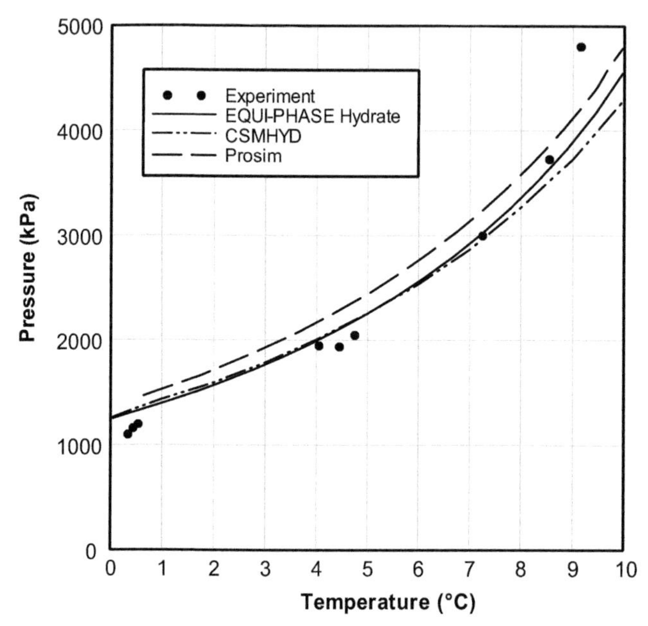

Figure 4.7 *Hydrate Locus for a Mixture of Carbon Dioxide (96.52 mol%) and Methane (3.48 mol%).*

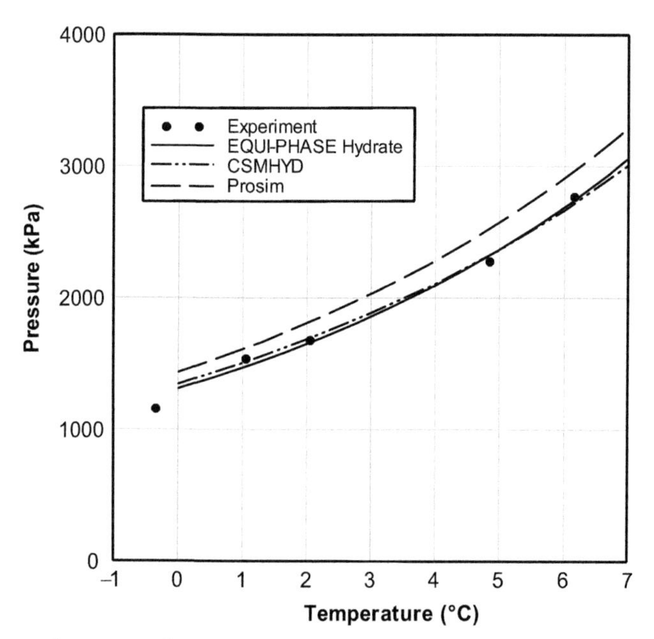

Figure 4.8 *Hydrate Locus for a Quaternary Mixture 88.53 mol% CO_2, 6.83% CH_4, 0.38% C_2H_6, and 4.26% N_2.*

this mixture. Prosim consistently predicts that the hydrate temperature is lower that the experimental data. Typically, the errors are less than 1 °C.

4.7.2.3 Data of Ng and Robinson

An important study is that of Ng and Robinson (1976, 1977). Because these data are so significant, it is highly likely that they were used in the development of the computer models. In the previous chapter, four mixtures of methane + *n*-butane were examined. Here, only a single mixture (CH_4 [96.09 mol%] + *n*-C_4H_{10} [3.91%]) will be presented. The main reason for this is clarity. In addition, as was mentioned, these data were almost certainly used to develop the model parameters and, thus, it should come as no surprise that these models accurately predict these data.

Figure 4.9 shows the hydrate curves for this mixture. Both CSMHYD and EQUI-PHASE Hydrate are excellent predictions of the data with errors less than 1 °C. Prosim consistently predicts that the hydrate temperature is less than the experimental data with errors larger than 2 °C.

Also, note that these calculated loci do not show the unusual behavior that the K-factor method did (Fig. 3.13). This unusual behavior was an artifact of the K-factor method.

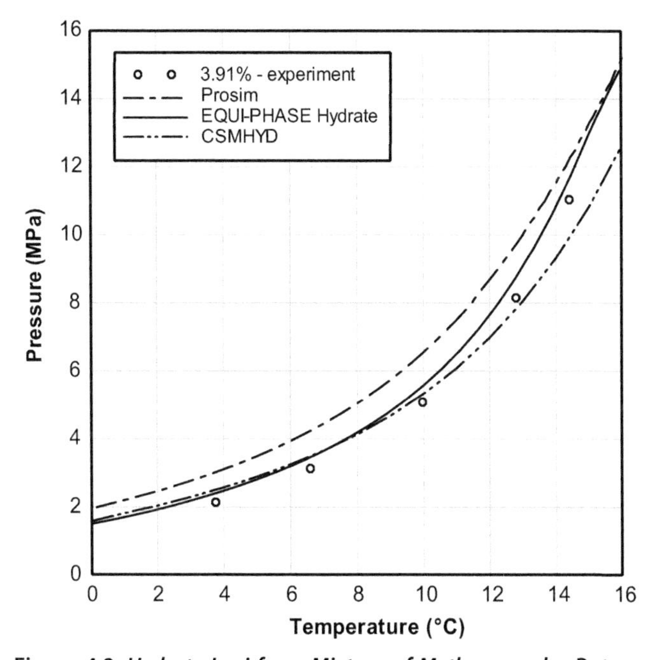

Figure 4.9 ***Hydrate Loci for a Mixture of Methane and n-Butane.***

4.7.2.4 Data of Wilcox et al

The data from Wilcox et al. (1941) were introduced in Chapter 3. These data are for sweet gas mixtures and the pressures range from 1.2 to 27.5 MPa (175–3963 psia), which corresponds to a temperature range of 3.6–25.2 °C (38–77 °F). There are three mixtures, therefore, the reader is referred to the previous chapter for more information about these mixtures.

The data were compared to predictions from CSMHYD, EQUI-PHASE Hydrate, and Hysys. The table shows the average error (AE) in the predicted temperature (simply the sum of the differences between the predicted value and the experimental value) and the average absolute error (AAE), the average of the absolute value of the temperature differences. These values are summarized in Table 4.1. The upper part of the table is for Celsius temperatures and the lower for Fahrenheit.

For all three packages, the maximum error in the predicted hydrate temperature is less than 2 °C (less than 3.5 °F).

This is only one set of data and it is difficult to draw general conclusions, but, for this set of data, CSMHYD appears to be the best. For CSMHYD, the AE is only +0.1 and the AAE is 0.81, both in Celsius degrees.

4.7.3 Sour Gas

From Chapter 2, we can see that of the components commonly found in natural gas, hydrogen sulfide forms a hydrate at the lowest pressure and it

Table 4.1 Comparison of Three Software Packages for Three Mixtures from Wilcox et al. (1941)

	CSMHYD		EQUI-PHASE Hydrate		Hysys	
	AE (°C)	AAE (°C)	AE (°C)	AAE (°C)	AE (°C)	AAE (°C)
Gas B	+0.24	0.48	+0.14	0.24	+0.12	0.23
Gas C	+0.26	0.47	+1.58	1.58	+1.57	1.57
Gas D	−0.34	0.40	−0.48	0.48	−0.49	0.49
Overall	+0.05	0.45	+0.53	0.88	+0.52	0.88
Maximum		1.00		1.93		1.92
	AE (°F)	AAE (°F)	AE (°F)	AAE (°F)	AE (°F)	AAE (°F)
Gas B	+0.43	0.86	+0.25	0.42	+0.21	0.42
Gas C	+0.46	0.84	+2.84	2.84	+2.83	2.83
Gas D	−0.61	0.72	−0.86	0.86	−0.89	0.89
Overall	+0.10	0.81	+0.96	1.58	+0.94	1.58
Maximum		1.80		3.47		3.45

AAE, average absolute error; AE, absolute error.

persists to the highest temperatures. It was also shown that mixtures of hydrogen sulfide and propane exhibit hydrae azeoptropy, thus, sour gas mixtures, natural gas containing H_2S, are an interesting class of mixtures.

Carroll (2004) did a thorough study of the hydrate formation in sour gas mixtures. A review of the literature established a database of approximately 125 points. The database was made from three studies of sour gas mixtures: Noaker and Katz (1954), Robinson and Hutton (1967), and Sun et al. (2003). The maximum H_2S concentration in the study of Noaker and Katz (1954) was 22 mol%. In their study, the temperature ranged from 38 to 66 °F (3.3–18.9 °C) and the pressure from 150 to 985 psia (1030–6800 kPa). Robinson and Hutton (1967) studied hydrates in ternary mixtures of methane, hydrogen sulfide, and carbon dioxide over a wide range of pressures (up to 2300 psia or 15,900 kPa) and temperatures (up to 76 °F or 24.4 °C). The hydrogen sulfide content of the gases in the study of Robinson and Hutton (1967) ranged from 5% to 15% and the carbon dioxide from 12% to 22%. Sun et al. (2003) also measured the hydrate conditions for the ternary mixture of CH_4, CO_2, and H_2S. This set covered a wide range of compositions (CO_2 about 7 mol% and H_2S from 5 to 27 mol%) for pressures up to 1260 psia (8700 kPa) and temperatures up to 80 °F (26.7 °C).

The AEs for both CSMHYD and EQUI-PHASE Hydrate are about 1.5 °F (0.8 °C). Typically, these methods are able to predict the hydrate temperature to within 3 °F (1.7 °C) 90% of the time. The hydrate prediction routine in Hysys, a general-purpose process simulator, was also quite accurate with an AE of 1.5 °F (0.8 °C). Hysys is able to predict the hydrate temperature to within 3 °F (1.7 °C) more than 90% of the time. On the other hand, Prosim, another general-purpose simulator program, was not as accurate. The AE for Prosim was about 2.3 °F (1.3 °C). It was able to predict the hydrate temperature to within 3 °F (1.7 °C) only about 65% of the time.

Although the averages noted above give an overall impression of the accuracy of these methods, the maximum errors reveal the potential for significantly larger errors. Even the computer methods have larger maximum errors, 6.0 °F (3.3 °C) for EQUI-PHASE Hydrate, 7.4 °F (4.1 °C) for CSMHYD, and 6.0 °F (3.3 °C) for Hysys, and 8.0 °F (4.4 °C) for Prosim.

Table 4.2 summarizes the results of this study, not only for the computer methods that are the subject of this chapter but also for some of the hand calculations presented in the previous chapter. The AE allows for positive

Table 4.2 Errors in Predicting the Hydrate Temperatures for Sour Gas Mixtures from the Overall Data Set

	Number of Points	Average Deviation (°F)	Average Absolute Deviation (°F)	Maximum Deviation (°F)	% Deviation Larger than 3 °F	% Deviation Larger than 5 °F
K-factor	123	+2.3	2.7	10.9	40	16
Baillie-Wichert	99	−0.6	2.0	5.8	19	3
Mann et al.	123	+0.5	1.5	7.0	9	3
CSMHYD	123	+0.7	1.5	7.4	11	5
EQUI-PHASE Hydrate	124	−0.1	1.5	6.0	7	2
Hysys	125	−0.1	1.5	6.0	8	2
Prosim	124	+2.2	2.3	8.0	36	7

errors and negative errors can cancel, whereas with the absolute value all errors are positive. The last two columns indicate the number of predictions with errors greater than 3 and 5 °F. For example, CSMHYD predicts the experimental temperatures to within 3 °F 89% of the time (11% have deviations greater than 3 °F).

4.7.4 Third Party Studies

In his development of CSMGEM, Ballard (2002) did comparisons among the new software and four other software packages: CSMHYD, *DBRHydrate* (v. 5), Multiflash, and *PVTsim*. These results are repeated here without verification. The database that Ballard used consisted of more than 1500 data points. The purpose of this section is to present results that do not reflect any bias from this author.

The errors are the difference between the experimentally measured temperature and the predicted temperature for a given pressure for the specified mixture. These differences are averaged to get the values given in the tables that follow. Overall, the error for CSMGEM was 0.40 °C and 0.66, 0.64, 0.54, and 0.54 °C for CSMHYD, DBRHydrate, Multiflash, and PVTSim, respectively. These are equal to 0.72, 1.19, 1.15, 0.97, and 0.97 °F for CSMGEM, CSMHYD, DBRHydrate, Multiflash, and PVTSim, respectively.

Table 4.3 gives the errors in the temperatures for nine pure components. Tables 4.4, 4.5, and 4.6 give the errors for binary mixtures with methane, ethane, and propane. Note these are AEs and, thus, negative values will cancel positive ones so, overall, the values look relatively small.

Table 4.3 Average Errors (Celsius Degrees) for the Hydrate Predictions from Five Software Packages for Pure Components

	CSMGEM	CSMHYD	DBRHydrate	Multiflash	PVTSim
Methane	−0.01	0.68	−0.45	0.09	−0.30
Ethylene	−0.15	0.25	−0.11	−0.04	–
Ethane	0.06	0.69	0.05	0.41	−0.32
Nitrogen	0.04	0.56	0.05	0.25	−0.09
H_2S	0.73	0.40	0.47	0.85	0.70
CO_2	−0.17	0.11	0.00	−0.68	0.04
Propylene	−0.04	–	−0.07	0.27	–
Propane	0.17	0.10	−0.01	−0.25	−0.01
Isobutane	0.14	–	−0.09	0.19	0.45

Based on Ballard (2002).

Table 4.4 Average Errors (Celsius Degrees) for the Hydrate Predictions from Five Software Packages for Binary Mixtures Containing Methane

	CSMGEM	CSMHYD	DBRHydrate	Multiflash	PVTSim
Ethane	0.21	0.46	−0.12	0.55	−0.19
Nitrogen	−0.44	−0.55	0.00	−0.35	0.05
H_2S	0.60	0.14	0.68	0.38	0.68
CO_2	0.10	0.14	0.16	−0.35	−0.15
Propane	0.29	0.61	−0.40	−0.15	−0.01
Isobutane	−0.50	−0.52	−0.40	−1.10	−0.62
n-Butane	0.08	0.12	−0.30	0.18	0.26

Based on Ballard (2002).

Table 4.5 Average Errors (Celsius Degrees) for the Hydrate Predictions from Five Software Packages for Binary Mixtures Containing Ethane

	CSMGEM	CSMHYD	DBRHydrate	Multiflash	PVTSim
CO_2	−0.22	0.65	0.18	0.30	0.10
Propane	−0.18	−0.05	−0.92	−0.28	0.68

Based on Ballard (2002).

Table 4.6 Average Errors (Celsius Degrees) for the Hydrate Predictions from Five Software Packages for Binary Mixtures Containing Propane

	CSMGEM	CSMHYD	DBRHydrate	Multiflash	PVTSim
Nitrogen	−0.20	1.45	−0.05	−0.01	−0.05
CO_2	0.99	−0.62	0.88	−0.05	0.50
Propylene	0.68	0.68	–	0.51	–
Isobutane	0.05	–	1.12	0.30	0.70
n-Butane	2.12	0.25	2.40	1.48	1.36

Based on Ballard (2002).

Although the Ballard (2002) study shows that CSMGEM is consistently more accurate than the other packages, it is probably fair to conclude that all give a prediction of acceptable accuracy.

4.8 DEHYDRATION

One of the criteria for hydrate formation is that a sufficient amount of water be present. The hand calculations presented in Chapter 3 all assumed that plenty of water was present. The calculations presented from the software

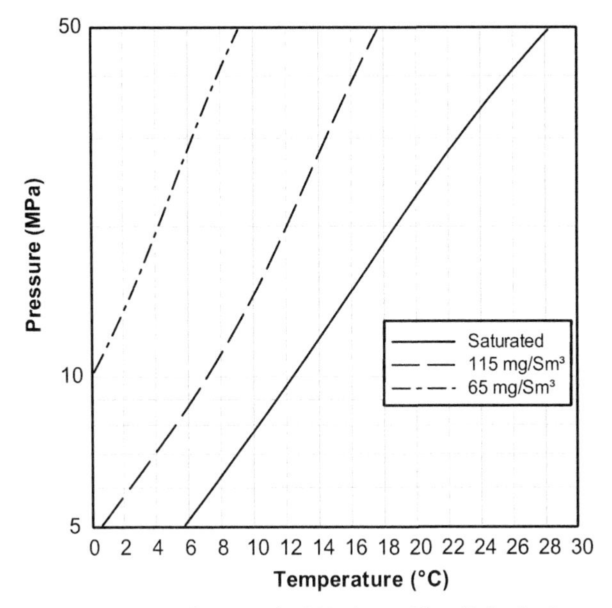

Figure 4.10 ***Hydrate Loci of Methane After Dehydration.***

packages presented so far, also assume that plenty of water was present. For pure components, the minimum amount of water to be in the saturation region was given in Chapter 2.

Methods for dehydration commonly used in the natural gas business are the topic of a subsequent chapter. In addition, further discussion of the water content of fluids in equilibrium with gases is presented later.

One of the advantages of a properly designed software package is that it can be used to predict the effect of dehydration on the hydrate formation conditions. Unfortunately, there is not a lot of experimental data available in the literature for building these models. More discussion of water content is given in Chapter 10 and the reader is referred to that chapter for more details.

Figure 4.10 shows some calculations for the effect of water content on the hydrate formation conditions. These calculations were performed using Prosim. The solid line on this plot labeled "saturation" is where plenty of water is present. This is the same curve as was plotted in Fig. 4.2. The other two plots on this figure are for a water content of 115 mg/Sm^3 (151 ppm or 7 lb/MMSCF) and the other is for 65 mg/Sm^3 (85.5 ppm or 4 lb/MMSCF).

Finally, it is probably fair to anticipate that the errors in the predictions shown on this plot are about ± 2 °C. This is based on the results for the systems presented earlier in this chapter.

The focus of Chapter 10 is the water content of natural gas but in equilibrium with liquid water and with solid phases. More details are presented in that chapter on the effect of water content.

4.9 MARGIN OF ERROR

One of the purposes of this chapter was to demonstrate that the software methods are not perfect. They are, however, very good. However, the design engineer should always build in a margin of error into the process designs. It is typical to have a safety factor of at least 3 °C (5 °F). That is, if you calculate a hydrate formation temperature of 10 °C, you should design to operate at 13 °C or more.

Examples

Example 4.1

Methane forms a type I hydrate at 0 °C and 2.60 MPa (see Table 2.2). Given that $c_{small} = 3.049/\mathrm{MPa}$ and $c_{large} = 13.941/\mathrm{MPa}$, calculate the saturation of the two cages for the methane hydrate using these values.

Answer: From Eqn (4.4) for the small cages:

$$Y_{small} = 3.049(2.60)/[1 + 3.049(2.60)] = 0.889$$
$$= 88.9$$

$$Y_{large} = 13.941(2.60)/[1 + 13.941(2.60)] = 0.973$$
$$= 97.3\%$$

Therefore, at 0 °C and 2.60 MPa, the small cages are just less than 90% filled and the large cages are slightly more than 97% filled.

Note, a Windows Excel spreadsheet is provided to perform such calculations. In addition, for higher accuracy, the fugacity should be used in place of the pressure.

Example 4.2

What would the *cs* in Eqn (4.4) be for propane in a type I hydrate?

Answer: To begin with, propane forms a type II hydrate. Furthermore, propane will never enter the cages of a type I hydrate. Therefore, the *cs* for propane in a type I hydrate are zero.

This may seem obvious, but it is not. For example, nitrogen also forms a type II hydrate. However, in the presence of a type I hydrate former, nitrogen can enter the type I lattice. Therefore, the cs for nitrogen in a type I hydrate are non-zero.

REFERENCES

Ballard, A.L., 2002. A Non-ideal Hydrate Solid Solution Model for a Multi-phase Equilibria Program (Ph.D. thesis). Colorado School of Mines, Golden, CO.

Ballard, L., Sloan, E.D., 2004. The next generation of hydrate prediction IV: a comparison of available hydrate prediction programs. Fluid Phase Equilib. 216, 257–270.

Carroll, J.J., May 19–21, 2004. An Examination of the Prediction of Hydrate Formation Conditions in Sour Natural Gas. GPA Europe, Dublin, Ireland.

Fan, S.-S., Guo, T.-M., 1999. Hydrate formation of CO_2-rich binary and quaternary gas mixtures in aqueous sodium chloride solutions. J. Chem. Eng. Data 44, 829–832.

Mei, D.-H., Liao, J., Yang, J.T., Guo, T.-M., 1998. Hydrate formation of a synthetic natural gas mixture in aqueous solutions containing electrolyte, methanol, and (electrolyte + methanol). J. Chem. Eng. Data 43, 178–182.

Ng, H.-J., Robinson, D.B., 1976. The role of n-butane in hydrate formation. AIChE J. 22, 656–661.

Ng, H.-J., Robinson, D.B., 1977. The prediction of hydrate formation in condensed systems. AIChE J. 23, 477–482.

Noaker, L.J., Katz, D.L., 1954. Gas hydrates of hydrogen sulfide-methane mixtures. Petro. Trans. AIME 201, 135–137.

Parrish, W.R., Prausnitz, J.M., 1972. Dissociation pressures of gas hydrates formed by gas mixtures. Ind. Eng. Chem. Process. Des. Dev. 11, 26–35.

Pedersen, K.S., Fredenslund, Aa, Thomassen, P., 1989. Properties of Oils and Natural Gases. Gulf Publishing, Houston, TX (Chapter 15).

Peng, D.-Y., Robinson, D.B., 1976. A new two-constant equation of state. Ind. Eng. Chem. Fundam. 15, 59–65.

Robinson, D.B., Hutton, J.M., 1967. Hydrate formation in systems containing methane, hydrogen sulphide and carbon dioxide. J. Can. Petro. Tech. 6, 1–4.

Soave, G., 1972. Equilibrium constants from a modified Redlich-Kwong equation of state. Chem. Eng. Sci. 27, 1197–1203.

Sun, C.-Y., Chen, G.-J., Lin, W., Guo, T.-M., 2003. Hydrate formation conditions of sour natural gases. J. Chem. Eng. Data 48, 600–603.

van der Waals, J.H., Platteeuw, J.C., 1959. Clathrate solutions. Adv. Chem. Phys. 2, 1–57.

Wilcox, W.I., Carson, D.B., Katz, D.L., 1941. Natural gas hydrates. Ind. Eng. Chem. 33, 662–671.

CHAPTER 5

Inhibiting Hydrate Formation with Chemicals

As has been stated earlier, hydrates are a significant problem in the natural gas industry. So what can be done when we encounter hydrates in our processes? What can we do to prevent them from forming in the first place? This chapter outlines some design information for battling hydrates using chemicals.

People who live in colder climates are well aware of methods for combating ice. In the winter, salt is often used to remove ice from roads and sidewalks. A glycol solution is sprayed on airplanes waiting for take-off in order to de-ice them. Similar techniques are used for combating hydrates.

Polar solvents, such as alcohol, glycol, and ionic salts (common table salt), are known to inhibit the formation of gas hydrates. It is important to note that they do not prevent hydrate formation, they inhibit it. That is, they reduce the temperature or increase the pressure at which a hydrate will form. The mere presence of an inhibitor does not mean that a hydrate will not form. The inhibitor must be present in some minimum concentration to avoid hydrate formation. The calculation of this minimum inhibitor concentration is addressed in detail in this chapter.

In the natural gas industry, the use of alcohols, particularly methanol, and glycols, EG, or triethylene glycol (TEG), is a common method for inhibiting hydrate formation. Figure 5.1 shows the inhibiting effect of methanol on the hydrate of hydrogen sulfide. The curve for pure H_2S is taken from Table 2.6 and the methanol data are from Ng et al. (1985). The curves for the methanol data are merely lines through the data points. They do not represent a prediction or fit of the data. This figure is presented as an example of the inhibiting effect. More such charts for other components will be presented later in this chapter.

Table 5.1 lists the properties of some common polar compounds that are used as inhibitors. Note that all of these compounds exhibit some degree of hydrogen bonding, and thus interfere with water's hydrogen bonds.

Ionic solids, such as sodium chloride (common table salt), also inhibit the formation of hydrates. This is similar to spreading salt on icy sidewalks or

Natural Gas Hydrates
ISBN 978-0-12-800074-8
http://dx.doi.org/10.1016/B978-0-12-800074-8.00005-3

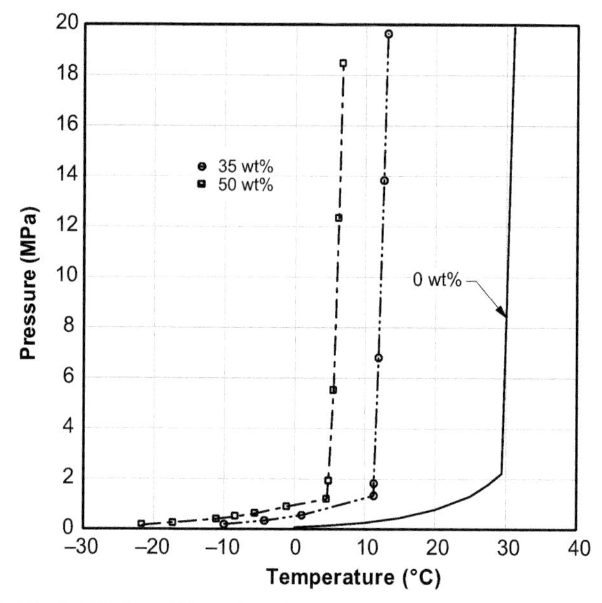

Figure 5.1 ***The Inhibiting Effect of Methanol on the Hydrate of Hydrogen Sulfide.***

Table 5.1 Properties of Some Hydrate Inhibitors

	Methanol	Ethanol	EG	TEG
Empirical formula	CH_4O	C_2H_6O	$C_2H_6O_2$	$C_6H_{14}O_4$
Molar mass, g/mol	32.042	46.07	62.07	150.17
Boiling point, °C	64.7	78.4	198	288
Vapor pressure (at 20 °C), kPa	12.5	5.7	0.011	<0.001
Melting point, °C	−98	−112	−13	−4.3
Density (at 20 °C), kg/m^3	792	789	1116	1126
Viscosity (at 20 °C), cp	0.59	1.2	21	49

EG = ethylene glycol, $HO{-}CH_2{-}CH_2{-}OH$.
TEG = triethylene glycol, $HO{-}CH_2{-}CH_2{-}O{-}CH_2{-}CH_2{-}O{-}CH_2{-}CH_2{-}OH$.

highways to melt the ice. It is unlikely that anyone would use salt as an inhibitor; the salt is almost always present in the produced water.

Although they would never be used as inhibitors per se, the alkanolamines, used for sweetening natural gas, also inhibit the formation of hydrates. The benefit of this is that the risk of hydrate formation in the amine unit is reduced because of the inhibiting effect of the amine.

5.1 FREEZING POINT DEPRESSION

The depression of the freezing point of a solvent by the presence of a small amount of solute is a fairly well-understood concept. In fact, the depression of the freezing point is commonly used to estimate the molar mass of a sample.

The theory behind freezing point depression can be found in any book on physical chemistry (for example, Laidler and Meiser, 1982). The derivation begins with the fundamental relationship for the equilibrium between a solid and a liquid and, after some simplifying assumptions, the resulting equation is:

$$x_{\mathrm{i}} = \frac{h_{\mathrm{sl}}\Delta T}{RT_{\mathrm{m}}^{2}} \tag{5.1}$$

where x_i is the mole fraction of the solute (inhibitor), ΔT is the temperature depression in °C, R is the universal gas constant (8.314 J/mol K), and T_m is the melting point of the pure solvent in K. Rearranging this equation slightly and converting from mole fraction to mass fraction gives:

$$\begin{aligned} \Delta T &= \frac{M_{\mathrm{s}}RT_{\mathrm{m}}^{2}}{h_{\mathrm{sl}}} \times \frac{W_{\mathrm{i}}}{(100 - W_{\mathrm{i}})M_{\mathrm{i}}} \\ &= K_{\mathrm{S}}\frac{W_{\mathrm{i}}}{(100 - W_{\mathrm{i}})M_{\mathrm{i}}} \end{aligned} \tag{5.2}$$

where M_s is the molar mass of the solvent, W_i is the weight percent solute (inhibitor), and M_i is the molar mass of the inhibitor. For water, it is $K_S = 1861$, when SI units are used. The leading term in this equation contains only constants, so the freezing point depression is a function of the concentration of the inhibitor and its molar mass.

It is worth noting that this equation is not applicable to ionic solutions, such as salt. This will be demonstrated later in this chapter.

To get a quick impression of the accuracy of Eqn (5.2), consider Fig. 5.2, which shows the freezing points of methanol + water and EG + water solutions. The freezing point depression for methanol is quite accurate up to concentrations of 30 wt%. For EG, the calculation is accurate only up to about 15 wt%. The fit for methanol + water is quite good because the solution is close to ideal. That is because the assumptions built into the derivation are applicable for methanol + water over a fairly wide range of concentration.

On the other hand, Fig. 5.3 shows the freezing point of sodium chloride (common salt) solutions. From this plot, it is clear that the true freezing

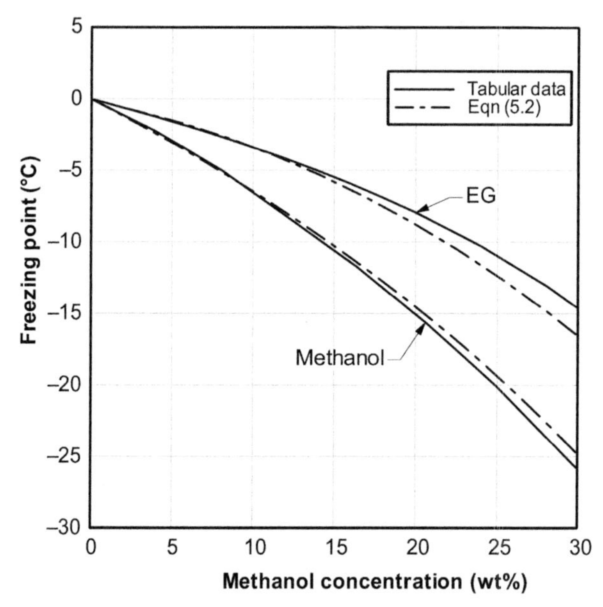

Figure 5.2 *Freezing Points of Methanol + Water and Ethylene Glycol (EG) + Water Mixtures.*

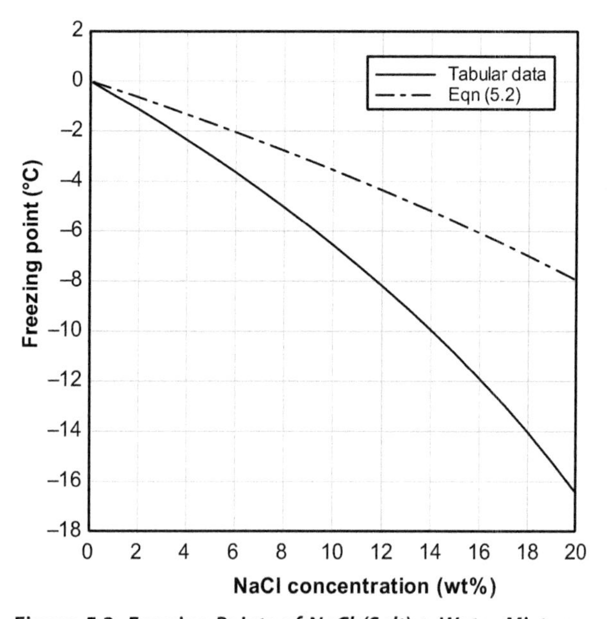

Figure 5.3 *Freezing Points of NaCl (Salt) + Water Mixtures.*

points of salt solutions are much lower than those predicted by the simple freezing point depression theory.

In both Figs 5.2 and 5.3, the curves labeled "Tabular Data" are from the *CRC Handbook* (Weast, 1978).

To use the freezing point depression method to find the molar mass of an unknown substance is relatively simple. It is quite straightforward to make a solution with a known weight fraction even though the nature of the solute is unknown. Adding 5 g of solute to 95 g of solvent makes a 5 wt% solution—it is that simple. The constant K_s in Eqn (5.2) is a property of the solvent only and values are readily available for any solvent that would be used for molar mass determination. Then the freezing point of the mixture is measured and this can be done to a high degree of accuracy. Equation (5.2) can be used to calculate the molar mass of the solute given the freezing point depression for a given concentration.

5.2 THE HAMMERSCHMIDT EQUATION

A relatively simple and widely used method to approximate the effect of chemicals on the hydrate forming temperature is the Hammerschmidt equation:

$$\Delta T = \frac{K_{\mathrm{H}} W}{M(100 - W)} \tag{5.3}$$

where ΔT is the temperature depression in °C, M is the molar mass of the inhibitor in g/mol, W is the concentration of the inhibitor in weight percent in the aqueous phase, and K_{H} is a constant with a value of 1297. To use this equation with American engineering units, then K_{H} is 2355 and ΔT is the temperature depression in °F. The units on the other two terms remain unchanged.

The concentration in this equation is on an inhibitor plus water basis (that is, it does not include the other components in the stream).

Note the similarity between this equation and the freezing point depression equation given earlier (Eqn (5.2)). Because of their similarity and their common origin, it is safe to assume that the Hammerschmidt equation is not applicable to ionic solids.

Equation (5.3) can be rearranged in order to calculate the concentration of the inhibitor required to yield the desired temperature depression, as:

$$W = \frac{100 M \Delta T}{K_{\mathrm{H}} + M \Delta T} \tag{5.4}$$

Table 5.2 Coefficients for the Hammerschmidt Equation, K_H (Eqn (5.3) in Text)

	Original	Ref. 1	Ref. 2	Ref. 3
Methanol	1297	1297	1297	1297
Ethanol	1297	–	1297	1297
Ethylene glycol	1297	2222	1222	1500
Diethylene glycol	1297	2222	2427	2222
Triethylene glycol	1297	2222	2472	3000

Ref. 1—*GPSA Engineering Data Book*.
Ref. 2—Arnold and Stewart (1989).
Ref. 3—Pedersen et al. (1989)—there is a mistake in their table, values for the constant are for degrees Fahrenheit, not Celsius.

To use the Hammerschmidt equation, you must first estimate the hydrate conditions without an inhibitor present. The Hammerschmidt equation only predicts the deviation from the temperature without an inhibitor present, not the hydrate forming conditions themselves.

Originally, the K_H in Eqns (5.3) and (5.4) was a constant, but, over the years, some have proposed making K_H a function of the inhibitor in order to improve the predictive capabilities of the equation. Some of these are listed in Table 5.2.

The value of 2222 for EG given in the *GPSA Engineering Data Book*, which they recommend for all glycols, is much too large. Better predictions are obtained using the original value of 1297. This will be demonstrated later in this chapter. On the other hand, this large value does improve the calculations for TEG.

The Hammerschmidt equation is limited to concentrations of about 30 wt% for methanol and EG and only to about 20 wt% for other glycols. The freezing point depression method, which was shown to bear a resemblance to the Hammerschmidt method, is only applicable to a few mole percent solute.

5.3 THE NIELSEN–BUCKLIN EQUATION

Nielsen and Bucklin (1983) first used principles to develop another equation for estimating hydrate inhibition of methanol solutions. Their equation is:

$$\Delta T = -72 \ln(1 - x_M) \tag{5.5}$$

where ΔT is in °C and x_M is the mole fraction of methanol. They claim that this equation is accurate up to mole fraction of 0.8 (about 88 wt%).

This equation can be rearranged to estimate the methanol concentration given the temperature depression:

$$x_M = 1 - \exp\left[\frac{-\Delta T}{72}\right] \tag{5.6}$$

and then to calculate the weight percent from this mole fraction, the following equation is used:

$$X_M = \frac{x_M M_M}{18.015 + x_M(M_M - 18.015)} \tag{5.7}$$

where X_M is the weight fraction methanol and M_M is the molar mass of methanol.

The Nielsen–Bucklin equation was developed for use with methanol; however, the equation is actually independent of the choice of inhibitor. The equation involves only the properties of water and the concentration of the inhibitor. Therefore, theoretically, it can be used for any inhibitor, where the molecular weight of the solvent is substituted for M_M in Eqn (5.7).

It is not clear when you compare Eqns (5.3) and (5.5) that these equations have similar limiting behavior. That is, at a low concentration, these two equations predict the same inhibiting effect for a given inhibitor solution. However, this is indeed the case.

Although this equation has a wider range of applicability than the Hammerschmidt equation, it has not gained wide acceptance. Most design engineers continue to use the simpler Hammerschmidt equation.

5.4 A NEW METHOD

The Hammerschmidt and Nielsen–Bucklin equations have some characteristics that make them very desirable. In addition to their simplicity, they exhibit the correct limiting behavior. In the limit, as the inhibitor approaches zero concentration, ΔT approaches zero. In the other limit, as one approaches pure inhibitor, the equation predicts infinite ΔT—no hydrate formation. The new equation should also have these limits.

In addition, Nielsen and Bucklin showed that the Hammerschmidt equation is a limiting case for their equation. Thus, the new equation should have the Nielsen–Bucklin (and hence the Hammerschmidt) equation as a low concentration limit.

Finally, the equation should have a firm basis in theory such that it can be extrapolated to conditions where no data exist.

With this in mind, the basis for the new equation is the same as that for the Nielsen–Bucklin equation. However, an activity coefficient is included to account for the concentration of the inhibitor. The starting equation is:

$$\Delta T = -72 \ln(\gamma_W x_W) \tag{5.8}$$

where γ_W is the activity coefficient of water and x_W is the mole fraction of water.

The next step is to find an activity coefficient model that is both realistic and simple. The simplest such model is the two-suffix Margules equation:

$$\ln \gamma_W = \frac{a}{RT} x_I^2 \tag{5.9}$$

To further simplify things, it will be assumed that the term a/RT is independent of the temperature and can be replaced by a more general constant that will be called A—the Margules coefficient. Thus, Eqn (5.8) becomes:

$$\Delta T = -72\left(A x_I^2 + \ln[1 - x_I]\right) \tag{5.10}$$

It turns out that this equation is sufficiently accurate over a wide range of inhibitor concentrations, which is what we desired.

The values for the Margules coefficients, A, were obtained by fitting experimental data from the literature and the values obtained are listed in Table 5.3.

Experimental data for methanol inhibition are relatively plentiful. In fact, measurements have been made up to concentrations of 85 wt%. Unfortunately, measurements for ethanol, which is not often used as an

Table 5.3 Margules Coefficients for Various Inhibitors and the Approximate Limits on the Correlation

			Limit	
Inhibitor	**Molar Mass (g/mol)**	**Margules Coefficient**	**Concentration (wt%)**	**ΔT (°C)**
Methanol	32.04	+0.21	<85	<94.3
Ethanol	46.07	+0.21	<35	<13.3
EG	62.07	−1.25	<50	<22.9
DEG	106.12	−8	<35	<10.3
TEG	150.17	−15	<50	<20.6

DEG, diethylene glycol; EG, ethylene glycol; TEG, triethylene glycol.

inhibitor, are relatively scarce. Thus, the Margules coefficient for ethanol was set equal to that for methanol.

Experimental data for EG and TEG are plentiful and are for concentrations up to 50 wt%. Data for diethylene glycol (DEG) are significantly less common. Fortunately, DEG is seldom used for this application. The value for the Margules coefficient used for DEG is the average of the values for EG and TEG.

5.4.1 A Chart

Admittedly, Eqn (5.10) is a little difficult to use, especially if the temperature depression is given and the required inhibitor concentration must be calculated. Therefore, a graphical version is presented in Fig. 5.4 in SI units and Fig. 5.5 in American engineering units. There are no experimental data for glycol concentrations greater than 50 wt%, therefore, beyond this concentration, the correlation is an extrapolation.

From this chart, it is quite easy to determine the temperature depression for a given inhibitor concentration and vice versa.

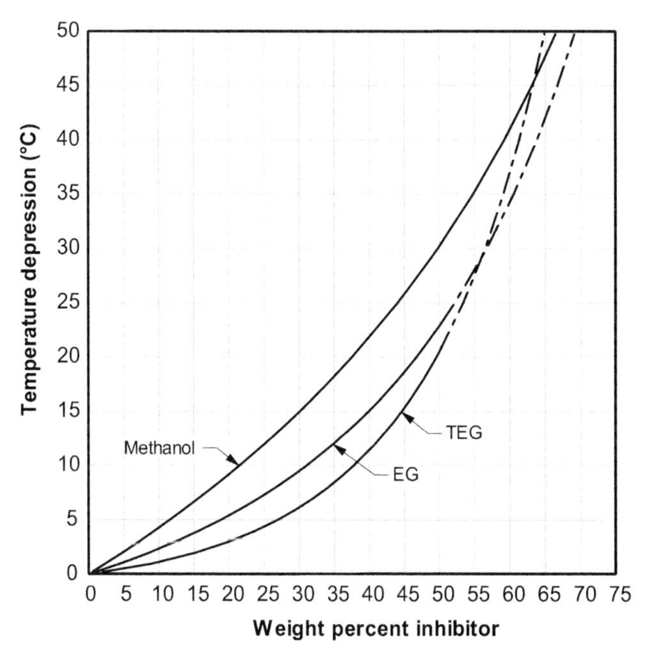

Figure 5.4 ***The Inhibiting Effect of Methanol, Ethylene Glycol (EG), and Triethylene Glycol (TEG)—SI Units.***

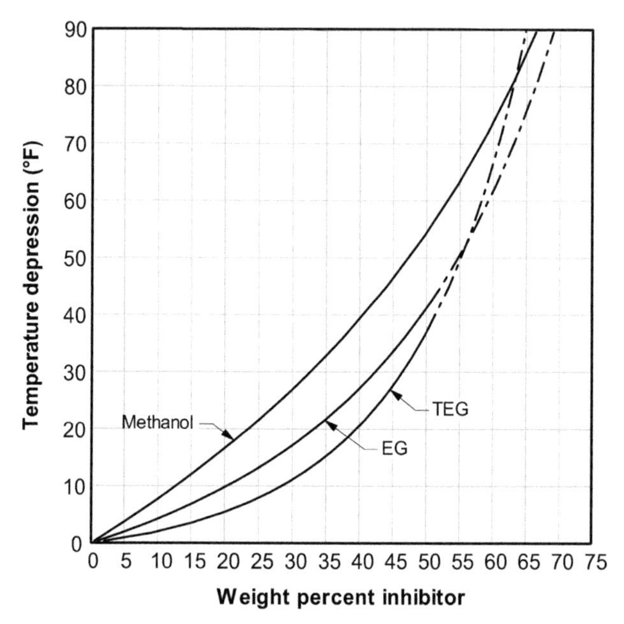

Figure 5.5 ***The Inhibiting Effect of Methanol, Ethylene Glycol (EG), and Triethylene Glycol (TEG)—American Engineering Units.***

5.4.2 Accuracy of the New Method

The new equation is compared with some data from the literature. For convenience, and with apologies to the original authors, the original sources of the data are not listed. The data were taken from the monograph of Sloan (1999).

Figure 5.6 shows the calculated depression for the methane hydrate using methanol. The curves are for 10, 20, 35, 50, 65, 73.7, and 85 wt% methanol. The new equation is shown to be very good even for these high methanol concentrations.

For comparison purposes, Fig. 5.7 shows only the 65 wt% curve. However, included in this plot are the predictions from the Hammerschmidt and Nielsen–Bucklin equations. At this methanol concentration, the Hammerschmidt equation predicts a temperature depression that is about 28 °C too large. The Nielsen–Bucklin equation is an improvement over the Hammerschmidt equation, but it is also overpredicts the inhibiting effect. The Nielsen–Bucklin equation is in error by about 4 °C. In a practical sense, this means that the methanol injection rate predicted using either the Hammerschmidt or the Nielsen–Bucklin equations would be too small.

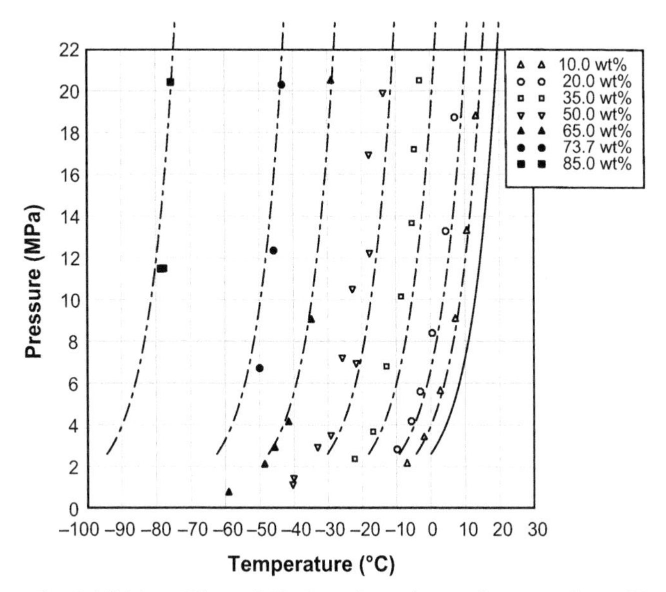

Figure 5.6 ***The Inhibiting Effect of Methanol on the Methane Hydrate.*** (Curves from Eqn (5.10).)

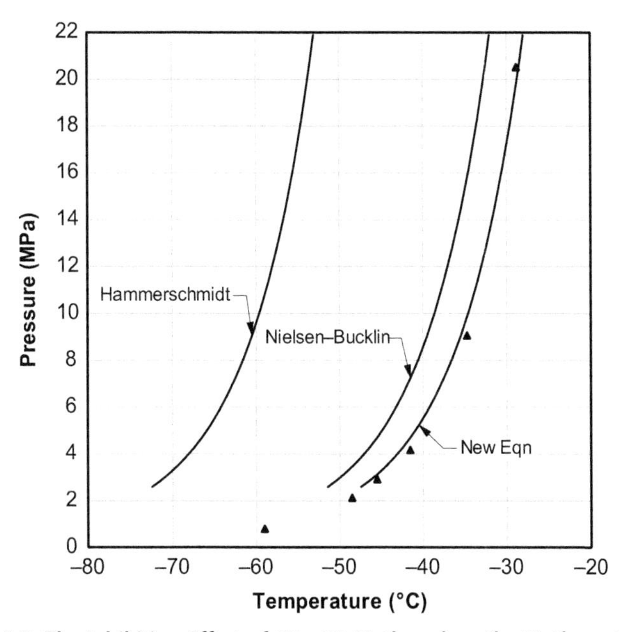

Figure 5.7 ***The Inhibiting Effect of 65 wt% Methanol on the Methane Hydrate.***

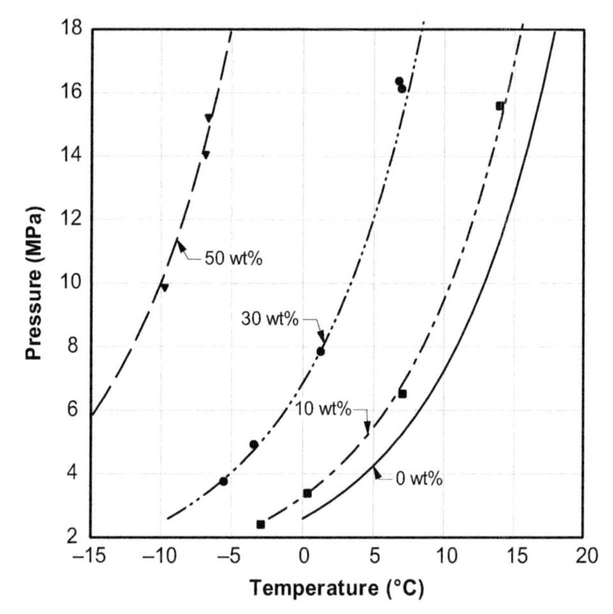

Figure 5.8 ***The Inhibiting Effect of Ethylene Glycol on the Methane Hydrate.***

Figure 5.8 shows the inhibiting effect of EG on the methane hydrate. Again, this figure demonstrates that the new equation is an excellent prediction of the experimental data.

Again, Fig. 5.9 is presented in order to compare the new equation with the correlations available in the literature. This figure is for 35 wt% EG. The original Hammerschmidt equation does a surprisingly good job. However, the Gas Processors Suppliers Association (GPSA) modification grossly overpredicts the temperature depression. The GPSA equation is in error by about 6 °C. Again, this translates to an EG injection rate that is too small for the desired inhibition.

5.5 BRINE SOLUTIONS

As was mentioned earlier, ionic solids also inhibit the formation of hydrate in much the same way that they inhibit the formation of ice. Some quick rules of thumb can be found in the observations based on experimental measurements.

Maekawa (2001) measured hydrate formation for methane and a mixture of methane and ethane, although rich in methane, in both pure water and 3 wt% NaCl for pressures from about 3 to 12 MPa. His data show

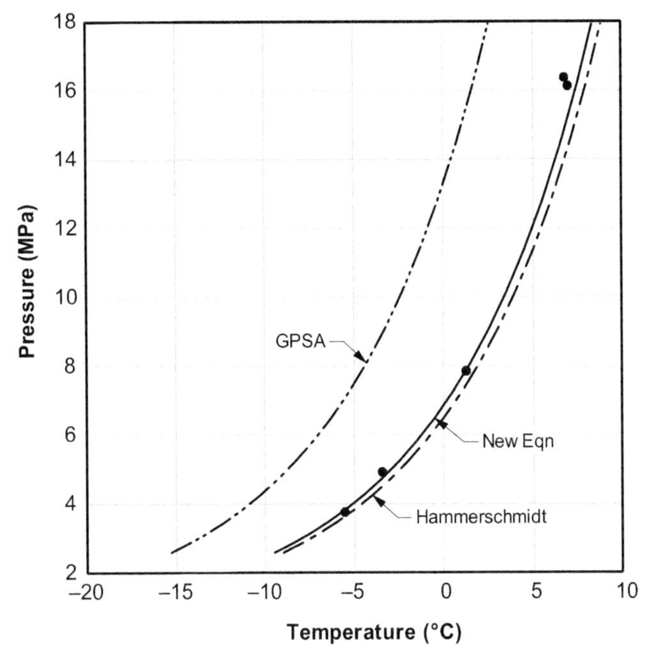

Figure 5.9 ***The Inhibiting Effect of 35 wt% Ethylene Glycol on the Methane Hydrate.***

that the hydrate formation temperature was reduced by about 1 °C for brine with this concentration. It is also demonstrated that the temperature depression is independent of the gas mixture studied.

Mei et al. (1998) measured the hydrate formation for a natural gas mixture in pure water and in solutions of various ionic salts and for pressures from about 0.6 to 2.5 MPa. Their data for pure water were presented earlier. For NaCl and KCl, their data show that ΔT is fairly constant over the range of temperatures shown. For $CaCl_2$, ΔT tends to increase with increasing temperature. The results of the study of Mei et al. (1998) are summarized in Tables 5.4 and 5.5.

The temperature depression for these data is independent of the concentration when expressed in either wt% or molality.

5.5.1 McCain Method

McCain (1990) provides the following correlation for estimating the effect of brine on the hydrate formation temperature:

$$\Delta T = AS + BS^2 + CS^3 \tag{5.11}$$

Table 5.4 Constants for the Østergaard et al. (2005) Correlation for the Inhibiting Effect of Ionic Solutions

Constant	NaCl	$CaCl_2$	NaBr
c_1	0.3534	0.194	0.419
c_2	1.375×10^{-3}	7.58×10^{-3}	6.5×10^{-3}
c_3	2.433×10^{-4}	1.953×10^{-4}	1.098×10^{-4}
c_4	4.056×10^{-2}	4.253×10^{-2}	2.529×10^{-2}
c_5	0.7994	1.023	0.303
c_6	2.25×10^{-5}	2.8×10^{-5}	2.46×10^{-5}
Max. con. (mass%)	26.5	40.6	38.8
Max. con. (mol%)	10	10	10

Constant	K_2CO_3	KBr	KCl
c_1	0.1837	0.3406	0.305
c_2	-5.7×10^{-3}	7.8×10^{-4}	6.77×10^{-4}
c_3	2.551×10^{-4}	8.22×10^{-5}	8.096×10^{-5}
c_4	6.917×10^{-2}	3.014×10^{-2}	3.858×10^{-2}
c_5	1.101	0.3486	0.714
c_6	2.71×10^{-5}	2.3×10^{-5}	2.2×10^{-5}
Max. con. (mass%)	40.0	36.5	31.5
Max. con. (mol%)	8	8	10

Table 5.5 Constants for the Østergaard et al. (2005) Correlation for the Inhibiting Effect Polar Inhibitors

Constant	Methanol	Ethanol	Glycerol
c_1	0.478	1.118	0.135
c_2	7.17×10^{-2}	-4.48×10^{-3}	8.846×10^{-3}
c_3	-1.44×10^{-5}	6.979×10^{-4}	-1.15×10^{-5}
c_4	2.947×10^{-2}	5.85×10^{-3}	1.335×10^{-2}
c_5	0.596	0.225	0.378
c_6	3.1×10^{-5}	3.4×10^{-5}	4.6×10^{-5}
Max. con. (mass%)	43.3	31.2	68.6
Max. con. (mol%)	30	15	30

Constant	EG	DEG	TEG
c_1	38.93	0.343	0.1964
c_2	-0.522	-3.47×10^{-3}	-5.81×10^{-3}
c_3	1.767×10^{-2}	2.044×10^{-4}	1.393×10^{-4}
c_4	3.503×10^{-4}	1.8×10^{-2}	2.855×10^{-2}
c_5	5.083×10^{-3}	0.3346	0.854
c_6	2.65×10^{-5}	2.74×10^{-5}	3.24×10^{-5}
Max. con. (mass%)	59.6	51	59.5
Max. con. (mol%)	30	15	15

DEG, diethylene glycol; EG, ethylene glycol; TEG, triethylene glycol.

where ΔT is the temperature depression in °F, S is the salinity in wt%, and the coefficients A, B, and C are functions of the gas gravity, γ, which is defined in Eqn (3.1), and are given below:

$$A = 2.20919 - 10.5746\gamma + 12.1601\gamma^2 \tag{5.12}$$

$$B = -0.106056 + 0.722692\gamma - 0.85093\gamma^2 \tag{5.13}$$

$$C = 0.00347221 - 0.0165564\gamma + 0.049764\gamma^2 \tag{5.14}$$

Equation (5.11) is limited to salt concentrations of 20 wt% and for gas gravities in the range $0.55 < \gamma < 0.68$. These equations were used to generate Figure 5.10 and 5.11, which is useful for rapid approximation of the inhibiting effect.

It is unlikely that anyone would add salt to the water in order to prevent hydrate formation. However, produced waters often contain brine and it is important to be able to estimate the effect of the brine in the produced water on the hydrate formation temperature.

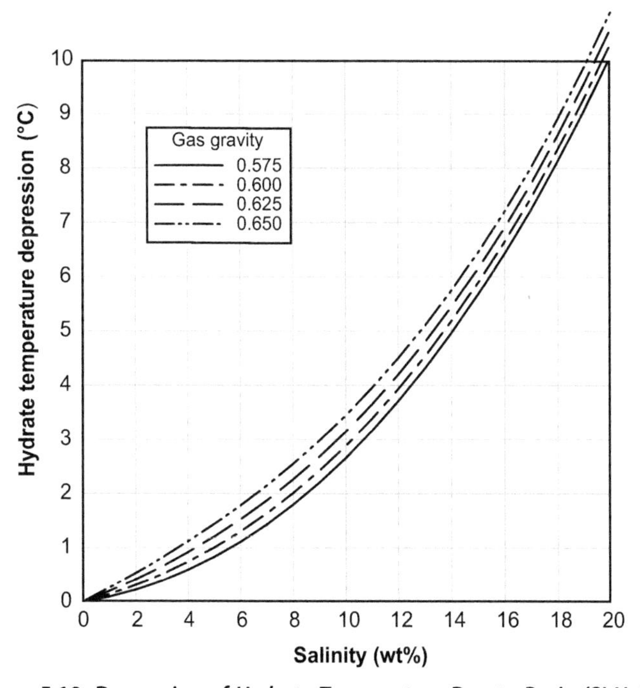

Figure 5.10 ***Depression of Hydrate Temperature Due to Brain (SI Units).***

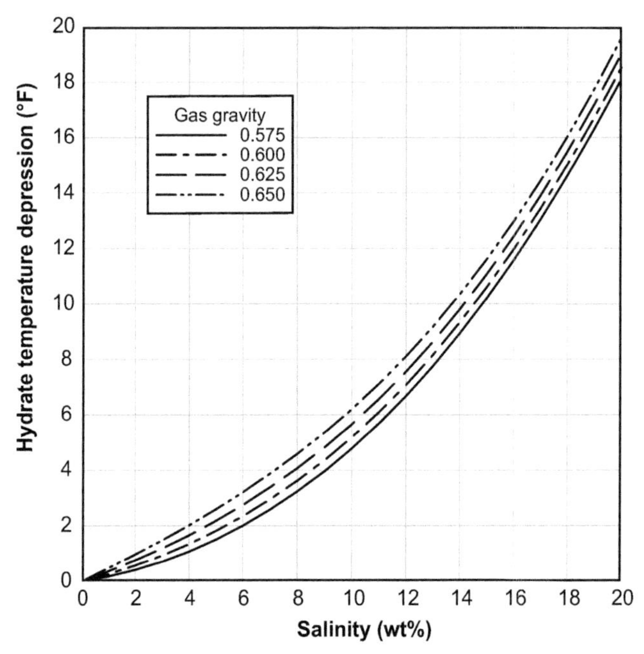

Figure 5.11 ***Depression of Hydrate Temperature Due to Brine (American Engineering Units).***

5.6 ØSTERGAARD ET AL.

Østergaard et al. (2005) presented a correlation that was significantly different from its predecessors. First, they built an equation applicable to both inorganic salts (such as NaCl) and organic compounds (such as alcohols and glycols). Second, it includes the effect of pressure.

The correlation has the form:

$$\Delta T = \left(c_1 W + c_2 W^2 + c_3 W^3\right)\left(c_4 \ln P + c_5\right)\left(c_6(P_0 - 1000) + 1\right) \quad (5.15)$$

where ΔT is the temperature depression in K or °C, P is the pressure of the system in kPa, W is the concentration of the inhibitor in liquid water phase in mass%, P_0 is the dissociation pressure of hydrocarbon fluid in pure water at 0 °C in kPa, and the c_is are parameters listed in Table 5.6. Also given in the tables is the stated range of applicability.

For solutions of ionic solids, the authors indicate that the average errors are less than 1.5% and for the polar inhibitors the average errors are slightly larger but still less than 2% for alcohols and glycols and about 5% for glycerol. On the other hand, the correlation is applicable over a wider range

Table 5.6 Hydrate Point Depression Observed by Mei et al. (1998) Due to Ionic Salts

	wt%	mol%	Molality	Average ΔT
NaCl	10	3.31	1.90	5
KCl	10	2.61	1.49	3
$CaCl_2$	10	1.77	1.00	4

of concentrations when applied to the polar solvent as compared with the ionic solids.

The equation is simple to use if you know the inhibitor concentration and you want to estimate the depression. It is more difficult to use if the temperature is specified and you wish to calculate the required inhibitor concentration. This requires an iterative solution.

5.7 COMMENT ON THE SIMPLE METHODS

These simple methods, Hammerschmidt, Nielsen–Bucklin, and the new method developed for this book, have several characteristics in common. At this point, it is time to review some of them.

All of these simple methods predict the depression of the hydrate temperature; they do not predict the actual hydrate formation conditions. First, you must begin with the methods presented in Chapters 3 and 4 to predict the hydrate formation in the absence of the inhibitor. Then, you can use the methods presented earlier to correct for the presence of the inhibitor. Therefore, if your method for estimating the hydrate formation condition in the absence of inhibitor is a poor model, then the corrections noted in the first part of this chapter will also be inaccurate. These correction factors cannot overcome a poor original prediction.

Note that these methods assume that the inhibition effect is independent of the pressure. Experimental work indicates that this is nearly the case. To within a good engineering estimate, it is safe to neglect the effect of pressure on the inhibiting effect.

Furthermore, these methods assume that the temperature depression is independent of both the nature of the hydrate former present and the type of hydrate formed. So the temperature depression in a 25 wt% methanol solution is the same for a methane hydrate (type I) as it is for a propane hydrate (type II).

There is a little bit of confusion about the units on the quantity calculated via these methods. The result obtained is a temperature difference, not an actual temperature. Therefore, if you calculate a ΔT of 5 °C, to convert this to Fahrenheit, simply multiply by 1.8 to get 9 °F. You do not convert the 5 °C to 41 °F as if it were a temperature reading.

5.8 ADVANCED CALCULATION METHODS

Just as there were more advanced methods for calculating hydrate formation conditions, there are more rigorous methods for estimating the effect of inhibitors (for example, see Anderson and Prausnitz (1986)). As with most rigorous methods, these methods are suited for computer calculations and not for hand calculations.

The van der Waals–Platteeuw-type models discussed in the previous chapter can be extended as follows:

$$\frac{\Delta\mu(T,P)}{RT} - \ln(x_W\gamma_W) = \frac{\Delta\mu(T_O,P_O)}{RT_O} - \int_{T_O}^{T}\frac{\Delta H}{RT^2}\mathrm{d}T + \int_{P_O}^{P}\frac{\Delta V}{R\overline{T}}\mathrm{d}P \tag{5.16}$$

where x_W is the mole fraction of water and γ_W is the activity coefficient of water in the aqueous solution. The bar over the temperature in the last term in the above equation indicates that this is an average temperature. The activity coefficient of water can be modeled using the equations commonly used for vapor–liquid equilibrium.

As a contrast to the simple model, these advanced models can and do account for the variables that are neglected in the simple models. For example, these models do include the effect of pressure and the type of hydrate.

5.9 A WORD OF CAUTION

Methanol is very useful for combating the formation of hydrates in pipelines and process equipment. However, methanol can have an adverse effect on subsequent processing of the hydrocarbon stream.

As an example of a processing problem that may arise with methanol use, it is possible that methanol will concentrate in the liquefied petroleum gas (LPG) stream. LPG is made up largely of propane and mixed butanes. It is known that propane + methanol and n-butane + methanol form azeotropes (Leu et al., 1992). These azeotropes mean that it is impossible to separate the systems using binary distillation. In a practical sense, this is why methanol may appear in unacceptable amounts in the LPG product.

In addition, methanol-hydrocarbon systems are notoriously difficult to model accurately. However, applied thermodynamics has made great strides in this area. The design engineer should be cautious about the models chosen to perform calculations on such systems.

Another interesting side-effect of methanol use is on corrosion inhibitors. At one site, methanol was injected into a pipeline to prevent hydrate formation. Inhibitor chemicals were also injected to prevent corrosion. The inhibitors were alcohol based and the methanol dissolved the inhibitor. This led to some unexpected corrosion problems.

Another potential corrosion problem related to methanol injection is dissolved air in the methanol. Methanol for inhibitor is usually stored on site in tanks that are open to the atmosphere. This allows some air to dissolve into the methanol. Then, upon injection into a process or pipeline, it may cause corrosion problems. Typically, the amount of dissolved oxygen is small, but over the long term this could cause problems.

5.10 AMMONIA

Ammonia was once suggested as an inhibitor for hydrate formation. It has a relatively low molar mass, 17.03 g/mol vs 32.04 g/mol for methanol, which as we have seen is advantageous for an inhibitor. Based on the Hammerschmidt equation, a 10 °C depression in the hydrate formation temperature requires an 11.6 wt% ammonia solution vs a 19.8 wt% methanol solution.

Ammonia may be more useful in thawing hydrate plugs in pipelines. Unlike liquid inhibitors, which require pressure drop in order to flow to reach a plug—typically not available in a plugged line—ammonia can diffuse through the gas phase to reach the hydrate plug.

Unfortunately, ammonia has several drawbacks as well. It is toxic and may be difficult to handle in oil filled applications. In addition, it reacts with both carbon dioxide and hydrogen sulfide in the aqueous phase.

Ammonia's volatility is a disadvantage as well as an advantage. Its high volatility translates into larger losses to the vapor.

The disadvantage outweighs any possible advantage and, therefore, ammonia is rarely (perhaps never) used as a hydrate inhibitor.

5.11 ACETONE

Nature is interesting in that she always seems to provide exceptions to the rules. In the field of hydrate inhibition, one exception is acetone. Acetone is a polar compound that is liquid at room temperature. It shares many characteristics in common with alcohols and would appear at first glance to be an excellent candidate for use a hydrate inhibitor. Closer scrutiny demonstrates that this is not the case.

In all honesty, no one would probably suggest using acetone as a hydrate inhibitor. Compared to methanol, it is more costly and is predicted to be less effective. For example, the Hammerschmidt equation indicates that, for a given weight fraction of inhibitor, the temperature depression is inversely proportional to the molar mass of the inhibitor. The molar mass of acetone (58.05 g/mol) is almost twice that of methanol. Therefore, for a given weight percent, the temperature depression from acetone would be approximately half that for methanol.

The reason that acetone was suggested for this application was that a company had a waste stream composed of methanol, ethanol, and acetone and wished to examine it as a potential hydrate inhibitor. It turned out that the blend performed much worse than anticipated and the culprit was the acetone. At low concentrations, the acetone enhanced the hydrate formation rather than depressed it. That is, for a given pressure, the hydrate in an acetone solution formed at a higher temperature than in pure water. Only at high concentrations did the acetone begin to inhibit the formation of the hydrate. More details of this work can be found in Ng and Robinson (1994). Confirmation of this unusual result was obtained by Mainusch et al. (1997). These experiments were conducted for the methane hydrate, but similar results could be anticipated for other hydrate formers.

Mainusch et al. (1997) were able to model the phenomenon but no detailed explanation was provided. That is, the role of the acetone in hydrate formation is unclear. Is acetone a hydrate former (i.e., does it enter the cages)? Is acetone a host (i.e., does it form part of the crystal lattice along with the water molecules)? Or does it play some other role?

5.12 INHIBITOR VAPORIZATION

Methanol is a volatile substance and, thus, when it is injected into natural gas and/or condensate some of the methanol will enter these phases. In practical terms, this means that more inhibitor must be injected than the amount predicted solely from the aqueous phase concentrations. If only that amount of inhibitor is injected, then not enough will be used.

The inhibitors are volatile and, thus, will evaporate from the aqueous liquid where they are required to inhibit the hydrate formation to the vapor.

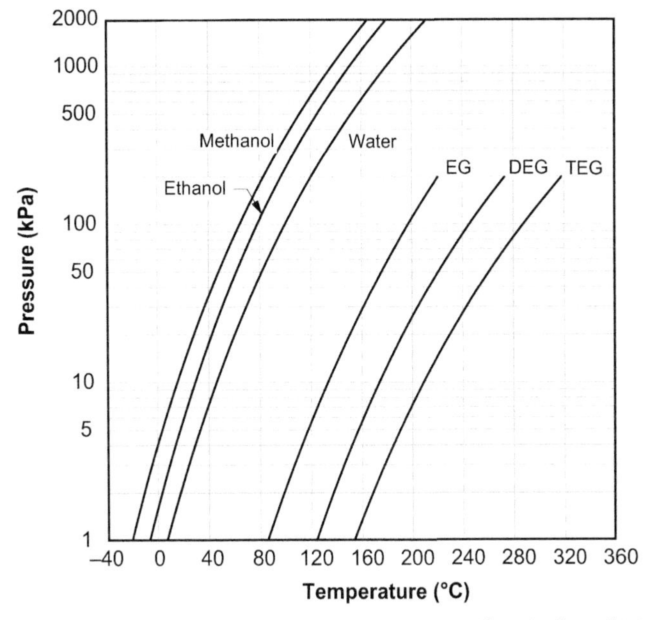

Figure 5.12 ***Vapor Pressures of Common Inhibitor Chemicals and Water.***

Some inhibitors are more volatile than others. The vapor pressures of some common inhibitors are plotted in Fig. 5.12.

Fortunately, there are charts available for estimating these losses. Figures 5.13 and 5.14 can be used to estimate the amount of methanol dissipated in the natural gas. These charts are approximations and should be used with care, especially if used for design.

The chart for the methanol losses to the vapor is a little confusing. The abscissa (x-axis) has units of $\frac{\text{kg MeOH}}{(10^6\ \text{Sm}^3)(\text{wt\% MeOH})}$ in SI units or $\frac{\text{lb MeOH}}{(\text{MMCF})(\text{wt\% MeOH})}$ in American engineering units. To calculate the methanol in the vapor locate the point that corresponds to the pressure (ordinate) and temperature (third parameter), and then read the value off the x-axis. The value on the x-axis is multiplied by the gas rate and the aqueous phase methanol concentration to get the methanol rate in the vapor. For example, at 9 °C and 5000 kPa, the value from the abscissa is 25. If the gas rate is 50×10^3 Sm^3/day and the aqueous phase concentration is 35 wt%, then the methanol in the gas is $25 \times (50 \times 10^3/10^6) \times 35 = 43.75$ kg/day.

From this chart, we can make some general observations. First, for a given temperature, the methanol losses increase with decreasing pressure. Second, at constant pressure, the losses increase with increasing temperature.

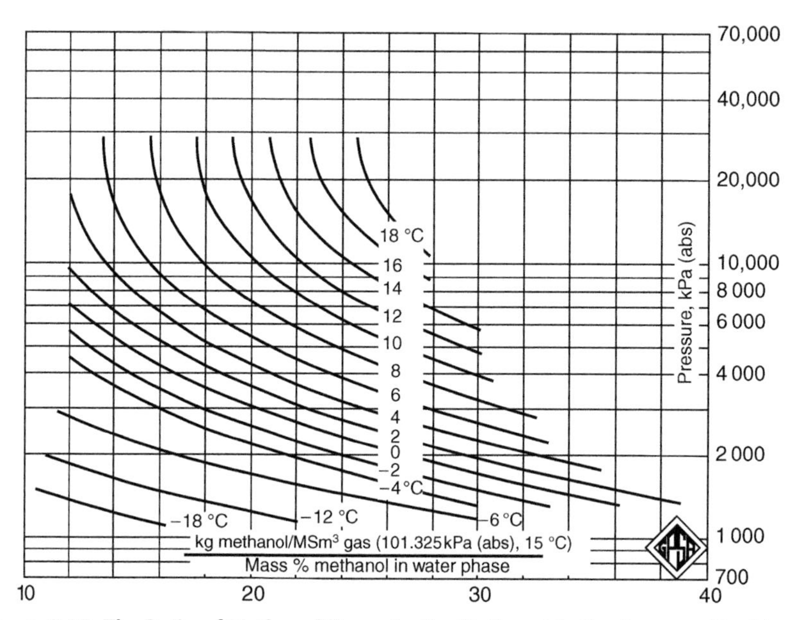

Figure 5.13 ***The Ratio of Methanol Vapor to the Methanol in the Aqueous Liquid as a Function of Pressure and Temperature in SI Units.*** *Reprinted from the* GPSA Engineering Data Book, *11th ed.—reproduced with permission.*

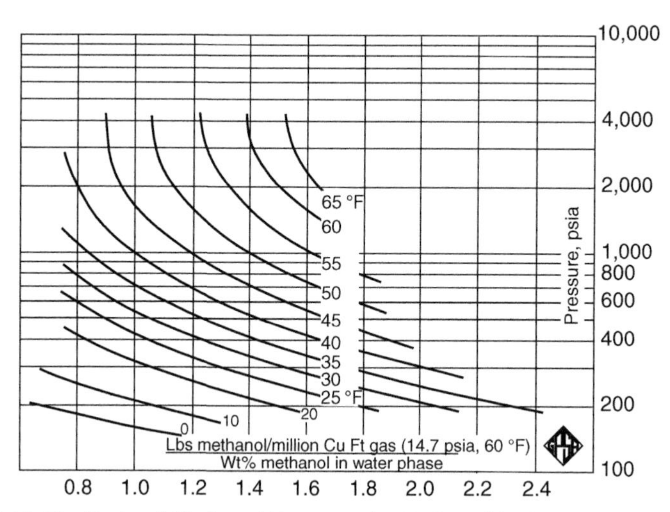

Figure 5.14 ***The Ratio of Methanol Vapor to the Methanol in the Aqueous Liquid as a Function of Pressure and Temperature in American Engineering Units.*** *Reprinted from the* GPSA Engineering Data Book, *11th ed.—reproduced with permission.*

In addition, the higher the gas rate, the higher the methanol losses to the vapor. Finally, the higher the methanol concentration in the aqueous phase, the greater the losses to the vapor. All of these are somewhat intuitive observations.

To use these charts to estimate the methanol in the nonaqueous phase, first one estimates the amount of methanol required using the methods outlined earlier. From that point, one can estimate the amount of methanol present in the other two phases. Only from this can the total amount of methanol injected can be calculated.

The glycols are much less volatile than methanol. In addition, glycols are usually used at low temperature. Thus, losses to the nonaqueous phases are less of a concern when using glycols.

Usually, the engineer will calculate the amount of inhibitor (usually methanol) that must be injected in order to avoid hydrate formation. This value is then passed along to field operations. The field operators will often adjust this value in order to minimize costs while still ensuring problem-free production.

5.12.1 A More Theoretical Approach

A crude estimate of the inhibitor losses to the vapor can be estimated assuming that Raoult's Law applies and that the nonidealities in the vapor phase can be neglected. This leads to the simple equation:

$$y_i = x_i\left(\frac{P_i^{sat}}{P}\right) \tag{5.17}$$

where y_i is the mole fraction of the inhibitor in the vapor phase, x_i is the mole fraction in the aqueous phase, P_i^{sat} is the vapor pressure of the inhibitor, and P is the total pressure. Rearranging this equation to more familiar units (and units that will be used later), we obtain the following equation in SI units:

$$Y_i = \left(\frac{760.4X_iM_i}{100M_i - [M_i - 18.015]X_i}\right)\left(\frac{P_i^{sat}}{P}\right) \tag{5.18}$$

where Y_i is the inhibitor in the vapor phase in kg/MSm^3, X_i is the weight percent inhibitor in the aqueous phase, and M_i is the molar mass of the inhibitor. In American engineering units, the equation is:

$$Y_i = \left(\frac{47484X_iM_i}{100M_i - [M_i - 18.015]X_i}\right)\left(\frac{P_i^{sat}}{P}\right) \tag{5.19}$$

where Y_i is the inhibitor in the vapor phase in pounds per million standard cubic feet (lb/MMCF), X_i is the weight percent inhibitor in the aqueous phase, and M_i is the molar mass of the inhibitor. With the appropriate properties, this equation can be used for any nonionic inhibitor.

Although this is an over simplified method, somewhat more accurate methods will be presented later, we can make some observations based on this equation. As the temperature increases, the vapor pressure increases and, thus, inhibitor losses increase.

Based on this simple analysis, the losses from methanol are about 2.5 times as large as ethanol, and about 200 times that for EG. TEG is even less volatile and losses are negligible.

Comparing this equation to the chart method reveals a potential correction. It can be seen that the error in the estimate from the equation increases with increasing pressure.

$$Y_i = C_f \left(\frac{760.4 X_i M_i}{100 M_i - [M_i - 18.015] X_i} \right) \left(\frac{P_i^{sat}}{P} \right) \tag{5.20}$$

or

$$Y_i = C_f \left(\frac{47484 X_i \ M_i}{100 M_i - [M_i - 18.015] X_i} \right) \left(\frac{P_i^{sat}}{P} \right) \tag{5.21}$$

where C_f is given by the following equation in SI units:

$$C_f = 1.1875 + 1.755 \times 10^{-4} P \tag{5.22}$$

where C_f is unitless and P is in kPa. In American engineering units, this equation becomes:

$$C_f = 1.1875 + 1.210 \times 10^{-3} P \tag{5.23}$$

where P is in psia. This is a purely empirical correction and there is no thermodynamic basis for making such a correction. Therefore, extrapolations outside the range of pressure and temperature should be done with caution.

5.12.2 Inhibitor Losses to the Hydrocarbon Liquid

In addition to the loss of inhibitor to the gas, if a liquid hydrocarbon is present, some of the inhibitor will enter that phase too. This section provides methods for estimating the loss of inhibitor to the condensate.

5.12.2.1 Methanol

The *GPSA Engineering Data Book* gives a chart for the distribution of methanol between a liquid hydrocarbon (condensate) and an aqueous solution reproduced here as Fig. 5.15. This chart is basically a plot of the raw experimental data. Figure 5.16 is a similar plot with some smoothing. Although the estimated values from the two charts are not the same, they are suitable for engineering calculations.

The chart is a plot of the mole fraction in the hydrocarbon liquid as a function of the temperature and the methanol concentration in the water rich phase. Using this chart for a given application requires the molar mass of the hydrocarbon liquid. Unfortunately, there is no typical value for the molar mass. For light condensate, it could be as low as 125 g/mol and for heavier oils it may be as large as 1000 g/mol.

For weight fractions between 20 and 70 wt%, it is sufficiently accurate to "eye ball" your estimate. For methanol concentrations less than 20 wt%, a linear approximation can be used, given the fact that at 0 wt% in the water the concentration in the hydrocarbon liquid is also 0. The resulting equation is:

$$x = \frac{x(20\ \text{wt\%})}{20} X \tag{5.24}$$

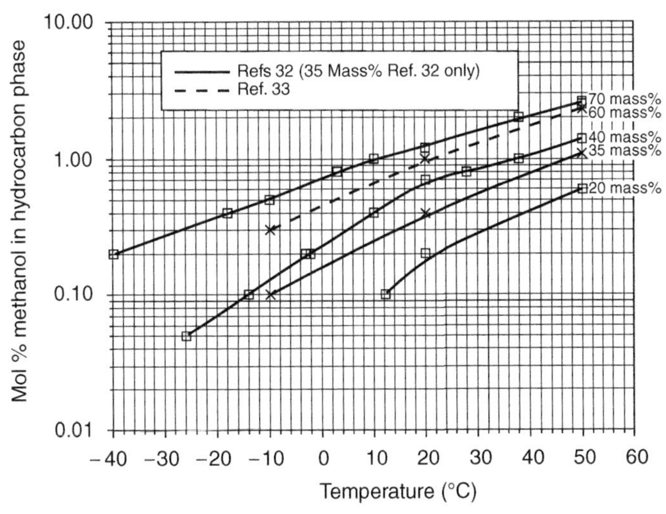

Figure 5.15 ***The Solubility of Methanol in Paraffinic Hydrocarbons as a Function of Aqueous Phase Composition, Pressure, and Temperature in SI Units.*** *Reprinted from the* GPSA Engineering Data Book, *11th ed.—reproduced with permission.*

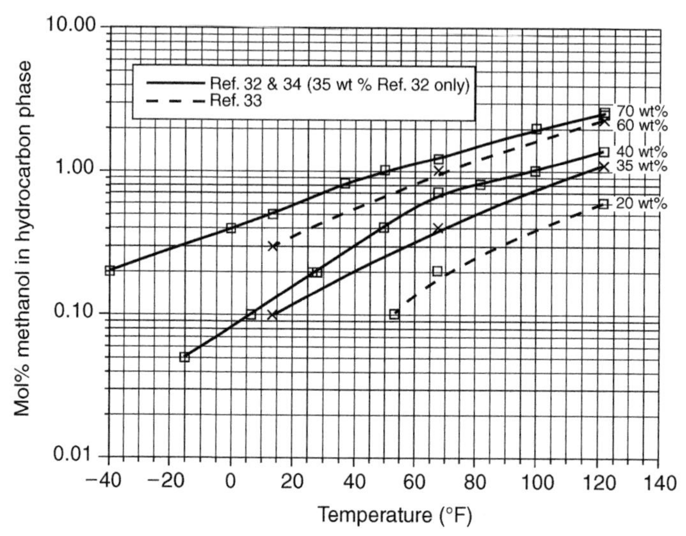

Figure 5.16 ***The Solubility of Methanol in Paraffinic Hydrocarbons as a Function of Aqueous Phase Composition, Pressure, and Temperature in American Engineering Units.*** *Reprinted from the* GPSA Engineering Data Book, *11th ed.—reproduced with permission.*

where X is the given weight percent methanol in the aqueous phase, x (20 wt%) is the mole percent methanol in the condensate at 20 wt% in the water, and x is the mole fraction in the hydrocarbon liquid for the given X. This equation applies to both SI and American engineering units.

The data of Chen et al. (1988) indicate that the methanol losses increase if the hydrocarbon liquid is aromatic. In an aromatic-rich condensate, the methanol losses could be as much as five times those in a paraffinic condensate.

Note the chart shows no effect of pressure on the distribution of methanol between the two liquid phases. This is common for liquid–liquid equilibrium and probably a good assumption in this case.

5.12.2.2 Glycol

A small amount of experimental data in the temperature range from −10 to 50 °C (14 to 122 °F) indicate that the EG in the hydrocarbon liquid is about at least 100 times less than the methanol, when expressed in terms of mole fraction. Based on these observations, it is probably safe to assume that EG losses to the condensate are negligibly small.

TEG losses to the condensate are just as small as those for EG.

5.13 A COMMENT ON INJECTION RATES

Methanol injection rates of 0.15–1.5 m^3/day (1–10 bpd) are common in the natural gas business. Occasionally, they may be more than this, but injection rates of more than 1.5 m^3/day become rather expensive. In oil field terms and from a cost point of view, these seem like relatively large numbers. In reality, they are very small rates. For example, 0.15 m^3/day converts to only 0.1 L/min (0.026 gpm) or 1.7 mL/s. A drop of liquid is approximately 0.5 mL in volume. Therefore, the injection rate is only three drops per second.

Another aspect of inhibitor injection is that often it is at high pressure. Injection pressures greater than 7000 kPa (1000 psia) are common.

Therefore, the injection pump must be designed to handle these conditions—low rates and high pressures. Fortunately, there are such pumps available. Two types are common (1) a diaphragm pump and (2) a piston pump. Both types work quite well for this application.

5.14 SAFETY CONSIDERATIONS

Table 5.7 lists some of the properties of methanol and the glycols from a safety point of view. Methanol is often used as a fuel and hence it is well known that this alcohol is flammable. This is reflected in the relatively high National Fire Protection Association (NFPA) rating. The glycols are significantly less flammable. Thus, from a fire prevention point of view, the storage of methanol is considerable more dangerous than that for glycols. The fire hazard of methanol vs glycol may be more significant in confirmed spaces such as on an offshore rig.

Both methanol and glycols are potential toxins if consumed in large enough quantities. Extrapolating the values for rats to a 70 kg (154 lb), for a human it indicates that about 670 g of pure methanol must be consumed for an LD_{50}. This is equal to slightly less than 1 L (about 1 US quart). This is strictly not how these values should be interpreted, but it gives an indication of the toxicity.

Another problem with the glycols is that they have a sweet flavor, which is attractive to children and some animals.

5.14.1 Diluted Methanol

For safety reasons, some companies do not store pure methanol at their well sites, batteries, plants, etc., choosing to store only diluted methanol. Typically, this would be less than 50 wt% and often much less. If the methanol

Table 5.7 Properties of Various Hydrate Inhibitor and Dehydration Chemicals from a Health and Safety Point of View

	Methanol	EG	DEG	TEG
Flammability				
NFPA[1] rating	3	1	1	1
Flash point	12 °C	111 °C	117 °C	163 °C
Method	Closed cup	Closed cup	Closed cup	Closed cup
UEL[2] in air	36.5%	15.3%	10.6%	–
LEL[3] in air	6.7%	3.2%	1.7%	–
Toxicity				
LD_{50}[4]	9.54 g/kg	4.70 g/kg	14.80 g/kg	17 g/kg
Method	Oral rat	Oral rat	Oral rat	Oral rat
Description	Practically non-toxic	Slightly toxic	Practically non-toxic	Practically non-toxic

DEG, diethylene glycol; EG, ethylene glycol; LD, lethal dose; TEG, triethylene glycol.
Notes on the above table:
[1]NFPA—National Fire Prevention Association (www.nfpa.org). In this rating scheme 4 is the highest fire hazard and 0 is the least.
[2]UEL—Upper Explosion Limit
[3]LEL—Lower Explosion Limit
[4]LD_{50}—LD stands for "Lethal Dose." LD_{50} is the amount of a material, given all at once, which causes the death of 50% (one half) of a group of test animals. The LD_{50} is one way to measure the short-term poisoning potential (acute toxicity) of a material. In general, the smaller the LD_{50} value, the more toxic the chemical is.

injected on-site is diluted, this will require additional methanol to be injected.

There is another problem with using diluted methanol. The concentration of the stored methanol puts an upper limit on the concentration attainable in the pipeline. If the stored methanol is 25 wt%, how can you achieve 50 wt% if that is required based on the hydrate considerations? No matter how much 25 wt% is injected, a concentration of 50 wt% will never be achieved.

5.15 PRICE FOR INHIBITOR CHEMICALS

Table 5.8 gives the approximate cost of chemicals that are used for inhibiting hydrate formation.

As can be seen from Table 5.8, the cost of glycol is significantly greater than methanol. Thus, it is usually cost-effective to recover the injected glycol. Usually, the methanol is injected and not recovered. Although it is

Table 5.8 Approximate Price for Hydrate Inhibitors

	Base Price		Adjusted Price	
	($US)	$US/gal	$(Canadian)/kg	$(Canadian)/L
Methanol[1]	$532/ton	1.60	0.53	0.42
Ethanol[2]	$3.30/US gal	3.30	1.10	0.87
EG[2]	$0.63/lb	5.87	1.39	1.55
DEG[2]	$0.32/lb	7.00	1.65	1.85
TEG[2]	$0.90/lb	8.44	1.98	2.23

DEG, diethylene glycol; EG, ethylene glycol; TEG, triethylene glycol.
Notes:
Base price is taken from original source.
$1 (US) ≈ $1.00 (Canadian) 2007.
Price is for pure ethanol.
Densities from Table 5.1, except DEG, $\rho = 1120$ kg/m^3.
References:
[1]Methanex, March 2008 (www.methanex.com).
[2]ICIS pricing, April 2008 (http://www.icispricing.com).

cost-effective to recover the injected glycol, one reason why this is usually not done is that the produced water is usually a brine. In the regeneration of the glycol, the salt is concentrated in the glycol, which ultimately causes processing problems.

5.16 LOW DOSAGE HYDRATE INHIBITORS

New hydrate inhibitors have entered the market that are markedly different from the thermodynamic inhibitors. The thermodynamic inhibitors are the alcohols, glycols, and ionic salts discussed earlier in this chapter. These basically inhibit the hydrate formation by depressing the freezing point—a thermodynamic effect.

The inhibition of hydrates using methanol or glycol is thermodynamic inhibition. The purpose of the addition of these chemicals is to shift the equilibrium to lower temperatures and higher pressures, thus reducing the region where hydrates can exist.

The new inhibitors, the so-called low-dosage hydrate inhibitors are in two classes: (1) kinetic inhibitors and (2) anticoagulants. They are called low dosage because, as we shall see, they can be used in significantly lower concentrations than the thermodynamic inhibitors.

Many of these new chemicals are patented. The appendix to this chapter gives a partial list of patents relating to low-dosage inhibitors. This is not meant to be an exhaustive list nor does it include any details from the patents.

5.16.1 Kinetic Inhibitors

Recently, there has been a class of chemicals developed called "kinetic inhibitors" or KHIs. Basically, the purpose of this type of inhibitor is to prevent the hydrate from crystallizing. In other words, its purpose is to slow the process by which the hydrate crystallizes. These chemicals delay hydrate formation and growth such there is sufficient time to transport the fluids to their destination. However, given sufficient time, hydrates will form even in the presence of the KHI.

Furthermore, it has been found that such chemicals can be used in very small concentrations. For example, according to Fu et al. (2001), these inhibitors can depress hydrate formation by 11 °C (20 °F) at concentrations less than 3000 ppm.

The chemicals that have been examined for this purpose are usually polymers and therefore are of high molar mass. In addition, they are water soluble (or at least have significant water solubility).

In recent years, there have been several successful field trials of these chemicals, examples include:

1. Fu et al. (2001) describe one trial for a 135 mile long, 4-in flow line transported between 11 and 15 million standard cubic feet per day (MMSCFD) of gas and 13 bpd of produced water, with a salinity of 6 wt% dissolved solids. Prior to the KHI test, methanol was injected at a rate of between 150 and 220 gallons per day. A new program was developed to inject a KHI at an effective concentration of 550 ppm polymer (an injection rate of 3 gallons per day of blended chemical). In the time period reported (more than 1 year), there were no hydrate incidents reported. It is reported that the switch from methanol to KHI results in a chemical cost savings of about 40%.
   Other field trials are described in Fu et al. (2001).
2. Notz et al. (1996) describe the use of a KHI in Texas and Wyoming.
3. Glenat et al. (2004) describe the application of KHI in the offshore flow lines operating at pressures greater than 1000 psia (7 MPa) South Pars field in Iran.
4. Thieu and Frostman (2005) describe the application of KHI at two unspecified locations.
5. Hagen (2010) describes three case studies using KHI in British Columbia and Alberta, Canada.
   a. This location in British Columbia is accessible by road only during the winter months. Thus to get methanol to the location during

other times it had to be flown by helicopter, which was expensive. A switch was made to a mixture of methanol and KHI. This change reduced the demand for chemicals and flying them in was no longer necessary. The chemical could be delivered during the winter months when the roads were passable.

b. This producer had a flow line transporting, on average, 10 m^3/day of oil and 400 × 10^3 Sm^3/day of sweet gas. The water rate was estimated to be 1 m^3/day but was not metered so the exact rate was not known. The line required approximately 600 L/day of methanol for hydrate prevention. After a review, the methanol was replaced by a KHI + methanol (10%) blend. The initial rate was set to 250 L/day and there were no issues with hydrates.

c. In this case, the water, liquid hydrocarbons, and gas were separated and the gas was transported via a pipeline. The gas entered the line at a rate of about 100 × 10^3 Sm^3/day. The gas is water saturated when it enters the line and is slightly sour containing about 2000 ppm of H_2S. Methanol was injected at a rate of about 64 L/day. Although there was no plugging because of hydrate formation, they were found during pigging operations. The treatment scheme was changed to a blend of a KHI and methanol. The initial rate for the blend was 10 L/day, but it was optimized to 6 L/day. Hydrates are no longer observed during pigging of the line.

5.16.2 Anticoagulants

Anticoagulant (also called anti-agglomeration) inhibitors work on a different level. They do not prevent hydrate formation, but they prevent the accumulation of hydrate into a plug. The hydrate stays in a slurry, which can still be transported and will not plug the line.

The use of an anticoagulants implies the transport of a slurry product. Sinquin et al. (2004) provided an interesting and thorough study of the rheology of hydrate slurries. The particles formed in the fluid modifies the flow properties. In the turbulent regime, pressure drop is controlled by the friction factor, whereas in the laminar regime, it is due to changes in the apparent viscosity.

Field application of anticoagulants have been described in the literature. Brief descriptions of a few applications are provided below:

1. Thieu and Frostman (2005) described an application at a field producing 8 MMSCFD of gas, 220 bpd of oil, and about 3 bpd of water. The

operator was injecting about 170 gal/day of methanol, but still had hydrate problems. They switched to a regime that included the injection of 4 gal/day of an anticoagulant and initially 27 gal/day of methanol. Subsequently, the methanol coinjection was terminated.

2. Cowie et al. (2003) described the application of an anticoagulant's inhibitor on an offshore project in Louisiana. Production includes 65,000 bpd of oil, 68 MMSCFD of gas, and water that is transported to the processing facilities via a 41-mile long, 12-inch diameter pipeline.
3. Klomp et al. (2004) described the application of an anticoagulant's inhibitor in a region of the Dutch North Sea. The field produces gas, a waxy crude oil, and water.
4. Hagen (2010) described two case studies using anticoagulant in Alberta, Canada.
   a. A well producing 9 m^3/day of brine, 36 m^3/day of oil, and 4×10^3 Sm^3/day of sour gas. The fluids were transported 9 km. Methanol injection was up to 6000 L/day and still hydrates formed. A program was started to inject AA and the rate was optimized to 90 L/day without hydrate plugging.
   b. At this site, the typical production was 2 m^3/day of water, 3.5 m^3/day of oil, and 0.5×10^3 Sm^3/day of gas. The injection rate for the AA was adjusted to about 15 L/day, which corresponds to a concentration of about 1750 ppm in the aqueous phase, and no hydrate blockages were noted through a winter season.

Examples

Example 5.1

Calculate the freezing point of a 10% solution of methanol in water.

Answer: Using Eqn (5.2):

$$\Delta T = \frac{M_s R T_m^2}{h_{sl}} \times \frac{W_i}{(100 - W_i) M_i}$$

The required input quantities are as follows:

$M_s = 18.015$ g/mol
$R = 8.314$ J/mol K
$T_m = 273.15$ K
$h_{sl} = 6006$ J/mol
$M_i = 32.042$ g/mol

Substituting these quantities into Eqn (5.2) yields:

$$\Delta T = \left[(18.015)(8.314)(273.15)^2/6006\right] \times [10/(100-10)(32.042)]$$
$$= 6.5\ ^\circ\mathrm{C}$$

The reader should verify that the units are correct. Therefore, the freezing point of the mixture is estimated to be −6.5 °C.

Example 5.2

The methane hydrate forms at 15 °C and 12.79 MPa (Table 2.2). Calculate the amount of methanol required to suppress this temperature by 10 °C using the Hammerschmidt equation.

Answer: The molar mass of methanol is 32.042 g/mol. From Eqn (5.4) we have:

$$W = 100(32.042)(10)/[1297 + (32.042)(10)]$$
$$= 32042/1617.42$$
$$= 19.8\ \mathrm{wt\%}$$

So it requires about 20 wt% methanol.

Example 5.3

Repeat the above problem using EG as the inhibitor.

Answer: The molar mass of EG is 62.07 g/mol. From Eqn (5.4) we have:

$$\mathrm{W} = 100(62.07)(10)/[1297 + (62.07)(10)]$$
$$= 62070/1917.7$$
$$= 32.3\ \mathrm{wt\%}$$

This is outside the range of this correlation, so this result should be taken with some trepidation. However, it demonstrates that significantly more EG is required, when expressed as weight fraction, because of its significantly higher molar mass.

Example 5.4

Repeat Examples 5.2 and 5.3 using Fig. 5.4.

Answer: From Fig. 5.4, to get a 10 °C depression in the hydrate formation temperature using methanol requires 21 wt% methanol and 31 wt% EG. In this case, there is good agreement between the chart and the Hammerschmidt equation.

Example 5.5

From Fig. 5.1, it can be estimated that the hydrate formation temperature of hydrogen sulfide is depressed by 18 °C in 35 wt% methanol and by 25 °C in 50 wt% methanol. Compare this to the values predicted by (1) the Hammerschmidt equation, (2) the Nielsen—Bucklin equation, and (3) Fig. 5.4.

Answer: (1) From the Hammerschmidt equation for 35 wt%, we get:

$$\begin{aligned}\Delta T &= 1297(35)/(34.042)(100-35)\\ &= 20.5\ ^\circ\text{C}\end{aligned}$$

and for 50 wt%:

$$\begin{aligned}\Delta T &= 1297(50)/(34.042)(100-50)\\ &= 38.1\ ^\circ\text{C}\end{aligned}$$

These are in error by 14% and 52%, respectively. Actually, the 35 wt% answer is closer than anticipated.

(2) To use the Nielsen–Bucklin equation, we must convert to mole fraction. For methanol, 35 wt% = 23.6 mol% and 50 wt% = 36.0 mol%. The readers can verify these conversions for themselves. Now, for the 35 wt% we get:

$$\begin{aligned}\Delta T &= -72\ln(1-0.236)\\ &= 19.4\ ^\circ\text{C}\end{aligned}$$

and for the 50 wt%:

$$\begin{aligned}\Delta T &= -72\ln(1-0.360)\\ &= 32.1\ ^\circ\text{C}\end{aligned}$$

These represent errors of 8% and 28%, respectively. Although the estimate for the 50 wt% mixture is an improvement over the Hammerschmidt equation, it is not as accurate as we would like considering the Nielsen–Bucklin equation is reputed to be accurate for greater concentrations than these.

(3) From Figure 5.4, you can read from the chart that 35 wt% gives a depression of 18 °C and 50 wt% is about 30 °C, which are errors of 0% and 20%, respectively. These values are improvements over both the Hammerschmidt and Nielsen–Bucklin equations.

Part of the reason for the inaccuracy of all of the methods may be the fact that we are dealing with hydrogen sulfide, which behaves significantly differently than sweet gas. For one thing, H_2S is more soluble in water than are the hydrocarbons. Furthermore, the solubility of H_2S in methanol, and hence aqueous solution, is even larger. This solubility has an effect on the hydrate depression.

Example 5.6

Natural gas is to be transported from a well site to a processing plant in a buried pipeline at a rate of 60×10^3 m^3[std]/day. The production also includes 0.1 m^3[std]/day of water, which is to be transported in the same pipeline. The gas enters the pipeline at 45 °C and 3500 kPa. The hydrate formation temperature of the gas is determined to be 40 °C at 3500 kPa. In the transportation through the pipeline, the gas is expected to cool to 8 °C. Calculate the amount of methanol that must be injected in order to prevent hydrate formation.

Answer: First, calculate the required temperature depression.

$$\Delta T = 40 - 8 = 32\ ^{\circ}\mathrm{C}$$

Include a 3 °C safety factor, the required temperature depression is 35 °C.

Now use Eqn (5.4) to estimate the concentration of methanol required.

$$\begin{aligned} W &= 100(32.042)(35)/[1297 + (32.042)(35)] \\ &= 32042/1617.42 \\ &= 46.4\ \mathrm{wt\%} \end{aligned}$$

This is outside the range of applicability of the Hammerschmidt equation. The value obtained from Fig. 5.4 is 55 wt%, which is both a better estimate and significantly larger than the Hammerschmidt estimate. We will use 55 wt% for the rest of this calculation.

Next, we calculate that 0.1 m^3 of water has a mass of about 100 kg. Therefore, to get a 55 wt% solution, based on water + methanol and not the entire stream, requires the injection of 122 kg/day of methanol. The density of methanol is 790 kg/m^3, so the methanol is injected at a rate of 0.155 m^3/day or 155 L/day.

If we used the values of 46.4 wt% from the Hammerschmidt equation, this converts to an injection rate of 86.6 kg/day.

Note, the amount of water in the natural gas (because it is being produced with free water, we can assume that it is saturated with water), has been neglected in this calculation. It is left to the reader to calculate the amount of water in the gas and determine if additional methanol is required to account for this water.

Example 5.7

Estimate the amount of methanol required to saturate the gas in Example 5.6, and from that determine the total amount of methanol that must be injected.

Answer: Using Fig. 5.12, we can read that at 8 °C and 3500 kPa, the methanol in the vapor is $27 \frac{\mathrm{kg\ MeOH}}{(10^6\ \mathrm{Sm^3})(\mathrm{wt\%\ MeOH})}$.

In the previous example, we determined that the aqueous phase concentration should be 55 wt%. Therefore, the amount of methanol in the vapor is:

$$27 \times \left(60 \times 10^3\ \mathrm{Sm^3}/10^6\right) \times (55\ \mathrm{wt\%}) = 89.1\ \mathrm{kg/day}$$

which converts to 0.113 m^3/day or 113 L/day.

Therefore, the total methanol requirement is 155 + 113 = 268 L/day. At this point, it is worth noting that 42% of the methanol injected will vaporized. Therefore, instead of the 155 L/day injection rate, which is based only on the aqueous phase, the actual injection rate that is required is more than 1.7 times that amount.

Example 5.8

Estimate the cost of the methanol injection from Example 5.7.

Answer: From the earlier calculation the injection rate was 268 L/day, so 268 L/day × \$0.27 (Canadian)/L = \$72 (Canadian)/day or about \$26,000 (Canadian)/year or \$17,300 (US)/year.

Example 5.9

Natural gas flowing in a pipeline exits the line at 45 °F and 700 psia and the flow rate of the gas is 5 MMSCFD. In order to prevent hydrate formation, it is estimated that there should be 25 wt% methanol in the aqueous phase. Calculate the methanol losses to the vapor phase (1) using the simple equation (Eqn (5.19)), (2) using the corrected equation (Eqn (5.21)), and (3) using the chart (Fig. 5.13). (4) Estimate the concentration of methanol in the condensate, if condensate were present.

Answer: (1) At 45 °F, the vapor pressure of methanol is 0.92 psia. From the simple equation:

$$Y_i = \left(\frac{47484 X_i M_i}{100 M_i - [M_i - 18.015] X_i}\right)\left(\frac{P_i^{sat}}{P}\right)$$

$$= \left(\frac{(47484)(25)(32.04)}{100(32.04) - [32.04 - 18.015](25)}\right)\left(\frac{0.92}{700}\right)$$

$$= 17.65 \text{ lb MeOH/MMCF}$$

Now multiply this value by the gas flow rate to get the total methanol loss.

$$(17.65)(5) = 88.25 \text{ lb/day} = \mathbf{3.7\ lb/h}$$

(2) The above value is corrected using the following equation:

$$C_f = 1.1875 + 1.210 \times 10^{-3} P = 1.1875 + 1.210 \times 10^{-3}(700)$$

$$= 2.035$$

Based on this correction factor, the simple equation underestimates the methanol losses by a factor of two. In other words, the actual losses will be twice those estimated using the simple equation.

$$Y_i = (2.035)(17.65) = 35.92 \text{ lb MeOH/MMCF}$$

and the total methanol loss is (35.95)(5) = 179.75 lb/day = **7.5 lb/h**.

(3) From the chart, $Y/X = 1.4$, therefore, $Y = 1.4(25) = 35$ lb MeOH/MMCF. Again multiplying by the gas flow rate gives $(35)(5) = 175$ lb/day = **7.3 lb/h**. The agreement between the corrected equation and the chart is very good in this case.

(4) Estimate the concentration of the methanol in the condensate at 30 °F if the methanol concentration in if the water phase is 15 wt%.

From the chart, the methanol concentration in the condensate at 20 wt% is 0.07 mol%. Then using the following equation:

$$x = \frac{x(20\ \text{wt\%})}{20} X = \frac{(0.07)(15)}{20} = \mathbf{0.053\ mol\%}$$

Example 5.10

Assuming a water flow rate of 1 bpd and methanol needs to be injected to have a 25 wt% solution in order to assure hydrates will not form in the pipeline.

1. How much pure methanol must be injected to achieve the goal of 25 wt%?
2. If the methanol is available with a concentration of 30 wt%, how much solution must be injected to achieve the goal of 25 wt%? How much of this solution is methanol?

Answer: First, 1 bpd of water is 42 gal/day. The density of water is 8.34 lb/gal and, thus $42 \times 8.34 = 350$ lb/day. This result is necessary for both parts of this question.

1. In order to have 25 wt%, which is 0.25 mass fraction, let x be the mass of methanol injected. Then from a simple ratio:

$$0.25 = x/(x + 350)$$

Solving for x gives **117 lb/day** of pure methanol.

2. In this case, let x be the amount of solution injected. This time the ratio is:

$$0.25 = 0.3x/(x + 350)$$

solving for x gives 1750 lb/day of 30 wt% methanol. This in turn means 0.3×1750 or **525 lb/day** of methanol are injected. Because the methanol is injected in dilute form, and water is added to the system, more total methanol needs to be injected.

Example 5.11

Assume a protein that inhibits the formation of ice is in the blood of an animal. Further assume that this inhibition is simply a freezing point depression. Estimate the concentration of the protein in order to achieve a 0.5 °C depression. Typically, these proteins are high molecular weight (greater than 2500 g/mol), so assume a value of 2500 g/mol.

Answer: Begin with the freezing point depression equation:

$$x_i = \frac{h_{sl} \Delta T}{R T_m^2}$$

$R = 8.314$ J/mol K
$T_m = 273.15$ K
$h_{sl} = 6006$ J/mol

$$x_i = \frac{h_{sl} \Delta T}{R T_m^2} = \frac{(6006)(0.5)}{(8.314)(273.15)^2}$$

$$= 4.84 \times 10^{-3}$$

Convert to weight fraction:

$$X_i = \frac{M_i x_i}{M_i x_i + M_w (1 - x_i)}$$

$$= \frac{(2500)\left(4.84 \times 10^{-3}\right)}{(2500)(4.84 \times 10^{-3}) + (18.015)(1 - 4.84 \times 10^{-3})}$$

$$= 0.403 = 40.3 \text{ wt\%}$$

It would seem highly unlikely that the protein would be 40 wt% of the blood, so freezing point depression is unlikely to account for the observed reduction in the freezing point.

REFERENCES

Anderson, F.E., Prausnitz, J.M., 1986. Inhibition of gas hydrates by methanol. AIChE J. 32, 1321–1333.

Arnold, K., Stewart, M., 1989. Surface Production Operations Vol. 2 Design of Gas-Handling Systems and Facilities. Gulf Publishing, Houston, TX.

Chen, C.J., Ng, H.-J., Robinson, D.B., 1988. The Solubility of Methanol or Glycol in Water-Hydrocarbon Systems. GPA Research Report RR-117. Gas Processors Association, Tulsa, OK.

Cowie, L., Shero, W., Singleton, N., Byrne, N., Kauffman, R.L., August 2003. New hydrate inhibitors enhance deepwater production at BP's Horn Mountain. Deepwater, 39–41.

Fu, S.B., Cenegy, L.M., Neff, C.S., 2001. A summary of successful field applications of a kinetic hydrate inhibitor. In: SPE Internat. Symp. Oil Field Chemicals, Paper No. SPE 65022, Houston, TX, Feb. 13–16.

Glenat, P., Peytatavy, J.L., Holland-Jones, N., Grainger, M., 2004. South-Pars phases 2 and 3: the kinetic hydrate inhibitor (KHI) experience at field start-up. In: 11th Abu Dhabi Petrol. Exhib. & Conf., Paper No. SPE 887751, Abu Dhabi, UAE, Oct. 10–13.

Hagen, S., 2010. Hydrate inhibitors: alternatives to straight methanol injection. In: NACE Northern Area Western Region Conf., Calgary, AB, Feb. 15–18.

Klomp, U., Le Clerq, M., Van Kins, S., 2004. The first use of a hydrate anti-agglomerant in a fresh water producing gas/condensate field. In: Petromin Deep Water Conf., Kuala Lumpur, Malaysia, August.
Laidler, K.J., Meiser, J.H., 1982. Physical Chemistry. Benjamin/Cummings Publishing Co., Menlo Park, CA.
Leu, A.-D., Robinson, D.B., Chung, S.Y.-K., Chen, C.-J., 1992. The equilibrium phase properties of the propane-methanol and *n*-butane-methanol binary systems. Can. J. Chem. Eng. 70, 330–334.
Maekawa, T., 2001. Equilibrium conditions for gas hydrates of methane and ethane mixtures in pure water and sodium chloride solution. Geochem. J. 35, 59–66.
Mainusch, S., Peters, C.J., de Swaan Arons, J., Javanmardi, J., Moshfeghian, M., 1997. Experimental determination and modelling of methane hydrates in mixtures of acetone and water. J. Chem. Eng. Data 42, 948–950.
McCain, W.D., 1990. The Properties of Petroleum Fluids, second ed. PennWell Publishing Co., Tulsa, OK.
Mei, D.-H., Liao, J., Yang, J.-T., Guo, T.-M., 1998. Hydrate formation of a synthetic natural Gas mixture in aqueous solutions containing electrolyte, methanol, and (electrolyte + methanol). J. Chem. Eng. Data 43, 178–182.
Ng, H.-J., Chen, C.-J., Robinson, D.B., 1985. Hydrate Formation and Equilibrium Phase Compositions in the Presence of Methanol: Selected Systems Containing Hydrogen Sulfide, Carbon Dioxide, Ethane, or Methane. GPA Research Report RR-87, Tulsa, OK.
Ng, H.-J., Robinson, D.B., 1994. New developments in the measurement and prediction of hydrate formation for processing needs. Ann. N.Y. Acad. Sci. 715, 450–462.
Nielsen, R.B., Bucklin, R.W., 1983. Why not use methanol for hydrate control? Hydro. Proc. 55 (4), 71–75.
Notz, P.K., Bumgardner, S.B., Schaneman, B.D., Todd, J.L., November 1996. Application of kinetic inhibitors to Gas hydrate problems. SPE Prod. Facil.
Østergaard, K.K., Masoudi, R., Tohidi, B., Danesh, A., Todd, A.C., 2005. A general correlation for predicting the suppression of hydrate dissociation temperature in the presence of thermodynamic inhibitors. J. Petrol. Sci. Eng. 48, 70–80.
Pedersen, K.S., Fredenslund, Aa, Thomassen, P., 1989. Properties of Oils and Natural Gases. Gulf Publishing, Houston, TX.
Sinquin, A., Palmero, T., Peysson, Y., 2004. Rheological and flow properties of Gas hydrate suspensions. Oil Gas. Sci. Tech. 59, 41–57.
Sloan, E.D., 1999. Clathrate Hydrates of Natural Gases, second ed. Marcel Dekker, New York, NY.
Thieu, V., Frostman, L.M., 2005. Use of low-dosage hydrate inhibitors in sour systems. In: SPE Internat. Symp. Oil Field Chemicals, Paper No. SPE 93450, Houston, TX, Feb. 2–4.
Weast, R.C. (Ed.), 1978. CRC Handbook of Chemistry and Physics. CRC Press, West Palm Beach, FL.

CHAPTER 6

Dehydration of Natural Gas

Dehydration is the process by which water is removed from natural gas. This is a common method used for preventing hydrate formation. If there is no water present, then it is impossible for a hydrate to form. If there is only a small amount of water present, then the formation of hydrate is less likely.

There are other reasons for dehydrating natural gas. The removal of water vapor reduces the risk of corrosion in transmission lines. Furthermore, dehydration improves the efficiency of pipelines by reducing the amount of liquid accumulating in the lines—or even eliminates it completely.

There are several methods of dehydrating natural gas. The most common of these are: (1) glycol dehydration (liquid desiccant), (2) molecular sieves (solid adsorbent), and (3) refrigeration. Each of these methods will be reviewed in this chapter.

There are several other dehydration methods that are less commonly used and they will not be discussed here.

6.1 WATER CONTENT SPECIFICATION

A typical water specification is 112 mg of water per standard cubic meter of gas (7 lb/MMCF) in many jurisdictions in the United States and 64 mg/Sm^3 (4 lb/MMCF) in many jurisdictions in Canada. In other jurisdictions, other specifications are used. Furthermore, the process may dictate that a lower water content is required. For example, cryogenic processes, those operating at very low temperatures, require that the gas be very dry.

It is also common to refer to the water content of a gas in terms of water dew point, with the dew point being the temperature at which the water just begins to condense. Thus another common specification is a −10 °C (14 °F) water dew point. However, this method must be used with some caution because dew points at temperatures below 0 °C (32 °F) represent a metastable condition. At temperatures below 0 °C, a true liquid dew point does not exist because the stable form of water at these temperatures is a solid phase, either ice or hydrate.

Natural Gas Hydrates
ISBN 978-0-12-800074-8
http://dx.doi.org/10.1016/B978-0-12-800074-8.00006-5

6.2 GLYCOL DEHYDRATION

The most common method for dehydration in the natural gas industry is the use of a liquid desiccant contactor–regeneration process. In this process, the wet gas is contacted with a lean solvent (containing only a small amount of water). The water in the gas is absorbed by the lean solvent, producing a rich solvent stream (one containing more water) and a dry gas.

6.2.1 Liquid Desiccants

Several liquids possess the ability to absorb water from a gas stream. Few liquids, however, meet the criteria for a suitable commercial application. Some of the criteria of commercial suitability are as follows:

1. The absorbing liquid should be highly hygroscopic, that is, it must have a strong affinity for water.
2. The hydrocarbon components of natural gas should have a low solubility in the solvent in order to minimize the loss of desired product and to reduce hydrocarbon emissions.
3. The desiccant should be easily regenerated to higher concentration for reuse, usually by the application of heat, which drives off the absorbed water.
4. The desiccant should have a very low vapor pressure. This will reduce the amount of solvent losses from vaporization.
5. The desiccant should exhibit thermal stability, particularly in the high temperature ranges found in the reboiler.
6. Suitable solutions should not solidify in the temperature ranges expected in the process of dehydration.
7. All liquids must be noncorrosive to the selected metallurgy of all dehydration equipment, especially the reboiler vapor space, the stripping column of the regenerator, and the bottom of the contactor.
8. The liquid desiccants should not chemically react with any of the natural gas constituents, including carbon dioxide and sulfur compounds.

6.2.1.1 Glycols

The organic compounds known as glycols approximate the properties that meet the commercial application criteria. Glycols have a higher boiling point than water and a low vapor pressure. Glycols will, however, decompose at elevated temperatures. The decomposition temperature limits the maximum temperature at which the process operates, particularly in the reboiler.

Several glycols have been found suitable for commercial application. Monoethylene glycol, which is commonly known as simply ethylene glycol (EG), diethylene glycol (DEG), triethylene glycol (TEG), and tetraethylene glycol (TREG) are the most common for dehydration applications. The glycols have the following chemical structures:

EG: $HO–CH_2–CH_2–OH$
DEG: $HO–CH_2–CH_2–O–CH_2–CH_2–OH$
TEG: $HO–CH_2–CH_2–O–CH_2–CH_2–O–CH_2–CH_2–OH$
TREG: $HO–CH_2–CH_2–O–CH_2–CH_2–O–CH_2–CH_2–O–CH_2–CH_2–OH$

TEG is by far the most used in natural gas dehydration. It exhibits most of the desirable characteristics listed previously and has other advantages compared with other glycols.

By comparison, DEG is marginally lower in cost than TEG. However, because DEG has a larger vapor pressure, it has larger losses. TEG has less affinity for water and thus has less dew point depression.

TREG is higher in cost and is more viscous than TEG. High viscosity translates into higher pumping costs. On the other hand, TREG has a lower vapor pressure that reduces losses.

6.2.2 Process Description

Basically, the liquid desiccant process is a two-step process. In the first step, the water is absorbed from the gas in a staged tower. The solvent is regenerated in a second column. The solvent is then returned to the first column to remove water from more feed gas. A simplified flow sheet for the glycol dehydration process is shown in Fig. 6.1.

The TEG natural gas dehydration unit operates at relatively high pressure on the contactor side and low pressure on the regeneration side. The high-pressure side comprises the glycol contactor and the inlet separator. The low-pressure side is made up of the regenerator and the flash tank and associated equipment.

The rest of this section discusses the individual components of the dehydration process in some detail.

6.2.2.1 Inlet Separator

The first step in the dehydration of natural gas is to remove any free liquids in the stream. A separator should be included upstream of the contactor to separate any hydrocarbon liquids and/or free water.

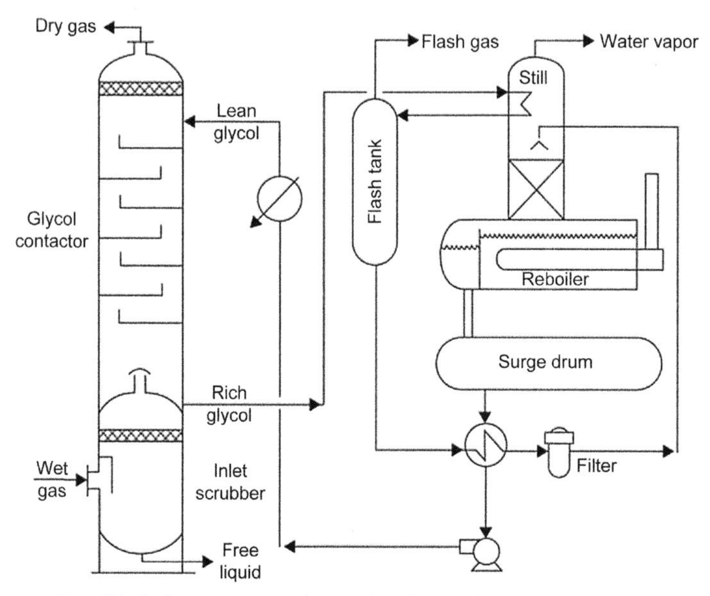

Figure 6.1 ***Simplified Flow Diagram for a Glycol Dehydration Unit.*** *Reprinted from the* GPSA Engineering Data Book, *11th ed.—reproduced with permission.*

This separator could be a two- or three-phase separator, depending on the amount of free water expected. The inlet separator can be freestanding with interconnecting piping to the contactor or it can be as an integral part of the contactor, usually at the base of the contactor with a chimney tray between the contactor bottom and the separator vessel.

The separator should be equipped with a high-efficiency wire mesh mist extractor in the top portion to remove any liquid entrainment and particulates from the gas stream before entering the absorber section. Integral separators are usually outfitted with a heating coil to prevent water from freezing. Hot solvent from the accumulator is circulated through this heating coil to provide the required heat.

6.2.2.2 The Contactor

The contactor (also called an absorber) is the workhorse of the dehydration unit. It is in the contactor that the gas and liquid are mixed and the actual water removal takes place.

The contactor is a typical absorber tower properly sized with the process objective in mind. The feed gas flow rate is the most significant factor in determining the diameter of the contactor. The outlet gas water content specification is the key to determine the contactor height, although other factors contribute as well. The contactor is made up of a number of equilibrium stages, enough to ensure mass transfer from the gas phase to the liquid is such that the outlet gas is at the desired water specification. The actual stages could be either (1) trays like bubble caps, valve trays, or sieve trays or (2) a suitable packing material. Structured packing is finding more acceptance in glycol contactors.

The contactor operates by the fundamental principles of an absorber. The flow of streams is countercurrent. Feed gas enters the bottom of the contactor and flows upward. Lean solvent enters the top of the contactor and flows downwards. The solvent absorbs water as it travels downward through the column and the gas transfers the water to the solvent as it travels upward.

The contactor pressure is set by the feed gas pressure, which is normally in the range of 4000–8500 kPa (600–1200 psia). The contactor is essentially isothermal; that is, the temperature profile is essentially uniform throughout the contactor.

6.2.2.3 The Flash Tank

The rich glycol is withdrawn from the bottom of the contactor, usually on level control. Typically the lean glycol is preheated, often by passing it through tubes in the overhead condenser at the top of the still column. Then it is flashed at low pressure in a flash tank where most of the volatile components (entrained and soluble) are vaporized. Flash tank pressures are typically in the range of 300–700 kPa (50–100 psia).

The glycol leaves the flash tank, again usually on level control, and then passes through a filter. Then the rich glycol enters the rich–lean heat exchanger.

6.2.2.4 Lean–Rich Exchanger

The basic purpose of the lean–rich exchanger is to conserve energy. In the lean–rich exchanger, hot, lean glycol from the regeneration is cooled with rich from the contactor. The lean glycol entering the contactor should be cool and the rich glycol to regeneration should be warm.

6.2.2.5 The Regenerator

A basic regeneration unit is made up of a combination of a fired reboiler, located at the lower section of a horizontal vessel with a vapor space above the tube bundle, a distillation column (still column) connected vertically to the vapor space of the reboiler vessel, and a surge tank located below the reboiler. Also included in the regeneration unit is a condensing coil added to the top of the still column to provide reflux to improve solvent/water separation. As was noted previously, often this coil performs the dual purpose of preheating the rich glycol ahead of the flash tank.

The size of the regenerator is determined by a balance between the solvent circulation rate, the amount of water vapor in the gas stream, and the reboiler temperature. The standard TEG dehydration unit operates effectively at a reboiler temperature around 175 °C (350 °F), or about 20 °C (30 °F) below the decomposition temperature of TEG.

In the regenerator, the separation of water from glycol takes place by fractionation. Water and glycol have widely varying boiling points 100 °C (212 °F) for water vs 288 °C (549 °F) for TEG. Furthermore, the two substances can be easily separated by fractional distillation. This is accomplished in the still column mounted directly on top of the reconcentration vessel. Within the column water, rich vapor rises in intimate contact with descending glycol-rich liquid. Between the two phases, there is a continuous exchange of material and heat. The temperature difference causes the glycol vapor (heavy component) to condense and liquid water (light component) to vaporize. At the top of the column, the vapor is virtually pure water, whereas there is very little water in the glycol in the bottom. A small portion of the vapor mixture (mainly water) at the top condenses at the overhead condenser to provide sufficient reflux that will aid in the process of fractionation. The main purpose of the still column is to effect final separation between the absorbed water and the absorbing TEG glycol, to vent the separated water to the atmosphere, and to recover the glycol vaporized by the reboiler.

The glycol-rich liquid, now becoming lean glycol, leaves the bottom of the packed still column and enters the reboiler vessel. Heat is applied in the reboiler, which is usually direct fired, to raise the temperature and cause partial vaporization. In a normal TEG dehydration unit, this temperature level has been found to cause no noticeable thermal decomposition of the TEG. The lean hot glycol leaves the reboiler vessel and overflows by gravity to the surge tank, a vessel normally located below the reboiler vessel.

The hot, lean glycol passes to the lean–rich exchanger where it is cooled. Ultimately it is returned to the contactor and the cycle is complete.

6.2.2.6 Glycol Pump

A circulation pump takes suction from the surge tank and raises the pressure of the concentrated glycol and delivers the glycol to the top tray of the glycol contactor. It is via the pump that the glycol circulation rate is established.

From the pump discharge, the glycol is passed through the lean–rich exchanger to experience some cooling and then through the lean gas exchanger to experience further cooling before entering the contactor.

6.2.3 Short Cut Design Method

The description of the process presented earlier in this chapter gives many of the process parameters. However, several parameters were not discussed and are reviewed in this section.

In this section, some methods are presented for the preliminary sizing of a TEG dehydration unit. Those interested in a more detailed design procedure are referred to the paper by Sivalls (1976).

The most important parameter in the design of a glycol dehydration unit is the glycol circulation rate. In general, the lower the circulation rate the lower the operating cost for the glycol dehydration unit. The circulation rate for the rich glycol is typically between 17 and 33 L of glycol per kilogram of water in the inlet gas (2–4 gal/lb of water). Circulation rates less than 17 L/kg (2 gal/lb) are not recommended.

Given the water content of the inlet gas and the specified circulation rate per mass of water removed allows for the calculation of the actual circulation rate.

$$L = \frac{L_W w_i G}{24} \tag{6.1}$$

A consistent set of units must be applied. For example, the circulation rate, L, will be in L/min if L_W is in L/kg, w_i is in kg/1000 Sm^3 (or g/Sm^3), and G is in 10^3 Sm^3/day. Alternatively, using American Engineering units: L will be in gpm if L_W is in gal/lb, w_i is in lb/MMCF, and G is in MMCFD.

Typically, the gas enters the dehydration unit saturated with water, although this is not always the case. Methods for calculating the water content of gas are presented in Chapter 10. The water content of the outlet gas is the design specification.

The amount of water removed can be calculated using the following expression:

$$W_R = \frac{(w_i - w_o)G}{24} \tag{6.2}$$

where W_R is the amount of water removed, w_i is the water content of the inlet gas, w_o is the water content of the exit gas, and G is the gas flow rate. As with the previous equation, Eqn (6.2) requires a consistent set of units.

The diameter of the contactor can be approximated using Fig. 6.2. This figure is only presented in American Engineering units. Even in countries where the metric system is commonly used, it is still typical to specify the column diameter in feet. Furthermore, it is common to round the approximate diameter to the nearest half-foot (6 in).

The duty for the regenerator reboiler can be estimated using the following equations:

$$Q = 560L \tag{6.3a}$$

where Q is the duty in kJ/h and L is the circulation rate, in L/min, calculated as specified previously. Or in American Engineering units:

$$Q = 2000L \tag{6.3b}$$

where Q is the duty in Btu/h and L is the circulation rate, gpm.

In addition to absorbing water vapor, some hydrocarbons are also absorbed. As an approximate rule of thumb, TEG typically absorbs about 0.007 m^3/L (1 SCF of sweet gas per gallon of glycol) at 7 MPa (1000 psia) and 38 °C (100 °F). The solubility will be significantly higher if the gas contains H_2S and CO_2.

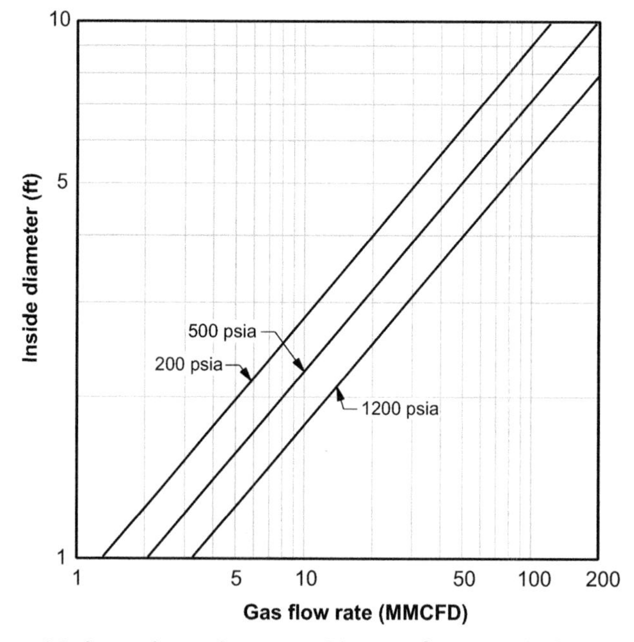

Figure 6.2 ***Approximate Contractor Diameter for TEG Dehydration Unit.***

This should give the reader sufficient information to perform preliminary sizing calculations for a TEG dehydration unit.

6.2.3.1 Benzene, Ethylbenzene, Toluene, and Xylenes

The solubility of paraffin hydrocarbons in TEG is very low. On the other hand, aromatic hydrocarbons (benzene, ethylbenzene, toluene, and xylenes, which are also called BTEX) have a significant solubility. About 25% of the aromatics in the gas stream could be absorbed by TEG under typical contactor conditions of temperature, pressure, and circulation rate. Aromatics absorption is higher for xylenes and lower for benzene. Aromatics absorption is a function of concentration in the gas steam, temperature and pressure of the contactor, and TEG circulation rate. The lower the circulation, the less is the absorption.

The absorption of aromatics by the desiccant and the subsequent liberation of these compounds during the regeneration step is a significant problem with the TEG dehydration process. The aromatic compounds are carcinogenic and therefore pose a health risk to those exposed to them.

6.2.3.2 Software

Process simulation has become very popular in the past few years. Powerful process simulation packages are available that easily fit onto the desktop computer of almost all design engineers.

The modern process simulation packages (*Hysys*, *Aspen*, *SimSci*, *Prosim*, and others) can be used to model glycol dehydration in all of its configurations. They can also be used for dehydration in a refrigeration process.

In addition, the *Gas Research Institute* sponsored a project to build a simulator specifically for glycol dehydration units. The resulting simulator is *GRI Glycalc*.

The design engineer should use these models with some caution. It was mentioned previously that hydrocarbon + methanol systems are difficult to model, and the same is true for glycol + hydrocarbon systems. When simulating these processes, the engineer should select a model suitable for these systems.

6.2.4 Approximate Capital Cost

Figure 6.3 is provided to make quick estimates (±30%) for the purchased cost of a glycol dehydration unit. These are based on information from the literature (Tannehill et al., 1994) and from the author's experience.

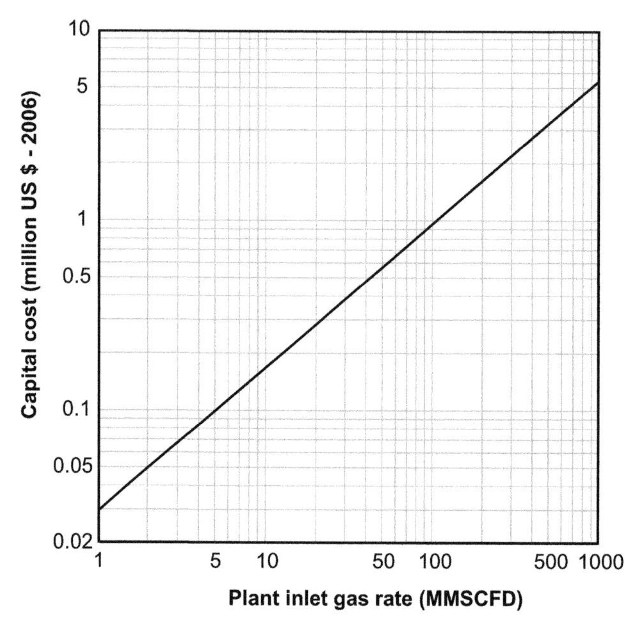

Figure 6.3 ***Approximate Capital Cost for TEG Dehydration Units.*** (Multiply by 1.25 for sour service.)

For both the dehydration units and the line heater, as an order of magnitude estimate, it can be assumed that the installation cost is between 0.8 and 1.2 times the purchased cost. In other words, the installation cost is approximately equal to the purchased cost.

Clearly these costs are approximate at best. Many factors have not been included, such as the operating pressure, TEG circulation rate, stripping gas, or no stripping gas, and so on. They should be used for rapid budget cost estimates only.

6.3 MOLE SIEVES

Unlike glycol dehydration, which is an *absorption* process, dehydration with molecular sieves is an *adsorption* process. Water in the gas adheres to the solid phase, the solid being the mole sieve and thus is removed from the natural gas. Molecular sieves are usually used when very dry gas is required (such as a cryogenic process). The mole sieve process can dry a gas to less than 1 ppm (1 mg/Sm^3 or 0.05 lb/MMCF).

In the mole sieve process, the wet gas enters a bed of adsorbent material. The water in the gas adsorbs onto the bed and a dry gas is produced. Once the bed becomes saturated with water, that is, when no more water can be adsorbed, the bed must be regenerated.

Often, to reduce the size of the mole sieve unit, a glycol dehydration unit is used for bulk water removal. This is followed by a mole sieve unit for the final drying.

6.3.1 Process Description

The adsorption of water vapor from a gas stream is a semibatch process; therefore, at least two beds are required. One bed is in the adsorption phase and the other is in the regeneration or cooling phase. As the bed adsorbs water it becomes saturated, and that portion of the bed can no longer adsorb water. Once the entire bed is saturated with water, the bed must be regenerated.

Regeneration is achieved via the application of heat. Thus a hot gas stream is passed through the bed to strip the water and thus regenerate the bed. Following the stripping stage, the bed must be cooled before it can be placed back in service.

A simplified flow sheet for a two-bed adsorption scheme is shown in Fig. 6.4.

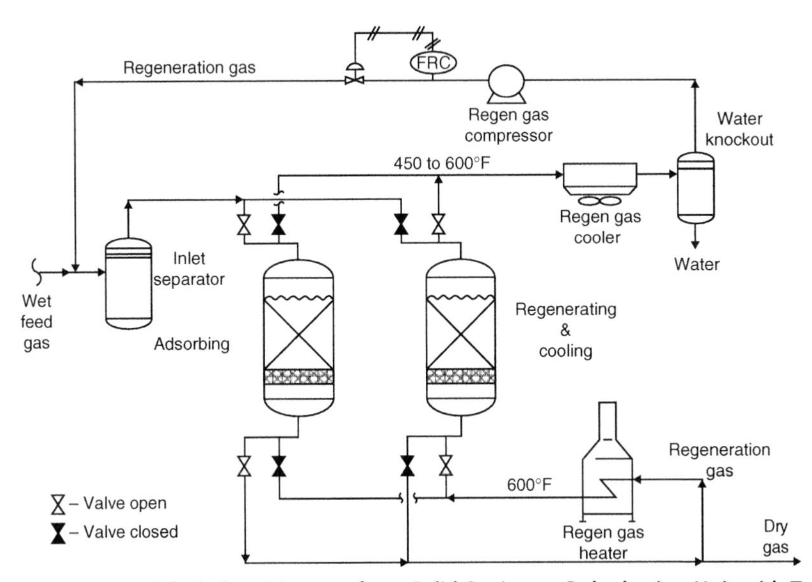

Figure 6.4 ***Simplified Flow Diagram for a Solid Desiccant Dehydration Unit with Two Towers.*** *Reprinted from the* GPSA Engineering Data Book, *11th ed.—reproduced with permission.*

6.3.2 Simplified Modeling

Mole sieves are more difficult to model and usually require specialized software. One reason that mole sieves are more difficult to simulate is that they are semibatch operations. Typical process simulators are steady state and cannot properly handle the batch mode.

However, the adsorption takes place in a packed bed and the usual methods for sizing packed beds are applicable. The reader should consult Trent (2001) for guidelines for designing mole sieves.

Presented in this chapter are some rules of thumb for performing preliminary design calculations for molecular sieve dehydration. For dehydration of natural gas, the most commonly used mole sieve is Type 4A, which comes in either granular or spherical pellets. A summary of the properties of Type 4A is listed in Table 6.1.

The first consideration in the design of a mole sieve unit is to determine the amount of water to be adsorbed. This is calculated using Eqn (6.2).

Next the amount of mole sieve required can be estimated. It is typical to use an adsorption cycle of between 8 and 24 h, with 8 being typical. For design purposes with the Type 4A mole sieve, it is common to use an adsorption of 1 kg of water per 10 kg of adsorbent (1 lb water/10 lb mole sieve). This results in the mass of adsorbent required in the bed. The mass of adsorbent can then be converted to a volume with the bulk density of the adsorbent.

The next consideration is the velocity through the bed. When dealing with packed beds, it is usual to refer to the superficial velocity rather than the actual velocity. The actual velocity through the bed is a complex function of bed geometry and is a very difficult parameter to deal with. On the other hand, the superficial velocity is a synthetic number but it is readily calculated.

The maximum superficial velocity is a function of the total pressure. Figures 6.5 and 6.6 show the relationship between these two parameters, the first in SI units and the second in American Engineering units. If the velocity is greater than this value, the bed will not function properly.

Table 6.1 Properties of Type 4A Mole Sieve Material

Regeneration temperature	450–550 °F	230–290 °C
Bulk density	40–45 lb/ft^3	640–720 kg/m^3
Heat capacity	0.25 Btu/lb °F	1.0 kJ/kg °F

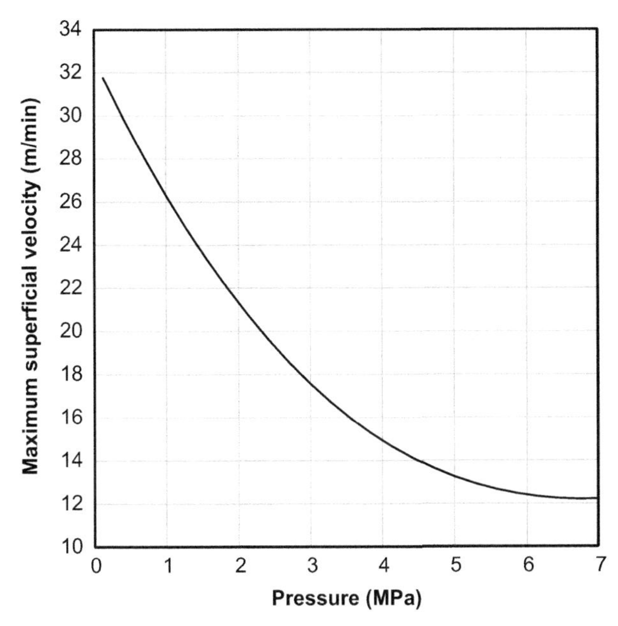

Figure 6.5 *Maximum Superficial Velocity through a Mole Sieve Bed (SI Units).*

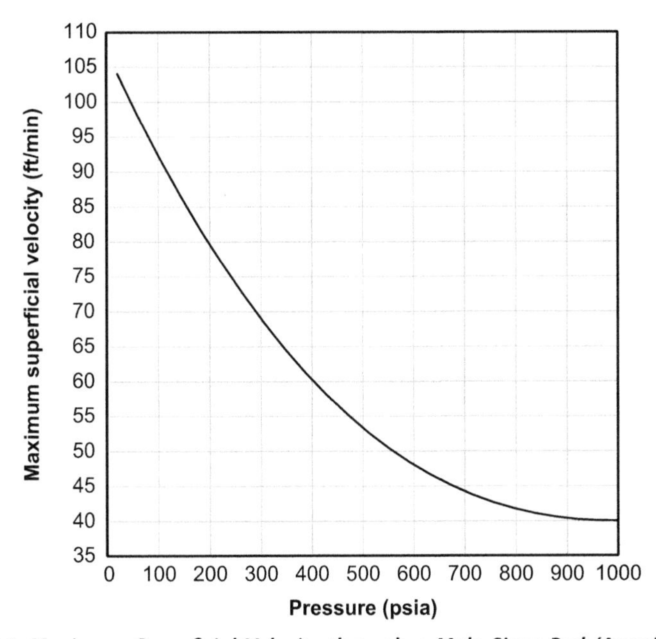

Figure 6.6 *Maximum Superficial Velocity through a Mole Sieve Bed (American Engineering Units).*

Next, the pressure drop through the bed must be estimated. Because the beds are uniform and usually spherical, the pressure drop per unit length through the bed can be estimated using the Ergun equation:

$$\frac{\Delta P}{L} = \left[\frac{\rho v_s^2(1-\varepsilon)}{D_p \varepsilon^3}\right]\left[\frac{150(1-\varepsilon)\mu}{D_p \rho v_s} + 1.75\right] \tag{6.4}$$

where ΔP is the pressure drop across the bed, L is the bed length, ρ is the gas density at process conditions, v_s is the superficial velocity, ε is the void fraction of the bed, D_p is the particle diameter, and μ is the gas viscosity at process conditions.

It is typical to simplify this equation to the following relation

$$\frac{\Delta P}{L} = A\mu v_s + B\rho v_s^2 \tag{6.5}$$

where the parameters A and B are given in Table 6.2 for mole sieve materials. The units for the coefficients in Table 6.2 are $\Delta P/L$ in psia/ft, μ in centipoise, v_s in ft/min, and ρ in lb/ft^3. An important observation from this equation is that the higher the superficial velocity, the greater the pressure drop across the bed.

The design velocity is a tradeoff between the maximum velocity and the acceptable pressure drop. Typically, the pressure drop through the bed should be less than 35 kPa (5 psia), although in some cases it may be as high as 55 kPa (8 psia).

After the superficial velocity is determined, then the diameter of the bed, d, can be calculated using the following equation:

$$d = \sqrt{\frac{4G_{act}}{\pi v_s}} \tag{6.6}$$

where d is in m, G_{act} is the actual volumetric flow rate in m^3/min, π is 3.14159..., and v_s is the superficial velocity through the bed in m/min. This equation can also be used with American Engineering units if appropriate

Table 6.2 Coefficients for Eqn (6.5)

	A	*B*
1/8-in beads	0.0560	8.89×10^{-5}
1/8-in extruded	0.0722	1.24×10^{-4}
1/16-in beads	0.152	1.36×10^{-4}
1/16-in extruded	0.238	2.10×10^{-4}

From the *GPSA Engineering Data Book*.

substitutions are made. The flow rate in the gas business is usually given in at standard conditions and these must be converted to the actual flow through the bed using the gas laws.

$$G_{act} = z\left(\frac{T}{T_{std}}\right)\left(\frac{P_{std}}{P}\right)G_{std} \tag{6.7}$$

where z is the compressibility factor of the gas at T and P, T is the process temperature expressed as an absolute temperature, T_{std} is the standard temperature, again expressed as absolute temperature (usually 520 R or 288.7 K), P_{std} is the standard pressure (usually 14.7 psia, 101.3 kPa or 1 atm), P is the process pressure, and G_{std} is the gas volumetric flow rate at standard conditions. The design engineer should be able to calculate the compressibility factor and thus this is not discussed further at this point.

Next, given the diameter of the bed and the volume of material required, the height can be calculated from simple geometry.

6.4 REFRIGERATION

Another method for dehydrating natural gas is to use refrigeration. That is, to cool the gas. As will be demonstrated in a subsequent chapter, cool gas holds less water than hot gas.

The usual purpose of a refrigeration plant is to remove heavy hydrocarbons from a natural gas stream to make hydrocarbon dew point specification. But this process also removes water. The cold temperatures in a refrigeration process result in water removal because the cold gas can carry less water than warm. To prevent the formation of ice and hydrates, the cold gas is mixed with a polar solvent, usually EG. A typical refrigeration process can easily reduce the water content of a gas stream down to the 1 lb/MMCF level.

6.4.1 Process Description

A simplified flow sheet for the refrigeration process is shown in Fig. 6.7.

Gas enters a gas–gas exchanger, where it is precooled. Further cooling is achieved via refrigeration. The gas enters the reboiler of a refrigeration unit (it is the refrigerant that is boiled and the process steam that is cooled), called a "chiller."

To prevent freezing (ice and/or hydrates) and to pick up condensing water, a glycol is sprayed into both the gas–gas exchanger and the chiller.

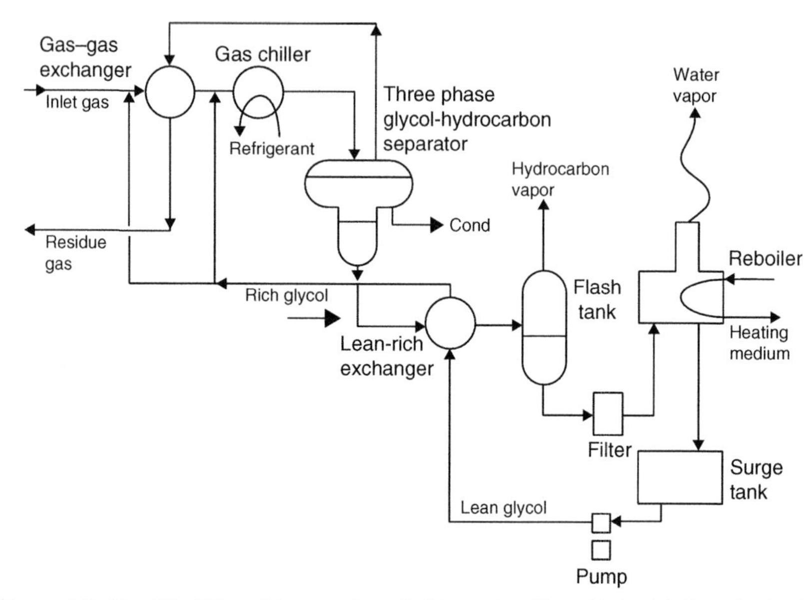

Figure 6.7 ***Simplified Flow Diagram for a Refrigeration Plant Unit with Glycol Injection and Recovery.*** *Reprinted from the* GPSA Engineering Data Book, *11th ed.—reproduced with permission.*

The glycol of choice for this application is EG because it has better low temperature properties than other glycols.

The mixture enters a low-temperature separator (LTS) where the gas comes off the top, liquid hydrocarbon comes out the middle, and a mixture of glycol and water come out the bottom (usually a boot). The hydrocarbon liquids are sent for further treating. The glycol–water mixture is sent to a regeneration still where the glycol is regenerated for reuse.

The sales gas, the gas off the top of the LTS, is very cold. Therefore it is sent back through the gas–gas exchanger to precool the feed gas and recover some of its energy.

With propane as the refrigerant, which is typical in the natural gas industry, the chiller temperature is usually in the range −10 to −40 °C (+15 to −40 °F).

One big advantage of this type of process is that it can produce a gas that meets both the hydrocarbon dew point and water content specifications. There is no need for separate a dehydration unit and a hydrocarbon dew-point control.

6.4.2 Glycol Injection

Ironically, the glycol injection rate and the concentration of the glycol are usually not dictated by the hydrate considerations discussed previously. The physical properties of the solution play a more important role in this design.

EG and water have a eutectic at approximately 80 wt% EG. A eutectic is approximately equivalent to an azeotrope in the boiling point. Mixtures of EG + water freeze at lower temperatures than the pure components. The freezing points of EG + water mixtures are shown in Figs 6.8 and 6.9. Freezing points of other glycol + water mixtures are also shown on these plots.

The minimum injection rate is about 2 L/min (0.5 gpm) because of mechanical and contact considerations. Injection rates less than this do not achieve the desired distribution in the exchangers.

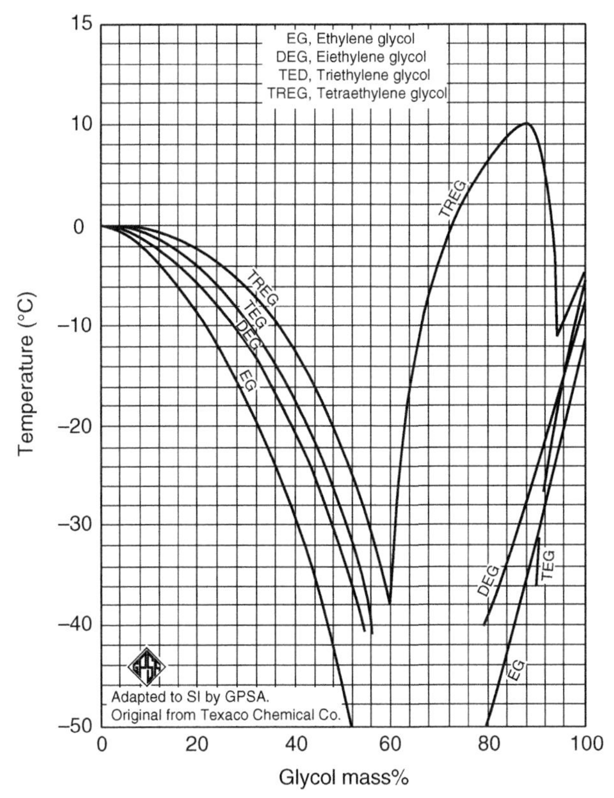

Figure 6.8 ***Freezing Points of Water + Glycol Mixtures for EG, DEG, TEG, and TREG in SI Units.*** *Reprinted from the* GPSA Engineering Data Book, *11th ed.—reproduced with permission.*

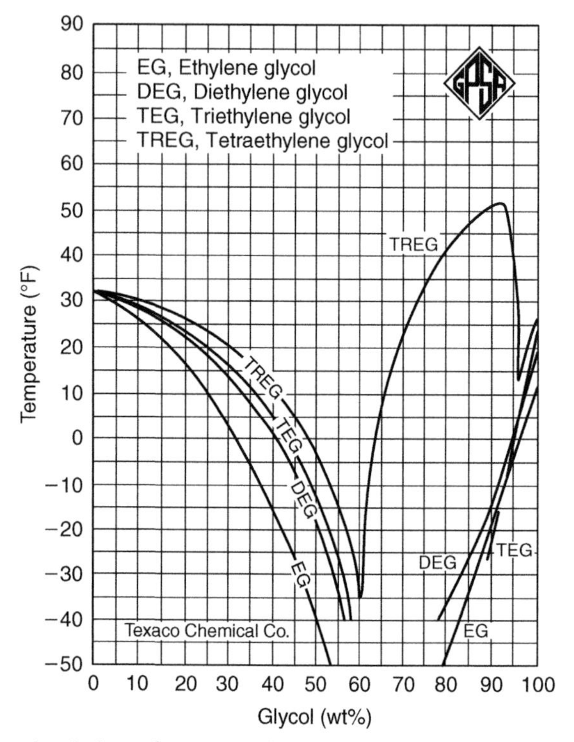

Figure 6.9 ***Freezing Points of Water + Glycol Mixtures for EG, DEG, TEG, and TREG in American Engineering Units.*** *Reprinted from the* GPSA Engineering Data Book, *11th ed.—reproduced with permission.*

Beyond the mechanical limits, typically the glycol is injected at 80 wt% and the minimum withdraw concentration is usually about 70 wt%.

Examples

Example 6.1

A wet natural gas stream is to be dehydrated using a TEG dehydration unit. The flow rate of the gas is 10 MMCF (283×10^3 Sm^3/day) and it is at 120 °F (48.9 °C) and 600 psia (4137 kPa). At these conditions, the gas is saturated with water, 150 lb/MMCF (2400 mg/Sm^3).

The gas is to be dried to 7 lb/MMCF (112 mg/Sm^3). Calculate the circulation rate, the diameter of the contactor, and the reboiler duty. Use a glycol rate of 2.5 gal of glycol per lb of water (0.3 kg/L).

Answer: From Eqn (6.1), the circulation rate can be calculated:

$$L = \frac{(2.5)(10)(150)}{24} = 156 \text{ gpm}$$

This converts to 590 L/min.

The contactor diameter can be estimated from Fig. 6.2. From that figure, the diameter is slightly more than 24 inches (61 cm).

Finally, the reboiler duty can be estimated from Eqn (6.3a):

$$Q = 2000(156) = 312,000 \text{ Btu/h}$$

this converts to 329,000 kJ/h.

Example 6.2

Estimate the capital cost for a TEG dehydration unit (sweet service) for a gas flow rate of 10 MMCFD flow.

Answer: From the chart this is slightly less than $100,000 (US). Reading the chart more closely reveals that the value is closer to $125,000 (US).

Example 6.3

Perform the sizing calculation for the mole sieve bed for dehydrating the gas in Example 6.1 to 1 ppm. The properties of the gas at 120 °F and 600 psia are as follows: density: 2.2 lb/ft^3, viscosity: 0.01 cp, z-factor: 0.90. This example will be done using only American Engineering units. The reader is invited to repeat the calculation using SI units.

Answer: The total water adsorbed is $150 \times 10 = 1500$ lb/day. Assuming a 12 h cycle, this means 750 lb of water are adsorbed per cycle. This requires 7500 lb of mole sieve. Given that the bulk density of the mole sieve is 42 lb/ft^3 means that 178.6 ft^3 of mole sieve in the bed.

Next convert the standard flow rate into an actual flow rate.

$$G_{act} = z\left(\frac{T}{T_{std}}\right)\left(\frac{P_{std}}{P}\right)G_{std}$$

$$= (0.90)\left(\frac{120+460}{60+460}\right)\left(\frac{14.7}{600}\right)(10)$$

$$= 0.246 \times 10^6 \text{act ft}^3/\text{day}$$

$$= 170.8 \text{ act ft}^3/\text{min}$$

From Fig. 6.6, the maximum allowable superficial velocity through the bed is 48 ft/min. Use this value as a starting point. Calculated the pressure drop per unit length:

$$\frac{\Delta P}{L} = A\mu v_s + B\rho v_s^2$$

$$= 0.056\,\mu\, v_s + 8.89 \times 10^{-5}\rho v_s^2$$

$$= 0.056(0.01)(48) + 8.89 \times 10^{-5}(2.2)(48)^2$$

$$= 0.477 \text{ psi/ft}$$

Next calculate the bed diameter:

$$d = \sqrt{\frac{4G_{\text{act}}}{\pi v_s}} = \sqrt{\frac{4(170.8)}{\pi(48)}}$$
$$= 2.13 \text{ ft}$$

Calculate the height of the bed from simple geometry:

$$L = \frac{V}{\pi d} = \frac{178.6}{\pi(2.13)} = 26.7 \text{ ft}$$

And finally the total pressure drop

$$\Delta P = (26.7)(0.477) = 12.7 \text{ psi}$$

This pressure drop is unacceptably large. Reduce the superficial velocity to 40 ft/min and repeat the calculation.

$$\frac{\Delta P}{L} = 0.056(0.01)(40) + 8.89 \times 10^{-5}(2.2)(40)^2$$
$$= 0.335 \text{ psi/ft}$$

$$d = \sqrt{\frac{4(170.8)}{\pi(40)}} = 2.33 \text{ ft}$$

$$L = \frac{178.6}{\pi(2.33)} = 24.4 \text{ ft}$$

$$\Delta P = (24.4)(0.335) = 8.17 \text{ psi}$$

This pressure drop is still a little large, so further reduce the velocity to 35 ft/min and once again repeat the procedure.

$$\frac{\Delta P}{L} = 0.056(0.01)(35) + 8.89 \times 10^{-5}(2.2)(35)^2$$
$$= 0.259 \text{ psi/ft}$$

$$d = \sqrt{\frac{4(170.8)}{\pi(35)}} = 2.49 \text{ ft}$$

$$L = \frac{178.6}{\pi(2.49)} = 22.8 \text{ ft}$$

$$\Delta P = (22.8)(0.259) = 5.91 \text{ psi}$$

This value is still marginally too large, but the procedure is well-established.

REFERENCES

Sivalls, C.R., 1976. Glycol dehydration design manual. In: Laurance Reid Gas Conditioning Conference, Norman, OK.

Tannehill, C.C., Echterhoff, L.W., Leppin, D., 1994. The cost of conditioning your natural gas for market. In: 73rd Annual GPA Convention, New Orleans, LA.

Trent, R.E., 2001. Dehydration with molecular sieves. In: Laurance Reid Gas Conditioning Conference, Norman, OK.

CHAPTER 7

Combating Hydrates Using Heat and Pressure

Much effort was expended in the earlier chapters of this book providing methods for determining the conditions of pressure and temperature at which hydrates would form for natural gas mixtures. Another means of combating hydrate formation is to avoid the regions of pressure and temperature where hydrates would form. This is the topic of this chapter.

The reason why a plug forms in a pipeline is because the three criteria for hydrate formation, given in Chapter 1, are present. There is water, gas, and the right combination of temperature and pressure. In this chapter, we are going to examine how we can take advantage of the third criteria in the battle against hydrates.

7.1 PLUGS

Perhaps the most significant problem with hydrates is the plugging of pipelines, and much of the focus of this chapter is on pipelines. The hydrates really become problematic when they block or severally restrict the flow in a pipeline. Often hydrates form but flow with the fluid in the line causing only minimal flow problems.

It is worth noting that hydrate plugs tend to be porous and permeable, especially in condensate lines (Austvik et al., 2000). However, this is not always true and should not be assumed a priori.

Another important consideration is that one should not assume that there is only a single hydrate plug blocking the line. One should prepare for the possibility of multiple plugs blocking the flow line.

When it comes to melting hydrate plugs, patience truly is a virtue. It may take several days to melt a large plug. In addition, the application of a remedial measure usually will not result in an immediate observable change in the situation.

7.1.2 Plug Formation

For pipe flow, the hydrates initially are as for small particles. As mentioned in Chapter 1, these particles are formed at a nucleation site. The particles accumulate and may eventually form a flow blocking plug.

Natural Gas Hydrates
ISBN 978-0-12-800074-8
http://dx.doi.org/10.1016/B978-0-12-800074-8.00007-7

Typically, the outset of hydrate plugging is associated with an increase in pressure drop. This initial accumulation is called bedding and the pressure drop associated with bedding is due to restriction in flow. The nature of bedding is a function of many variables, including the velocity and the water cut. At high velocity, the hydrate particles are carried with the flow and thus there is less tendency to bed.

The pressure drop is erratic because the bedding is not yet a plug and quickly changes in size. As the hydrate accumulates into a plug, the pressure drop can increase dramatically. Eventually the hydrate accumulates into a flow blocking plug.

Another reason for the changes in pressure drop in hydrates systems is due to changes in viscosity. The formation of hydrate in a flow system will tend to increase the effective viscosity of the liquid phase, which in turn will increase the pressure drop.

Operators who observe these changes in the flow can assume that a flow blocking plug is inevitable and should try to take remedial action before the flow stops. This may include increasing the rate at which inhibitor is being injected or the use of some of the concepts presented in this chapter.

This description of the formation of hydrate plugs in pipe flow is based on research conducted at the Center for Hydrate Research at the Colorado School of Mines in Golden, Colorado.

7.2 THE USE OF HEAT

We have already discussed the range of temperature and pressure where hydrates may form. To prevent the formation of hydrates, one merely has to keep the fluid warmer than the hydrate-forming conditions (with the inclusion of a suitable margin for safety). Alternatively, it may be possible to operate at a pressure less than the hydrate formation pressure.

With a buried pipeline, which loses heat to the surroundings as the fluid flows, the temperature must be such so that no point in the pipeline is in the region where a hydrate will form. This heating is usually accomplished by two means, either by using line heaters or heat tracing.

A heater can be used to warm the fluid. Because this is a single-point injection of energy, the amount of energy must be such that the fluid remains above the hydrate point until the next point where heat is added is reached. This means that the fluid entering the pipeline must be well above the hydrate temperature.

Another method to add heat to a system is to use heat tracing. In this method, the heat is injected continuously along a line. Thus the fluid temperature does not need to be very high at any single point, but with the continuous injection of energy the temperature of the fluid needs to be only above the hydrate temperature and not warmer.

Heat tracing can be electrical or a fluid medium (hot oil or glycol, for example). In either case, the heat trace is placed adjacent to the line that is to be heated.

Heat tracing is especially useful on valves. Valves are notorious for freezing because of cooling from the Joule–Thomson effect.

Another important tool in the fight against hydrate formation is the use of insulation. An insulated pipeline will lose heat at a slower rate than an un-insulated one. This translates into a lower temperature requirement for the outlet of the heater and ultimately a lower heater duty. And a lower duty translates into lower operating costs. As a matter of fact, the proper use of insulation may, in some case, negate the requirement for a heater altogether.

7.2.1 Heat Loss from a Buried Pipeline

Heat loss from a buried pipeline can be estimated using the fundamental principles of heat transfer. You begin with the basic heat transfer equation:

$$Q = UA\Delta T_{\mathrm{lm}} \tag{7.1}$$

where Q is the heat transfer rate, U is the overall heat transfer coefficient, A is the area available for heat transfer, and ΔT_{lm} is the logarithmic mean temperature, which is difference given by:

$$\Delta T_{\mathrm{lm}} = \frac{T_{\mathrm{fluid,in}} - T_{\mathrm{fluid,out}}}{\ln\left[\frac{T_{\mathrm{fluid,in}} - T_{\mathrm{soil}}}{T_{\mathrm{fluid,out}} - T_{\mathrm{soil}}}\right]} \tag{7.2}$$

where T is the temperature and the subscripts are sufficiently descriptive.

The overall heat transfer coefficient, U, is the sum of four terms: (1) convection from the fluid flowing in the line, (2) conduction through the pipe, (3) conduction through insulation (if present), and (4) resistance from the soil. The U is obtained from the following equation:

$$\frac{1}{UA} = \frac{1}{h_{\mathrm{i}}A_{\mathrm{i}}} + \frac{\ln(d_{\mathrm{o}}/d_{\mathrm{i}})}{2\pi k_{\mathrm{p}}L} + \frac{\ln[(d_{\mathrm{o}} + 2t_{\mathrm{ins}})/d_{\mathrm{o}}]}{2\pi k_{\mathrm{ins}}L} + \frac{1}{k_{\mathrm{s}}S} \tag{7.3}$$

where h_i is the convective heat transfer coefficient for the fluid in the pipe, A_i is the inner surface area of the pipe, d_o and d_i are the outside and inside diameters of the pipe, k_p is the thermal conductivity of steel used to construct the pipe, L is the length of a pipe segment, t_{ins} is the thickness of the insulation, k_{ins} is the thermal conductivity of the insulation, k_s is the thermal conductivity of the soil, and S is the shape factor for the buried pipe.

7.2.1.1 Fluid Contribution

There are many correlations for estimating heat transfer coefficients. These are based on the properties of the fluid and flow considerations. However, almost all of them can be expressed in the dimensionless form:

$$\mathrm{Nu} = f(\mathrm{Re}, \mathrm{Pr}) \tag{7.4}$$

where Nu is the Nusselt number, which is a dimensionless heat transfer coefficient; Re is the Reynolds number, which is a combination of the flow conditions and fluid properties and it is also dimensionless; and Pr is the Prandtl number, which is the dimensionless description of the fluid properties.

An example of such a correlation is the Dittus–Boelter equation (Holman, 1981):

$$\mathrm{Nu} = 0.023\mathrm{Re}^{0.8}\mathrm{Pr}^{n} \tag{7.5}$$

where $n = 0.4$ for heating or $n = 0.3$ for cooling, which is the case for heat loss from a pipeline. Substituting the properties for the dimensionless groups yields:

$$\frac{h_i d_i}{k} = 0.023\left(\frac{\rho v d_i}{\mu}\right)^{0.8}\left(\frac{C\mu}{k}\right)^{n} \tag{7.5a}$$

where ρ is the density of the fluid, v is the fluid velocity, μ is the viscosity of the fluid, C is the heat capacity of the fluid, and k is the thermal conductivity of the fluid (all other symbols were defined earlier). Rearranging this slightly yields:

$$h_i = 0.023\frac{\rho^{0.8} v^{0.8} k^{1-n} C^{n}}{d_i^{0.2}\mu^{0.8-n}} \tag{7.5b}$$

The Dittus–Boelter equation is limited to Reynolds numbers in the range $5000 < \mathrm{Re} < 500{,}000$ and for Prandtl numbers from 0.6 to 1000.

It is also worth noting that there are several other correlations available for estimating the heat transfer coefficient in tube flow. The reader should consult any textbook on the subject of heat transfer, such as Holman (1981), for more such correlations.

7.2.1.2 Pipe Contribution

The second term on the right hand side of Eqn (7.3) is the contribution from the conduction through the metal of the pipe.

Typically pipe comes in standard sizes that have descriptions via their nominal diameter in inches. Even in countries where the metric system of units is used, nominal pipe sizes are usually given in inches. For example, there is 3-in schedule 80 pipe, where the schedule number reflects the wall thickness of the pipe. Although this pipe is called 3-in, the outside diameter of the pipe is 3.500 in (88.90 mm) and the inside diameter is 2.900 in (73.66 mm). The physical sizes of standard pipe are readily available.

The thermal conductivity of carbon steel is about 40 W/m °C (25 Btu/ft h °F), whereas for stainless steel it can be as low as 10 W/m °C (6 Btu/ft h °F). The design engineer should consult standard references for the thermal conductivity of the material used in their particular application. On the other hand, the resistance from the pipe is often negligibly small.

7.2.1.3 Soil Contribution

The final contribution is the resistance from the soil. The shape factor for a buried pipe can be calculated from the following formula (Holman, 1981):

$$S = \frac{2\pi L}{\ln[4D/(d_o + 2t_{ins})]} \tag{7.6}$$

where most of the symbols are as defined previously and D is the depth to which the pipe is buried.

Although soil temperatures vary to some degree with ambient temperature, time of the year (and hence the season), and other factors, these effects are usually neglected. In addition, the soil temperature also varies with location. That is, the soil temperature is different in Texas than it is in Alberta, Canada. In fact, different locations within a state or province may have different soil temperatures. Even the pipeline itself has an effect on the soil temperature.

In the design of a pipeline, it is typical to assume that the soil temperature is a constant. The value used varies from location to location, but for a given pipeline design, it is assumed to be constant.

7.2.1.4 Overall

An approximate value for the overall heat transfer coefficient for an uninsulated pipe is 5 W/m^2 °C (1 Btu/ft^2 h °F). For an insulated pipe the heat transfer coefficient is reduced.

For an uninsulated pipeline, the resistance from the soil dominates. Typically, this resistance accounts for more than 90% of the total resistance.

7.2.1.5 Heat Transferred

The energy lost by the fluid to the surroundings can be calculated from the change in enthalpy using the following equation:

$$Q = \dot{m}_{\text{fluid}}\left(H_{\text{fluid,in}} - H_{\text{fluid,out}}\right) \tag{7.7}$$

where $\dot{m}$ is the mass flow rate and H is the specific enthalpy.

If there is no phase change, then the enthalpies can be replaced with the product of the temperature and the heat capacity. Equation (7.7) becomes:

$$Q = \dot{m}_{\text{fluid}} C_{\text{fluid}}\left(T_{\text{fluid,in}} - T_{\text{fluid,out}}\right) \tag{7.8}$$

If there is a phase change, then Eqn (7.7) must be used because it accounts for the enthalpy associated with the phase change, whereas Eqn (7.8) does not.

7.2.1.6 Additional Comments

The temperature profile is only one of the many factors that go into the proper design of a pipeline. The optimum design of a pipeline should include all of these considerations.

On the disk accompanying this book is a simple program for calculating the temperature loss from a buried pipeline. This program follows the procedure outlined here. However, it is useful only for performing approximate calculations and should be used with caution. It does not include factors such as the effect of temperature on the properties of the fluid (in the program, they are assumed to be constant). Nor does it account for the possibility of a phase change in the line.

For detailed design, engineers should use software packages that account for all of these effects.

If the pipeline is not buried but exposed to the air or to water, then there is a slight change in the calculation. The term representing the thermal resistance from the soil is replaced with a term for the thermal resistance from the ambient air or water. In still air or water, this convection term is free convection. If the fluid moves, in the case of air by winds and for water by currents, then the convection is forced convection.

7.2.2 Line Heater Design

Figure 7.1 shows a schematic diagram of a line heater. The fire tube is a large diameter U-tube, which is the source of energy for the heater. The fuel gas and air enter the fire tube and are burned to produce the heat.

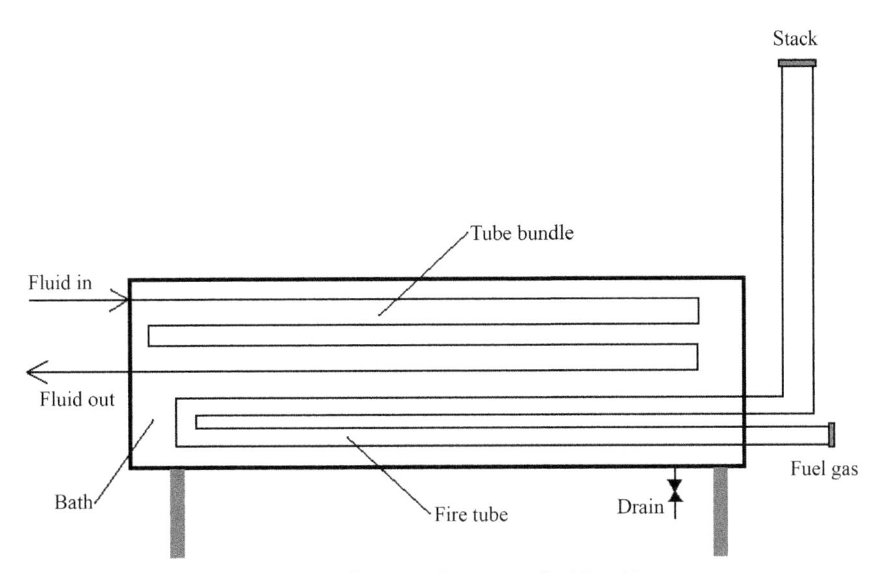

Figure 7.1 A Schematic Diagram of a Line Heater.

The tube bundle shown in the schematic shows only four tubes but there are usually more than four tubes in the bundle. The tubes in the bundle are typically 50.8–101.6 mm (2–4 in) in diameter. The diameter of the tubes is dictated by the pressure drop and heat transfer considerations. Smaller diameter tubes result in high velocities that in turn result in high pressure drops but also higher heat transfer rates. The wall thickness of the tubes is dependent upon the fluid pressure.

The schematic does not show a choke. With high-pressure gas wells, it is common to produce at high pressure directly into the heater. Then the gas is choked down to pipeline pressure. This pressure reduction will result in cooling of the gas according to the Joule–Thomson effect. The gas then enters a second stage of heating. This is often referred to as "preheat-reheat." Both the preheat and the reheat take place in the same heater. The amount of heat in the preheat must be such that hydrates do not form after the choke.

The design for the size of the tube bundle begins with the fundamental heat transfer equation (Eqn (7.1)). However, in this case, the log mean temperature difference is calculated as follows:

$$\Delta T_{\mathrm{lm}} = \frac{\left(T_{\mathrm{fluid,in}} - T_{\mathrm{bath}}\right) - \left(T_{\mathrm{fluid,out}} - T_{\mathrm{bath}}\right)}{\ln\left[\frac{T_{\mathrm{fluid,in}} - T_{\mathrm{bath}}}{T_{\mathrm{fluid,out}} - T_{\mathrm{bath}}}\right]} = \frac{T_{\mathrm{fluid,in}} - T_{\mathrm{fluid,out}}}{\ln\left[\frac{T_{\mathrm{fluid,in}} - T_{\mathrm{bath}}}{T_{\mathrm{fluid,out}} - T_{\mathrm{bath}}}\right]} \tag{7.9}$$

The subscripts in this equation should be completely descriptive.

A typical bath temperature is about 90 °C (200 °F); however, the bath temperature depends upon the fluid used for the bath. Typically this fluid is water, a glycol, a glycol–water mix, or a specialized heat transfer medium. Clearly, if water is used for the bath, the temperature must be less than 100 °C to prevent all of the water from boiling away.

The overall heat transfer coefficient, U_o, is the sum of four terms: (1) free convection from the bath to the tube bundle, (2) conduction through the pipe, (3) fouling on both sides of the tube, and (4) convection from the fluid flowing in the tube bundle. The U_o is obtained from the following equation:

$$\frac{1}{U_o A_o} = \frac{1}{h_o A_o} + \frac{\ln(d_o/d_i)}{2\pi k_p L} + R_{\text{foul}} + \frac{1}{h_i A_i} \tag{7.10}$$

The reader should note a similarity between this equation and that for the buried pipeline. The difference between the two equations is the final term, where the term for the resistance from the soil is replaced by the resistance due to the fluid bath.

7.2.2.1 Bath

The tube bundle is in a fluid bath and the heat transfer from the fluid bath to the tubes is via free convection. That is, the flow of the fluid is due to temperature gradients. There is no pump or stirrer to force the flow of the fluid.

The outside heat transfer can be estimated using a correlation for the free convection. There are many correlations for estimating heat transfer coefficients and most all of them can be expressed in the following dimensionless form:

$$\text{Nu} = f(\text{Gr}, \text{Pr}) \tag{7.11}$$

where Nu is the Nusselt number; Gr is the Grashof number, which is a dimensionless number roughly equivalent to the Reynolds number; and Pr is the Prandtl number.

$$\text{Nu} = C\text{Gr}^m\text{Pr}^m \tag{7.12}$$

values for C and m are given in Table 7.1. Substituting for the dimensionless numbers gives:

$$\frac{h_o d_o}{k} = C\left(\frac{g\beta\Delta T d_o^3 \rho^2}{\mu^2}\right)^m \left(\frac{C\mu}{k}\right)^m \tag{7.13}$$

Table 7.1 Parameters for Eqn (7.12) for Ranges of Grashof Number Times Prandtl Number

Gr·Pr	C	m
10^4 to 10^9	0.53	1/4
10^9 to 10^{12}	0.13	1/3

From Holman (1981).

The properties in this equation are those of the bath fluid and are evaluated at the film temperature (the average between the bath temperature and the temperature of the tube wall). Moreover, d_o is the outside diameter of the tubes, g is the acceleration from gravity (9.81 m/s^2 or 32.2 ft/s^2), β is coefficient of expansion and has units of reciprocal temperature, and all of the other symbols are the same as those given previously. At room temperature, for water, $\beta = 0.2 \times 10^{-3}$/K and for glycol $\beta = 0.65 \times 10^{-3}$/K (Holman, 1981).

7.2.2.2 Tube Bundle

The heat transfer coefficient for the fluid inside the tubes can be calculated using the convection equations discussed previously, which were applicable to the fluid flowing inside of a pipeline.

7.2.2.3 Fire Tube

The area of the fire tube is calculated assuming a flux of 31.5 kW/m^2 (10,000 Btu/h ft^2). Some authors recommend other values (some as low as 25 kW/m^2), but 31.5 kW/m^2 is recommended here.

The minimum fire tube diameter can be estimated as follows (Arnold and Stewart, 1989):

$$d = \sqrt{\frac{Q}{16,500}} \tag{7.14}$$

In this equation, the heat duty, Q, must be in Btu/h and the tube diameter, d, in inches. From the flux, which is the heat transfer rate per unit area, and the diameter, the tube length can be calculated. The fire tube is a U-tube, so the calculated tube length is divided by two and rounded to the nearest foot to obtain the length.

Ultimately, the design of the fire tube is a tradeoff between the tube diameter and length. Smaller diameters result in longer tubes. The design engineer should attempt to fit his or her design to the standard fire tube dimensions. However, if necessary, it is possible to have a custom fire tube constructed.

7.2.2.4 Other Considerations

In the design of any heat exchanger, it is customary to attempt to account for the aging of the exchanger and the buildup of corrosion products, etc., on the exchanger. This is done through the use of a fouling factor, R_{foul}. Typically the fouling factor for line heaters is approximately 0.0002 m^2 °C/W (0.001 h ft^2 °F/Btu).

Finally, the design engineer is wise to confirm all of these values for the various resistances either through the use of heat transfer software, personal experience, or from information provided by vendors.

7.2.2.5 Heat Transfer

The rate of heat transfer to the process fluid can be calculated using Eqn (7.7) or Eqn (7.8).

There is an interesting design twist in the case of preheat-reheat heats. Because the choke valve is an isenthalpic process (constant enthalpy), then the overall energy balance is unaffected. Therefore the heat duty can be estimated based on inlet and outlet conditions only.

7.2.3 Two-Phase Heater Transfer

It is common in both pipeline flow and line heaters that the fluid is not single phase. For a two-phase system, the overall heat transfer coefficient is estimated as a weighted average of those for the two phases (here we assume that the phases are water, W, and gas, G).

$$U_o = U_G(1 - X) + U_W X \tag{7.15}$$

where X is the mass fraction of the phase denoted W. The two overall heat transfer coefficients, U_G and U_W, can be calculated using Eqn (7.3) for buried pipelines and Eqn (7.10) for a line heater.

The total heat transfer rate is:

$$Q = \dot{m}_G C_G \left(T_{G,in} - T_{G,out}\right) + \dot{m}_W C_W \left(T_{W,in} - T_{W,out}\right) \tag{7.16}$$

However, because the two fluids are at the same temperature, this becomes:

$$Q = \left(\dot{m}_G C_G + \dot{m}_W C_W\right)\left(T_{fluid,in} - T_{fluid,out}\right) \tag{7.17}$$

Even though these equations seem rather complex, they do not account for exchange between the two phases and thus do not include any latent heat effects. Thus they too should be used with some caution.

7.3 DEPRESSURIZATION

Another method that is used to rid hydrates once they have formed is to reduce the pressure. Based on the information presented earlier, when the pressure is reduced the hydrate is no longer the stable phase. This is different from ice. Depressurizing would have little effect on the freezing point of ice.

Theoretically, this should work, but the process is not instantaneous. It takes some time to melt the hydrate. There are many horror stories about people who depressurized a line and then uncoupled a connection only to have a hydrate projectile shot at them.

Both theory and experiment indicate that the plugs tend to melt radially (from the pipe wall into the center of the pipe) (Peters et al., 2000). The plug shrinks inward but tends to settle to the bottom of the line because of gravity. This forms a flow path for communication between the two sides of the plug, which is typically established quite quickly. In contrast, if the plug melted linearly a flow link would not be established until almost the entire plug had melted.

The porous-permeable nature of most hydrate plugs means that there can be some flow communication through the plug and this tends to equalize pressures on both sides. However, it is unwise to assume that this will be the case, because it the rare case the plug is not very permeable.

Typically, the lower the pressure the faster the plug melts. However, when the pressure is reduced, there is a cooling resulting from the Joule–Thomson effect (see Chapter 11). If the depressurizing is done quickly, there is no time to equilibrate with the surroundings and one must wait for the system to warm.

A potentially dangerous scenario for the melting of a hydrate plug using pressure reduction is presented in Fig. 7.2. As shown in the figure, an attempt is made to melt the hydrate blockage by bleeding the pressure off the line.

However, in this case, the pressure is only bled off one side of the hydrate. The plug can loosen and will be projected along the line at high velocity. The hydrate can accelerate along the line like a bullet in a rifle barrel. The speed of the plug can often be enhanced by a fine film of water that acts as a lubricant.

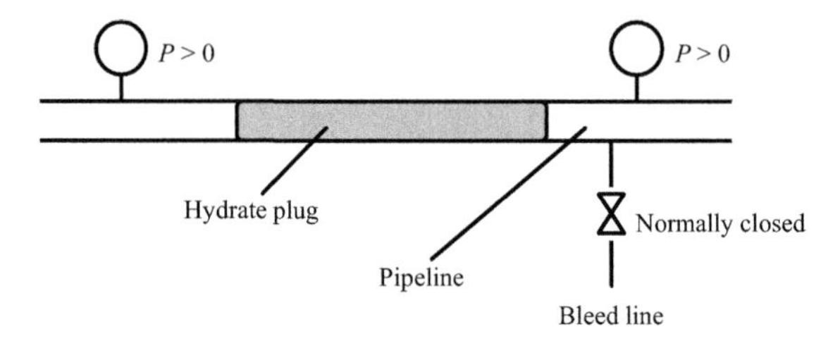

A hydrate formed under pressure is blocking a pipeline and preventing flow.

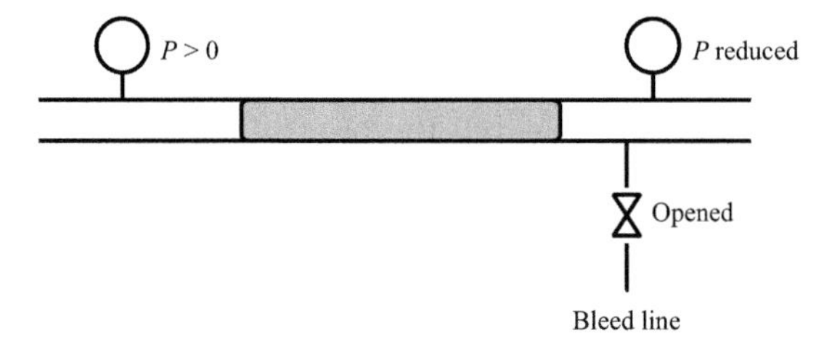

The valve on the bleed line is opened to reduce the pressure in an attempt to melt the hydrate.

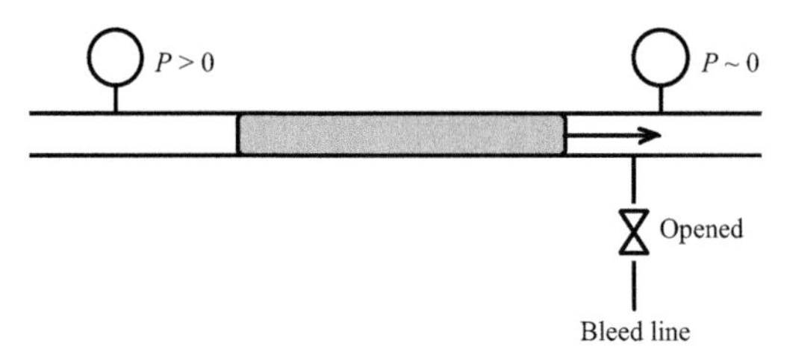

The hydrate plug begins to move potentially at high velocity.

Figure 7.2 Improper Depressurization of Hydrate Plug Resulting in the Hydrate Plug Being Launched Like a Projectile.

In this situation, it is better if the pressure is bled off both sides of the blockage. And if possible an attempt should be made to keep the pressure nearly equal on both sides of the plug. Maintaining equal pressure on both sides will prevent significant movement of the plug. Based on experience, a maximum of 10% difference in the pressure across a hydrate plug is commonly specified, but it would not be classified as an "industry standard."

If it is not possible to bleed off both sides of the line, then the pressure should bled off one side of the plug in a step-like manner. First, release some pressure and allow the plug to melt, which will increase the pressure. Then more pressure is released. Continue to step down in pressure until the plug melts. The problem with this method is that sufficient pressure must be bled off that hydrate melting will occur but not result in the plug becoming a projectile. If the pressure in the line is well above the hydrate formation pressure, then bleeding off some pressure will not melt the hydrate. If the line pressure is high, even bleeding of 1400–2000 kPa (200–300 psia), which is sufficient to create a dangerous projectile, may not be enough to melt the blockage. Therefore, it is important to know the hydrate formation pressure in order to know how much pressure to bleed.

Standard procedures are in place in most jurisdictions that are designed to prevent accidents during the maintenance of oil field equipment. Often, operating companies have even more stringent procedures for contractors working in their employ. It is wise to follow such procedures.

7.4 MELTING A PLUG WITH HEAT

Another method to remove a hydrate plug is to melt it with the application of heat. Heat can be applied by spraying steam on the line or electrical resistance heating or the like. However, this method should also be used with caution. A potentially dangerous situation is depicted in Fig. 7.3.

The melted plug will release gas and produce liquid water. As will be demonstrated in Chapter 8, 1 m^3 [act] of methane hydrate releases 170 Sm^3 of gas. The melting also produces 51.45 kmol of liquid water, which occupies 0.927 m^3 as a liquid. This means that if the hydrate is melted in a confined space, there is only 0.073 m^3 available for the 170 Sm^3 of released gas.

The pressure of the released gas can be crudely estimated using the ideal gas law:

$$P_1 V_1 = P_2 V_2 \quad \text{or} \quad P_2 = P_1 V_1 / V_2$$

$$\begin{aligned} P_2 &= (170)(101.325)/0.073 \\ &= 236{,}000 \text{ kPa} \\ &= 236 \text{ MPa} = 34{,}000 \text{ psia} \end{aligned}$$

Although there are some errors in this analysis, it provides better than an order of magnitude estimate of the pressure buildup. And as can be seen, the pressure is very large and capable of bursting most pipes.

If you redo the calculation, you will see that the calculated pressure is independent of the volume melted as long as the melting occurs in a confined space. Therefore do not think the problem is reduced if a volume

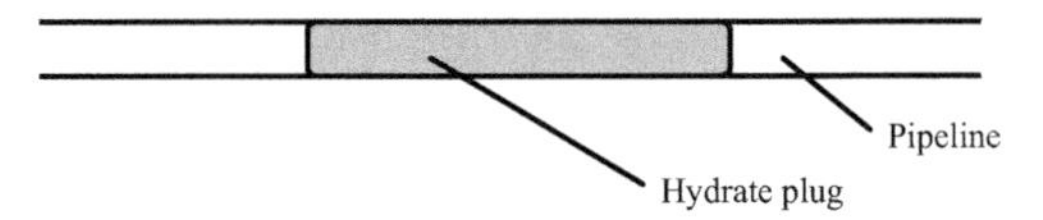

A hydrate formed under pressure is blocking a pipeline and preventing flow.

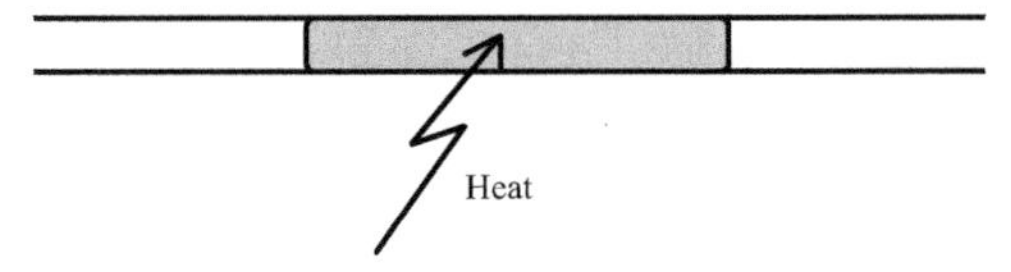

Heat is applied to the plug in order to melt it. In this case the heat is applied near the center of the plug.

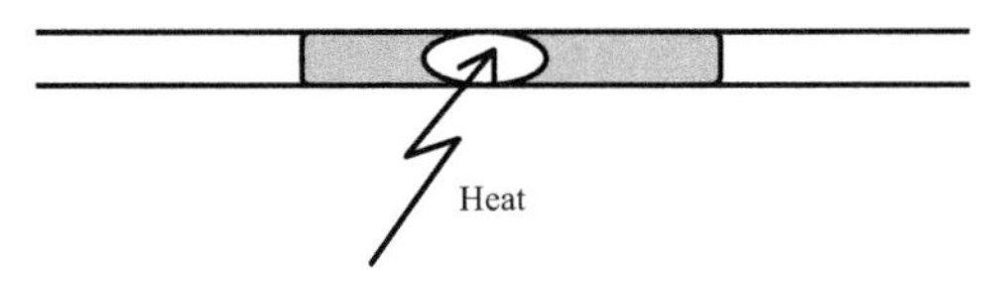

The plug begins to melt and the pressure within the plug rises.

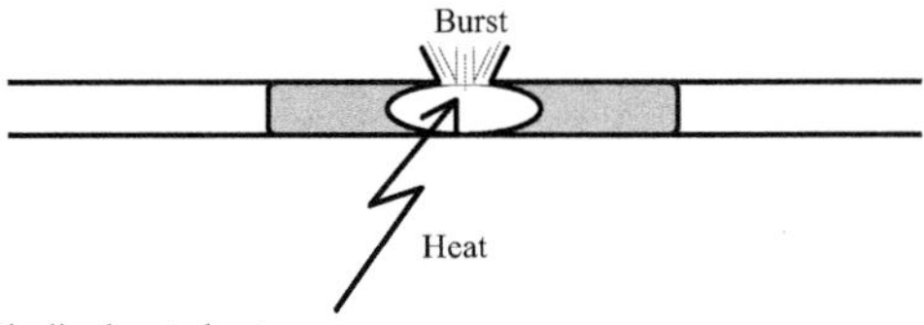

The line bursts due to over pressure.

Figure 7.3 Scenario for Pipeline Rupture during the Melting of a Gas Hydrate Plug.

less than 1 m^3 is melted—this is not the case. Even melting a very small volume will result in the same dangerous over pressurization.

Those more familiar with American Engineering Units can redo the calculation and see that the result is also independent of the set of units used. Melting a hydrate in a confined space results in a large over pressurization regardless of the set of units used for the calculation.

On the other hand, if the hydrate plug can move, then this becomes similar to the scenario shown in Fig. 7.2. The pressure buildup in the melted section can result in a hydrate projectile. Again the hydrate is projected like a bullet with a significant potential for causing damage.

To avoid these dangerous situations, it is important to heat the entire hydrate plug. When heating externally, this may be difficult because it requires locating the entire plug within the pipeline. It is wise to heat more of the line rather than less. Err on the side of caution. Heat more line than you believe is occupied by the gas hydrate plug.

7.5 HYDRATE PLUG LOCATION

As was demonstrated in the examples previously, it is important to do your best to locate the hydrate plug. It is difficult if not impossible to determine the exact location of a hydrate plug. However, with some engineering tools and Sherlock Holmes-like deduction, a best estimate of the plug location can be determined. Some guidelines are presented in this section.

If you have access to the pipe, a temperature survey may reveal the location of the hydrate. However, if the pipeline is buried, this maybe a difficult problem. Thus some preliminary analysis of the situation may lead to possible locations for the hydrate plug. Flow modeling should indicate where the flow conditions cross the hydrate curve. However, the point when hydrates begin to form is not where they will accumulate. Likely locations for the hydrates to accumulate are low points in the pipe or constriction from fitting or valves. These locations should be checked first.

If the line is exposed, the temperature can be measured using an infrared temperature "gun." One simply points the gun at the pipeline and obtains a surface temperature for the pipe. A cold spot in the line would be indicative of a hydrate plug.

7.6 BUILDINGS

In colder climates, such as Western Canada, it is common to house process equipment, field batteries, and even wells along with some of their associated facilities in heated buildings. The buildings not only provide comfort to operators during inclement weather, they also can prevent freezing of the equipment.

In the early gas industry in Western Canada, the plants were constructed in the style of warmer climates. That is, they were built without building. Freezing was a serious problem. This freezing was not all from hydrate formation, but some of it almost certainly was. Since that time, most gas plants in Western Canada are constructed with buildings and the interior of the buildings is heated.

7.7 CAPITAL COSTS

Figure 7.4 gives quick estimates (±30%) for the purchased cost of a line heater. This chart is based partially on information from the literature, but largely from the experience of the author.

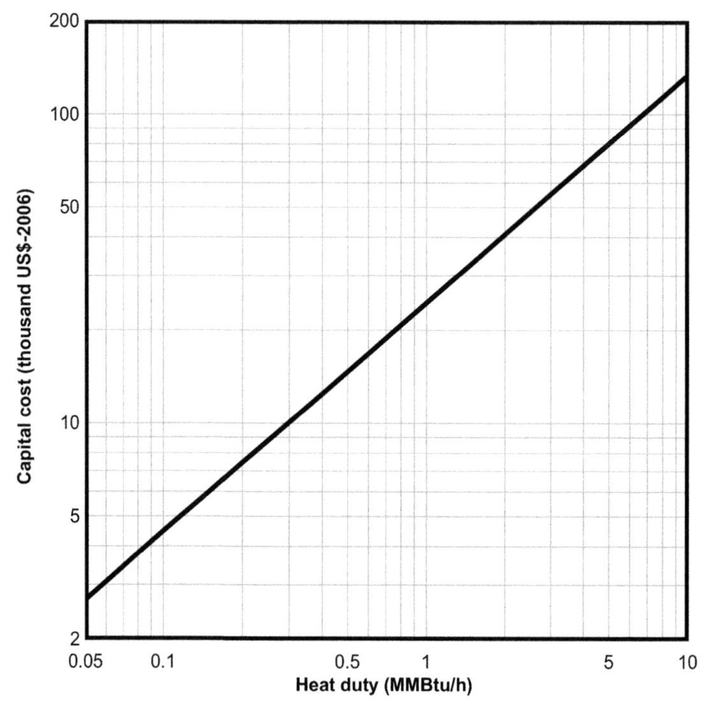

Figure 7.4 Approximate Capital Cost for Line Heaters.

As an order of magnitude estimate, it can be assumed that the installation cost is between 0.8 and 1.2 times the purchased cost. In other words, the installation cost is approximately equal to the purchased cost.

Clearly, these costs are approximations at best. Many factors have not been included such as the operating pressure, gas flow rate, etc. They should be used for rapid budget cost estimates only.

7.8 CASE STUDIES

7.8.1 Case 1

The National Energy Board of Canada (2004) reported an incident that occurred in British Columbia. An 18-in pipeline (DN 450) ruptured, and the force from the blast was sufficient to knock down a nearby worker who was unhurt. The pipeline was transporting sour gas (0.4% H_2S), which was released to the environment. An emergency situation was triggered and nearby residences were evacuated. Fortunately no one was adversely affected by the release.

The investigation revealed that the likely cause was due to a shock wave after a hydrate plug was released by high differential pressure. It also concluded that the point where the pipeline burst was probably a weak point from a manufacturer defect.

The Board recommends that companies develop procedures for melting hydrate plugs and that the procedures are followed.

The hydrate plug must have been of significant size to block an 18-in line.

7.8.2 Case 2

The Edmonton Journal (Sands, 2011) reported an incident near Fox Creek, Alberta. According to an article, a crew was sent to a pipeline to melt an "ice plug" inside the pipe. They used steam wands to melt the "ice" (i.e., steam was directly applied to the pipe at the suspected location of the plug), which resulted in a release of sour gas. The leak was deadly—one of the workers died as a result of H_2S poisoning. Others, including a policeman investigating the situation, were sent to hospital.

The newspaper article is intended for a general audience and requires some interpretation. First, because this was a natural gas pipeline, it is safe to assume that it was not ice but hydrate that blocked the line. Other details in the article are a little thin, but this appears to be a case of improperly locating the hydrate and melting from a more or less central point. The overpressure resulted in a pipeline leak.

Furthermore, although the article does not mention defects in the pipe, it is possible that a manufacturer's defect contributed to the failure, as was the case in the previous incident.

However, this incident demonstrates the need for locating the hydrate plug before melting it. Also care should be taken not to melt the plug too fast, which would result in an over pressurization of the line.

Examples

Example 7.1

Gas is to flow at a rate of 195.7×10^3 Sm^3/day (6.91 MMCFD) in a buried pipeline from a well site to a gas plant 7.5 km (4.66 miles) away. From other considerations (pressure drop, wall thickness, etc.), it is decided to use 4-in schedule 80 pipe ($k = 48.5$ W/m K). The gas at the wellhead is at 48.89 °C (120 °F) and 7500 kPa (1087 psia).

From the methods presented earlier, it is determined that at the pressure in the pipeline a hydrate will form at 15 °C (59 °F).

Table 7.2 Properties of the Natural Gas for Example 7.1

	SI Unit	Engineering Units
Density	68.1 kg/m^3	4.25 lb/ft^3
Molar Mass	23.62 kg/kmol	23.62 lb/lb mol
Heat Capacity	2.76 kJ/kg K	0.66 Btu/lb °F
Viscosity	0.013 mPa s	0.013 cP
Thermal Conductivity	0.040 W/m K	0.023 Btu/h ft °F

If the pipeline is uninsulated, estimate the temperature of the gas arriving at the plant. Furthermore, estimate the thickness of insulation ($k = 0.173$ W/m K) required such that the gas will arrive at the plant 5 °C (9 °F) above the hydrate temperature. That is, such that the gas arrives at the plant at 20 °C.

The physical properties of the gas are given in Table 7.2. The soil temperature is 1.67 °C (35 °F) and the thermal conductivity of the soil is 1.3 W/m K.

Answer: First, convert the flow rate from volumetric to a mass flow.

$$\dot{n} = \frac{Pv}{RT} = \frac{(101.325)(195.7 \times 10^3)}{(8.314)(15.55 + 273.15)} = 8261.3 \text{ kmol/day}$$

$$\dot{m} = \frac{(8260.3)(24)}{20.363} = 9737 \text{ kg/h}$$

From standard tables for pipe properties, a 4-in schedule 80 pipe has an outside diameter of 11.430 cm (4.50 in) and an inside diameter of 9.718 cm (3.826 in).

All of the other information required for the program is given in the statement of the problem or in Table 7.2. In general, the design engineer would be required to find these values in reference books or calculate the properties.

The program on the enclosed disk can be used to perform this calculation. The output from this run is appended to this chapter. From the output the exit temperature is estimated to be 8.1 °C, well below the hydrate formation temperature.

Three cases of insulated pipe were also examined: 2.54 cm (1 in) of insulation, 5.08 cm (2 in), and 7.62 cm (3 in). The enclosed program was used for these cases as well. The complete output is also in the appendix and the outlet temperatures are summarized below:

No Insulation	8.1 °C
2.54 cm	15.0 °C
5.08 cm	19.0 °C
7.62 cm	21.7 °C

Therefore 5.08 cm (2 in) of insulation is not quite enough and 7.62 cm (3 in) is too much.

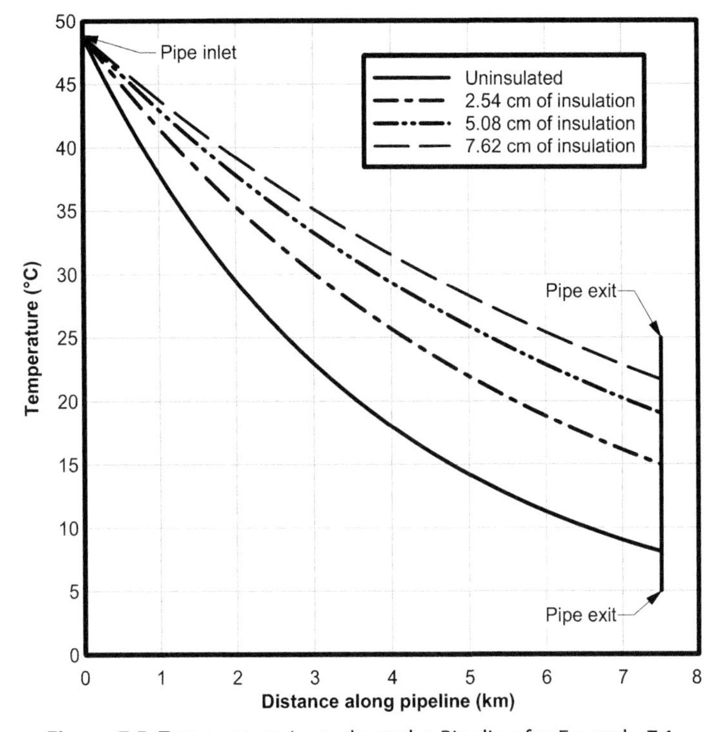

Figure 7.5 Temperature Loss along the Pipeline for Example 7.1.

Finally, Fig. 7.5 is a plot showing the temperature along the pipeline for the four cases. In addition, the reader is invited to review the complete output listed in the Appendix to obtain further insights into the heat loss from pipelines in general and the results for this specific case.

Example 7.2

As an alternative to insulating the line, it is suggested that a heater be used. How hot should the inlet gas be heated to in order to arrive at the plant 5 °C above the expected hydrate formation temperature?

Answer: Again the program on the accompanying disk is used for this calculation. However, this requires an iterative solution.

Inlet Temperature	Outlet Temperature
48.89	8.13—This is from Example 7.1.
100	15.15
200	28.82
135	9.92—This value of 135 °C obtained by linearly interpolating the previous two values.

At this point, it is worth commenting that there are limits to the temperature that the gas can enter a pipeline. One concern is that high temperatures damage the yellow jacket. Furthermore, to obtain a fluid temperature of 135 °C would require a bath temperature of at least 140 °C (an approach temperature of 5 °C). Clearly, to obtain this temperature, the bath fluid cannot be water.

Example 7.3

Estimate the duty required to heat the gas from 48.89 to 135 °C. Furthermore, estimate the capital cost of the line heater required.

Answer: First calculate the heat duty using Eqn (7.8):

$$Q = (9737\ \text{kg/h})(2.76\ \text{kJ/kg K})\ (135 - 48.89) = 2,314,000\ \text{kJ/h}$$

$$= 643,000\ \text{W} = 643\ \text{kW}$$

Convert to engineering units:

$$Q = 2.192\ \text{MMBtu/h}$$

From Fig. 7.2 the approximate capital cost is $35,000 (US) (about $52,500 Cdn).

APPENDIX 7A OUTPUT FROM PIPE HEAT LOSS PROGRAM FOR THE EXAMPLES IN THE TEXT

```
*********************************************
**     Buried Pipeline Heat Loss Calculation     **
**     Vers. 1.1                     Sept. 2000     **
**     ----------- BETA RELEASE-----------     **
*********************************************
Project: Field Book Example
Job Number: 01234     Date: 07-11-2001     Time: 09:01:03

INPUT PARAMETERS:
------------------

Fluid Properties:
------------------

Heat Capacity (kJ/kg-K)                          2.76
Viscosity (cp)                                   0.013
Density (kg/m3)                                  68.1
Thermal Conductivity (W/m-K)                     0.04
Mass Flow Rate (kg/hr)                           9737
Fluid Temperature (deg C)                        48.89
```

```
Pipe Properties:
-----------------

Inside Diameter (cm)                          9.718
Outside Diameter (cm)                         11.43
Thermal Conductivity (W/m-K)                  48.5
Buried Depth (m)                              1.5
Length (km)                                   7.5

Insulation Properties:
----------------------

***  Pipe Uninsulated  ***

Yellow Jacket:
--------------

Pipe coated with Yellow Jacket

Soil Properties:
----------------

Thermal Conductivity (W/m-K)                  1.3
Temperature (deg C)                           1.67

CALCULATED RESULTS:
-------------------

Fluid Exit Temperature                  8.13 deg C
Temperature Change                      40.76 deg C
Log Mean Temperature Change             20.50 deg C


Reynolds Number        2.726E+06
Prandtl Number         8.970E-01
Nusselt Number         3.133E+03


Fluid Velocity         5.355 m/s
Pressure Drop          1.726E+02 Pa/m     ***Approximate***
Total Pressure Drop    1.295E+03 kPa      ***Approximate***


Inside Heat Transfer Coeff             1.290E+03 W/m2-K
Inside Overall Heat Transfer Coeff     6.483E+00 W/m2-K
Inside Surface Area                    2.290E+03 m2
Outside Overall Heat Transfer Coeff    5.392E+00 W/m2-K
Outside Surface Area                   2.753E+03 m2
Total Heat Transfer                    3.042E+02 kW
```

Pipeline Profile

Distance (km)	Fluid Temperature (deg C)	Heat Loss (kW)
0.000	48.89	—
0.188	46.60	17.09
0.375	44.42	16.27
0.563	42.35	15.48
0.750	40.38	14.73
0.938	38.50	14.01
1.125	36.71	13.33
1.313	35.01	12.69
1.500	33.40	12.07
1.688	31.86	11.49
1.875	30.39	10.93
2.063	29.00	10.40
2.250	27.67	9.89
2.438	26.41	9.41
2.625	25.21	8.96
2.813	24.07	8.52
3.000	22.99	8.11
3.188	21.95	7.72
3.375	20.97	7.34
3.563	20.03	6.99
3.750	19.14	6.65
3.938	18.29	6.33
4.125	17.49	6.02
4.313	16.72	5.73
4.500	15.99	5.45
4.688	15.30	5.18
4.875	14.64	4.93
5.063	14.01	4.69
5.250	13.41	4.47
5.438	12.84	4.25
5.625	12.30	4.04
5.813	11.78	3.85
6.000	11.29	3.66
6.188	10.83	3.48
6.375	10.38	3.31
6.563	9.96	3.15
6.750	9.56	3.00
6.938	9.17	2.86
7.125	8.81	2.72
7.313	8.46	2.59
7.500	8.13	2.46

```
Contributions to Overall Heat Transfer Coefficient:
-----------------------------------------------------

Resistance Due to Fluid                           0.50%
Resistance Due to Pipe                            0.11%
Resistance Due to Insulation                      0.00%
Resistance Due to Yellow Jacket                   3.96%
Resistance Due to Soil                           95.44%

   ************************************************
   **     Buried Pipeline Heat Loss Calculation   **
   **     Vers. 1.1               Sept. 2000      **
   **     ----------- BETA RELEASE ------------   **
   ************************************************
Project: Field Book Example
Job Number: 01234     Date: 07-11-2001     Time: 09:10:07

INPUT PARAMETERS:
-----------------

Fluid Properties:
-----------------

Heat Capacity (kJ/kg-K)                           2.76
Viscosity (cp)                                    0.013
Density (kg/m3)                                   68.1
Thermal Conductivity (W/m-K)                      0.04
Mass Flow Rate (kg/hr)                            9737
Fluid Temperature (deg C)                         48.89

Pipe Properties:
----------------

Inside Diameter (cm)                              9.718
Outside Diameter (cm)                             11.43
Thermal Conductivity (W/m-K)                      48.5
Buried Depth (m)                                  1.5
Length (km)                                       7.5

Insulation Properties:
----------------------

Thermal Conductivity (W/m-K)                      0.173
Insulation Thickness (cm)                         2.54

Yellow Jacket:
--------------

Pipe coated with Yellow Jacket

Soil Properties:
----------------

Thermal Conductivity (W/m-K)                      1.3
Temperature (deg C)                               1.67
```

CALCULATED RESULTS:

Fluid Exit Temperature	14.98 deg C
Temperature Change	33.91 deg C
Log Mean Temperature Change	26.78 deg C

Reynolds Number	2.726E+06
Prandtl Number	8.970E−01
Nusselt Number	3.133E+03

Fluid Velocity	5.355 m/s	
Pressure Drop	1.726E+02 Pa/m	***Approximate***
Total Pressure Drop	1.295E+03 kPa	***Approximate***

Inside Heat Transfer Coeff	1.290E+03 W/m^2-K
Inside Overall Heat Transfer Coeff	4.129E+00 W/m^2-K
Inside Surface Area	2.290E+03 m^2
Outside Overall Heat Transfer Coeff	2.394E+00 W/m^2-K
Outside Surface Area	3.950E+03 m^2
Total Heat Transfer	2.532E+02 kW

Pipeline Profile

Distance (km)	Fluid Temperature (deg C)	Heat Loss (kW)
0.000	48.89	—
0.188	47.42	10.99
0.375	45.99	10.64
0.563	44.61	10.31
0.750	43.27	9.99
0.938	41.98	9.68
1.125	40.72	9.38
1.313	39.50	9.09
1.500	38.32	8.80
1.688	37.18	8.53
1.875	36.07	8.26
2.063	35.00	8.00
2.250	33.96	7.76
2.438	32.96	7.51
2.625	31.98	7.28
2.813	31.04	7.05
3.000	30.12	6.83

```
3.188          29.23               6.62
3.375          28.38               6.41
3.563          27.54               6.21
3.750          26.74               6.02
3.938          25.96               5.83
4.125          25.20               5.65
4.313          24.47               5.47
4.500          23.76               5.30
4.688          23.07               5.14
4.875          22.40               4.98
5.063          21.75               4.82
5.250          21.13               4.67
5.438          20.52               4.53
5.625          19.93               4.39
5.813          19.36               4.25
6.000          18.81               4.12
6.188          18.28               3.99
6.375          17.76               3.86
6.563          17.26               3.74
6.750          16.77               3.63
6.938          16.30               3.51
7.125          15.85               3.40
7.313          15.41               3.30
7.500          14.98               3.20

Contributions to Overall Heat Transfer Coefficient:
----------------------------------------------------

Resistance Due to Fluid                    0.32%
Resistance Due to Pipe                     0.07%
Resistance Due to Insulation               42.65%
Resistance Due to Yellow Jacket            1.75%
Resistance Due to Soil                     55.22%

   *************************************************
   **     Buried Pipeline Heat Loss Calculation   **
   **     Vers. 1.1                Sept. 2000     **
   **     ----------- BETA RELEASE ------------   **
   *************************************************

Project: Field Book Example
Job Number: 01234     Date: 07-11-2001     Time: 09:13:24

INPUT PARAMETERS:
-----------------
```

```
Fluid Properties:
-----------------

Heat Capacity (kJ/kg-K)                          2.76
Viscosity (cp)                                   0.013
Density (kg/m3)                                  68.1
Thermal Conductivity (W/m-K)                     0.04
Mass Flow Rate (kg/hr)                           9737
Fluid Temperature (deg C)                        48.89

Pipe Properties:
----------------

Inside Diameter (cm)                             9.718
Outside Diameter (cm)                            11.43
Thermal Conductivity (W/m-K)                     48.5
Buried Depth (m)                                 1.5
Length (km)                                      7.5

Insulation Properties:
----------------------

Thermal Conductivity (W/m-K)                     0.173
Insulation Thickness (cm)                        5.08

Yellow Jacket:
--------------

Pipe coated with Yellow Jacket

Soil Properties:
----------------

Thermal Conductivity (W/m-K)                     1.3
Temperature (deg C)                              1.67

CALCULATED RESULTS:
-------------------

Fluid Exit Temperature                     19.04 deg C
Temperature Change                         29.85 deg C
Log Mean Temperature Change                29.85 deg C


Reynolds Number       2.726E+06
Prandtl Number        8.970E-01
Nusselt Number        3.133E+03


Fluid Velocity        5.355 m/s
Pressure Drop         1.726E+02 Pa/m    ***Approximate***
Total Pressure Drop   1.295E+03 kPa     ***Approximate***
```

Inside Heat Transfer Coeff	1.290E+03 W/m^2-K
Inside Overall Heat Transfer Coeff	3.261E+00 W/m^2-K
Inside Surface Area	2.290E+03 m^2
Outside Overall Heat Transfer Coeff	1.451E+00 W/m^2-K
Outside Surface Area	5.147E+03 m^2
Total Heat Transfer	2.229E+02 kW

Pipeline Profile.

Distance (km)	Fluid Temperature (deg C)	Heat Loss (kW)
0.000	48.89	—
0.188	47.72	8.71
0.375	46.59	8.49
0.563	45.48	8.28
0.750	44.40	8.08
0.938	43.34	7.88
1.125	42.31	7.68
1.313	41.31	7.49
1.500	40.33	7.31
1.688	39.37	7.13
1.875	38.44	6.95
2.063	37.53	6.78
2.250	36.65	6.61
2.438	35.78	6.45
2.625	34.94	6.29
2.813	34.12	6.13
3.000	33.32	5.98
3.188	32.54	5.83
3.375	31.77	5.69
3.563	31.03	5.55
3.750	30.31	5.41
3.938	29.60	5.28
4.125	28.91	5.15
4.313	28.24	5.02
4.500	27.58	4.90
4.688	26.94	4.78
4.875	26.32	4.66
5.063	25.71	4.54
5.250	25.11	4.43
5.438	24.54	4.32
5.625	23.97	4.22
5.813	23.42	4.11
6.000	22.88	4.01

```
6.188        22.36           3.91
6.375        21.85           3.81
6.563        21.35           3.72
6.750        20.86           3.63
6.938        20.39           3.54
7.125        19.93           3.45
7.313        19.48           3.37
7.500        19.04           3.28

Contributions to Overall Heat Transfer Coefficient:
------------------------------------------------------

Resistance Due to Fluid             0.25%
Resistance Due to Pipe              0.05%
Resistance Due to Insulation        58.25%
Resistance Due to Yellow Jacket     1.06%
Resistance Due to Soil              40.38%

   *************************************************
   **   Buried Pipeline Heat Loss Calculation     **
   **   Vers. 1.1                  Sept. 2000     **
   **   ----------- BETA RELEASE ------------     **
   *************************************************

Project: Field Book Example
Job Number: 01234     Date: 07-11-2001     Time: 09:26:12

INPUT PARAMETERS:
-----------------

Fluid Properties:
-----------------

Heat Capacity (kJ/kg-K)                          2.76
Viscosity (cp)                                   0.013
Density (kg/m^3)                                 68.1
Thermal Conductivity (W/m-K)                     0.04
Mass Flow Rate (kg/hr)                           9737
Fluid Temperature (deg C)                        48.89

Pipe Properties:
----------------

Inside Diameter (cm)                             9.718
Outside Diameter (cm)                            11.43
Thermal Conductivity (W/m-K)                     48.5
Buried Depth (m)                                 1.5
Length (km)                                      7.5

Insulation Properties:
----------------------

Thermal Conductivity (W/m-K)                     0.173
Insulation Thickness (cm)                        7.62
```

```
Yellow Jacket:
--------------
Pipe coated with Yellow Jacket

Soil Properties:
----------------
Thermal Conductivity (W/m-K)              1.3
Temperature (deg C)                       1.67

CALCULATED RESULTS:
-------------------
Fluid Exit Temperature                    21.69 deg C
Temperature Change                        27.20 deg C
Log Mean Temperature Change               31.70 deg C


Reynolds Number        2.726E+06
Prandtl Number         8.970E-01
Nusselt Number         3.133E+03


Fluid Velocity         5.355 m/s
Pressure Drop          1.726E+02 Pa/m    ***Approximate***
Total Pressure Drop    1.295E+03 kPa     ***Spproximate***


Inside Heat Transfer Coeff               1.290E+03 W/m²-K
Inside Overall Heat TransferCoeff        2.797E+00 W/m²-K
Inside Surface Area                      2.290E+03 m²
Outside Overall Heat TransferCoeff       1.009E+00 W/m²-K
Outside Surface Area                     6.344E+03 m²
Total Heat Transfer                      2.030E+02 kW

Pipeline Profile
----------------
```

Distance (km)	Fluid Temperature (deg C)	Heat Loss (kW)
0.000	48.89	—
0.188	47.89	7.48
0.375	46.91	7.32
0.563	45.95	7.17
0.750	45.01	7.01
0.938	44.09	6.86
1.125	43.19	6.72
1.313	42.31	6.58

```
1.500        41.45          6.44
1.688        40.60          6.30
1.875        39.78          6.17
2.063        38.97          6.04
2.250        38.18          5.91
2.438        37.40          5.78
2.625        36.64          5.66
2.813        35.90          5.54
3.000        35.17          5.42
3.188        34.46          5.31
3.375        33.77          5.19
3.563        33.09          5.08
3.750        32.42          4.98
3.938        31.77          4.87
4.125        31.13          4.77
4.313        30.50          4.67
4.500        29.89          4.57
4.688        29.29          4.47
4.875        28.71          4.38
5.063        28.13          4.28
5.250        27.57          4.19
5.438        27.02          4.10
5.625        26.48          4.02
5.813        25.96          3.93
6.000        25.44          3.85
6.188        24.94          3.77
6.375        24.44          3.69
6.563        23.96          3.61
6.750        23.49          3.53
6.938        23.03          3.46
7.125        22.57          3.38
7.313        22.13          3.31
7.500        21.69          3.24

Contributions to Overall Heat Transfer Coefficient:
----------------------------------------------------

Resistance Due to Fluid              0.22%
Resistance Due to Pipe               0.05%
Resistance Due to Insulation        66.56%
Resistance Due to Yellow Jacket      0.74%
Resistance Due to Soil              32.45%
```

```
*********************************************
**    Buried Pipeline Heat Loss Calculation    **
**    Vers. 1.1                  Sept. 2000    **
**    ----------- BETA RELEASE ------------    **
*********************************************
Project: Field Book Example
Job Number: 01234     Date: 07-11-2001     Time: 10:42:23

INPUT PARAMETERS:
-----------------
Fluid Properties:
-----------------

Heat Capacity (kJ/kg-K)                        2.76
Viscosity (cp)                                 0.013
Density (kg/m3)                                68.1
Thermal Conductivity (W/m-K)                   0.04
Mass Flow Rate (kg/hr)                         9737
Fluid Temperature (deg C)                      135

Pipe Properties:
----------------

Inside Diameter (cm)                           9.718
Outside Diameter (cm)                          11.43
Thermal Conductivity (W/m-K)                   48.5
Buried Depth (m)                               1.5
Length (km)                                    7.5

Insulation Properties:
----------------------
*** Pipe Uninsulated ***

Yellow Jacket:
--------------
Pipe coated with Yellow Jacket

Soil Properties:
----------------

Thermal Conductivity (W/m-K)                   1.3
Temperature (deg C)                            1.67

CALCULATED RESULTS:
-------------------

Fluid Exit Temperature                   19.92deg C
Temperature Change                       115.08 deg C
Log Mean Temperature Change              57.87deg C


Reynolds Number        2.726E+06
Prandtl Number         8.970E-01
Nusselt Number         3.133E+03
```

```
Fluid Velocity        5.355 m/s
Pressure Drop         1.726E+02 Pa/m   ***Approximate***
Total Pressure Drop   1.295E+03 kPa    ***Approximate***

Inside Heat Transfer Coeff               1.290E+03 W/m2-K
Inside Overall Heat Transfer Coeff       6.483E+00 W/m2-K
Inside Surface Area                      2.290E+03 m2
Outside Overall Heat Transfer Coeff      5.392E+00 W/m2-K
Outside Surface Area                     2.753E+03 m2
Total Heat Transfer                      8.590E+02 kW
```

Pipeline Profile

Distance (km)	Fluid Temperature (deg C)	Heat Loss (kW)
0.000	135.00	–
0.188	128.53	48.27
0.375	122.38	45.93
0.563	116.53	43.70
0.750	110.96	41.58
0.938	105.66	39.56
1.125	100.61	37.65
1.313	95.82	35.82
1.500	91.25	34.08
1.688	86.91	32.43
1.875	82.77	30.86
2.063	78.84	29.36
2.250	75.10	27.94
2.438	71.54	26.58
2.625	68.15	25.29
2.813	64.92	24.07
3.000	61.86	22.90
3.188	58.94	21.79
3.375	56.16	20.73
3.563	53.52	19.73
3.750	51.00	18.77
3.938	48.61	17.86
4.125	46.34	16.99
4.313	44.17	16.17
4.500	42.11	15.39
4.688	40.15	14.64
4.875	38.28	13.93
5.063	36.51	13.25
5.250	34.82	12.61

```
5.438      33.21          12.00
5.625      31.68          11.42
5.813      30.22          10.86
6.000      28.84          10.34
6.188      27.52          9.84
6.375      26.27          9.36
6.563      25.08          8.90
6.750      23.94          8.47
6.938      22.86          8.06
7.125      21.83          7.67
7.313      20.85          7.30
7.500      19.92          6.95

Contributions to Overall Heat Transfer Coefficient:
--------------------------------------------------

Resistance Due to Fluid            0.50%
Resistance Due to Pipe             0.11%
Resistance Due to Insulation       0.00%
Resistance Due to Yellow Jacket    3.96%
Resistance Due to Soil             95.44%
```

REFERENCES

Arnold, K., Stewart, M., 1989. Surface Production Operations. Design of Gas-Handling Systems and Facilities. vol. 2. Gulf Publishing, Houston, TX.

Austvik, T., Li, X., Gjertsen, L.H., 2000. Hydrate plug properties. Formation and removal of plugs. Ann. N.Y. Acad. Sci. 912, 294–303.

Holman, J.P., 1981. Heat Transfer, fifth ed. McGraw-Hill, New York.

National Energy Board of Canada, 2004. Safety Advisory – NEB SA 2004-01. www.neb-one.gc.ca.

Peters, D., Selim, S., Sloan, E.D., 2000. Hydrate dissociation in pipelines by two-sided depressurization: experiment and model. Ann. N.Y. Acad. Sci. 912, 304–313.

Sands, S., May 1, 2011. Alberta worker dies after hydrogen sulphide leak. Edmonton J.

CHAPTER 8

Physical Properties of Hydrates

In the design of processes, the physical properties are important. This is no less true when the processes involve hydrates. Some of the properties of hydrates were reviewed in the previous chapters. This chapter focuses on properties that were not covered earlier.

The estimation of the properties of hydrates is complicated by the fact that the properties depend on: (1) the type of the hydrate, (2) the guest molecule encaged in the hydrate, and (3) the degree of saturation (remember that hydrates are nonstoichiometric). It is unfortunate, but most hydrate programs do not give the saturation numbers as a part of their calculations. An exception is *CSMHYD*, which does give saturation values. The newer *CSMGEM* gives composition of the hydrate phase, but not specifically the cell saturation.

The heat capacity, electrical, and mechanical properties of hydrates are similar to those for ice. The thermal conductivity is unique because it is significantly different from that of ice (Handa and Cook, 1987), as we shall see.

8.1 MOLAR MASS

The molar mass (molecular weight) of a hydrate can be determined from its crystal structure and the degree of saturation. The molar mass of the hydrate, M, is given by:

$$M = \frac{N_W M_W + \sum_{j=1}^{c} \sum_{i=1}^{n} Y_{ij} \nu_i M_j}{N_W + \sum_{j=1}^{c} \sum_{i=1}^{n} Y_{ij} \nu_i} \tag{8.1}$$

where N_W is the number of water molecules per unit cell (46 for Type I, 136 for Type II, and 34 for Type H), M_W is the molar mass of water, Y_{ij} is the fractional occupancy of cavities of type i by component j, ν_i is the number of type i cavities, n is the number of cavities of type i (two for both Type I and II, but is three for Type H), and c is the number of components in the cell.

Although this equation looks fairly complicated, it is just accounting for all of the molecules present and then using a number average to get the molar mass.

Natural Gas Hydrates
ISBN 978-0-12-800074-8
http://dx.doi.org/10.1016/B978-0-12-800074-8.00008-9

Table 8.1 Molar Masses of Some Hydrates at 0 °C

	Hydrate Type	Saturation		Molar Mass (g/mol)
		Small	Large	
Methane	I	0.8723	0.9730	17.74
Ethane	I	0.0000	0.9864	19.39
Propane	II	0.0000	0.9987	19.46
Isobutane	II	0.0000	0.9987	20.24
CO_2	I	0.7295	0.9813	21.59
H_2S	I	0.9075	0.9707	20.87

Note: calculated using Eqn (8.1). The saturation values were calculated using *CSMHYD*.

Table 8.1 summarizes the molar masses of a few hydrate formers. It is a little surprising that the molar masses of all six components are approximately equal (~20 g/mol). This is because the hydrate is composed mostly of water (18.015 g/mol).

It is interesting that the molar masses of hydrates are a function of the temperature and the pressure, since the degree of saturation is a function of these variables. We usually think of molar masses as being constants for a given substance.

8.2 DENSITY

The density of a hydrate, ρ, can be calculated using the following formula:

$$\rho = \frac{N_W M_W + \sum_{j=1}^{c} \sum_{i=1}^{n} Y_{ij} \nu_i M_j}{N_A V_{cell}} \tag{8.2}$$

where N_W is the number of water molecules per unit cell (46 for Type I, 136 for Type II, and 34 for Type H), N_A is Avogadro's number (6.023×10^{23} molecules/mole), M_W is the molar mass of water, Y_{ij} is the fractional occupancy of cavities of type i by component j, ν_i is the number of type i cavities, V_{cell} is the volume of the unit cell (see Table 2.1), n is the number of cavities types (two for both Type I and II, but is three for Type H), and c is the number of components in the cell.

Equation (8.2) can be reduced for a single component in either a Type I or Type II hydrate to:

$$\rho = \frac{N_W M_W + (Y_1 \nu_1 + Y_2 \nu_2) M_j}{N_A V_{cell}} \tag{8.3}$$

Again, although Eqns (8.2) and (8.3) look complicated, they are just accounting for the number of molecules in a unit cell of hydrate. The mass

Table 8.2 Densities of Some Hydrates at 0 °C

	Hydrate Type	Density (g/cm^3)	Density (lb/ft^3)
Methane	I	0.913	57.0
Ethane	I	0.967	60.3
Propane	II	0.899	56.1
Isobutane	II	0.934	58.3
CO_2	I	1.107	69.1
H_2S	I	1.046	65.3
Ice	–	0.917	57.2
Water	–	1.000	62.4

Note: calculated using Eqn (8.3). The saturation values were calculated using *CSMHYD*. Properties of ice and water from Keenan et al. (1978).

of all of these molecules divided by the unit volume of the crystal gives the density of the hydrate.

When using these equations, be careful with the units. Follow the examples at the end of the chapter closely.

As noted earlier, most hydrate software packages do not give the degree of saturation, making it difficult to calculate the density of the hydrate. The K-factors from the Katz method (Chapter 3) do not give the saturation, even though they have the appearance of doing so. Remember, the compositions thus calculated are on a water-free basis.

The densities of some pure hydrates at 0 °C are given in Table 8.2. Note that the densities of the hydrates of the hydrocarbons are similar to ice. The hydrates of carbon dioxide and hydrogen sulfide are significantly denser. In fact, they are denser than water.

8.3 ENTHALPY OF FUSION

Another useful property is the enthalpy of fusion of the hydrate (sometimes called the heat of formation). From this, the amount of heat required to melt a hydrate can be estimated. Table 8.3 lists some enthalpies of fusion for a few hydrates. Ice is included for comparison.

These values represent the formation of a hydrate from liquid water and a gaseous guest molecule. This explains why they are significantly larger than the heat of fusion of water. For pure water, the ice is becoming liquid. When a hydrate melts, it forms a liquid and a gas and the gas is a more highly energetic state.

On the other hand, the enthalpies of fusion are comparable to the enthalpy of sublimation of ice (the phase change going from a solid directly to a gas).

Table 8.3 Enthalpies of Fusion for Some Gas Hydrates

	Hydrate Type	Enthalpy of Fusion (kJ/g)	Enthalpy of Fusion (kJ/mol)	Enthalpy of Fusion (Btu/lb)
Methane	I	3.06	54.2	1320
Ethane	I	3.70	71.8	1590
Propane	II	6.64	129.2	2850
Isobutane	II	6.58	133.2	2830
Ice	–	0.333	6.01	143

Note: Molar enthalpies of fusion converted to specific values (i.e., per unit mass) by using the molar masses from Table 8.1.
Original values from Sloan (1998). Properties of ice and water from Keenan et al. (1978).

For water, this is 2.83 kJ/g or 51.0 kJ/mol. This process is probably more comparable to the formation of a hydrate than is the simple melting of ice.

One method for estimating the effect of temperature on the heat of fusion is the so-called Clapeyron approach. A Clapeyron-type equation is applied to the three-phase locus. The Clapeyron-type equation used in this application is:

$$\frac{\mathrm{d}\ln P}{\mathrm{d}1/T} = -\frac{\Delta H}{zR} \tag{8.4}$$

where ΔH is the enthalpy of fusion, z is the compressibility factor of the gas at the conditions of interest, and R is the universal gas constant. Inherent in this equation is the assumption that the molar volume of the liquid and the hydrate are insignificantly small in comparison to that of the gas; also, this is the only assumption in Eqn (8.4).

From the correlation provided in Chapter 2, the derivative required for Eqn (8.4) is obtained as follows:

$$\frac{\mathrm{d}\ln P}{\mathrm{d}1/T} = \mathrm{B}T^2 - \mathrm{C} + \mathrm{D}T \tag{8.5}$$

Therefore, to calculate the heat of fusion, an analytical expression is required for the three-phase locus. This expression is then differentiated and the enthalpy of fusion is calculated.

8.4 HEAT CAPACITY

There are limited experimental data for the heat capacity of hydrates. Table 8.4 lists some values. For comparison, ice is also included in this table. Over the narrow range of temperatures that hydrates can exist, it is probably safe to assume that these values are constants.

Table 8.4 Heat Capacities for Some Gas Hydrates

	Hydrate Type	Heat Capacity (J/g °C)	Heat Capacity (J/mol °C)	Heat Capacity (Btu/lb °F)
Methane	I	2.25	40	0.54
Ethane	I	2.2	43	0.53
Propane	II	2.2	43	0.53
Isobutane	II	2.2	45	0.53
Ice	—	2.06	37.1	0.492

Original values from Makogon (1997). Properties of ice and water from Keenan et al. (1978).

8.5 THERMAL CONDUCTIVITY

There have been limited studies into the thermal conductivity of hydrates. However, they show that hydrates are much less conductive than ice. The thermal conductivity of ice is 2.2 W/m K, whereas the thermal conductivities of hydrates of hydrocarbons are in the range 0.50 ± 0.01 W/m K.

The thermal conductivity is a key parameter in the process to melt hydrates. This relatively small value is one of the reasons why hydrates take a long time to melt.

8.6 MECHANICAL PROPERTIES

In general, the mechanical properties of hydrates are comparable to those of ice. In the absence of additional information, it is safe to assume that the mechanical properties of the hydrate equal those of ice.

One should not assume that hydrates are soft, slushy material. Hydrate blocks can be as hard as ice. When projected from a pipe under high velocity they can do significant damage.

8.7 VOLUME OF GAS IN HYDRATE

The purpose of this section is to demonstrate the volume of gas encaged in a hydrate. For the purposes of this section, we will examine only the methane hydrate.

The following are the properties of the methane hydrate at 0 °C: the density is 913 kg/m^3, the molar mass (molecular weight) is 17.74 kg/kmol, and methane concentration is 14.1 mol percent —this means there are 141 molecules of methane per 859 molecules of water in the methane hydrate. The density and the molar mass are from earlier in this chapter and the concentration is from Chapter 2.

This information can be used to determine the volume of gas in the methane hydrate. From the density, 1 m^3 of hydrate has a mass of 913 kg. Converting this to moles, 913/17.74 = 51.45 kmol of hydrate, of which 7.257 kmol are methane.

The ideal gas law can be used to calculate the volume of gas when expanded to standard conditions (15 °C and 1 atm or 101.325 kPa)

$$V = nRT/P = (7.257)(8.314)(15 + 273)/101.325$$
$$= 171.5 \text{ Sm}^3$$

Therefore, 1 m^3 of hydrate contains about 170 Sm^3 of methane gas.

Or, in American Engineering Units, this converts to 1 ft^3 of hydrate contains 170 SCF of gas—not a difficult conversion. And 1 ft^3 of hydrate weighs about 14.6 lb, so 1 lb of hydrate contains 11.6 SCF of methane.

By comparison, 1 m^3 of liquid methane (at its boiling point 111.7 K or −161.5 °C) contains 26.33 kmol, which converts to 622 m^3 of gas at standard conditions.

Alternatively, 1 m^3 compressed methane at 7 MPa and 300 K (27 °C) (1015 psia and 80 °F) contains 3.15 kmol or 74.4 Sm^3 of methane gas. The properties of pure methane are from Wagner and de Reuck (1996).

To look at this another way, to store 25,000 Sm^3 (0.88 MMSCF) of methane requires about 150 m^3 (5300 ft^3) of hydrates. This compares with 40 m^3 (1400 ft^3) of liquefied methane, or 335 m^3 (11,900 ft^3) of compressed methane.

8.8 ICE VERSUS HYDRATE

The properties of hydrates over wide ranges of conditions are not readily available. In the absence of additional information, one might be tempted to assume that the property of the hydrate is the same as that for ice. However, as some of the examples have shown, this may lead to significant errors.

Examples

Example 8.1

Calculate the molar mass of a hydrate from an equimolar mixture of hydrogen sulfide and propane at 0 °C. This is a Type II hydrate and the saturations are given in the table below, which are from *CSMHYD*.

	Feed (mol%)	Molar Mass (g/mol)	Cell 1 Satur'n (%)	Cell 2 Satur'n (%)
C_3H_8	50.0	44.094	0.00	97.26
H_2S	50.0	34.080	63.33	1.81

Answer: Calculate the contributions from the propane and hydrogen sulfide first:

$$C_3H_8: \quad [(0.0000)16 + (0.9726)8]44.094 = 214.43$$

$$H_2S: \quad [(0.6333)16 + (0.0181)8]34.080 = 350.26$$

Substitute into Eqn (8.1):

$$M = \frac{136(18.015) + 214.43 + 350.26}{136 + (0.0000 + 0.6333)(16) + (0.9726 + 0.0181)(8)}$$

$$= \frac{3014.7}{154.06} = 19.56 \text{ g/mol}$$

Again, the resulting molar mass is approximately 20 g/mol.

Example 8.2

From *CSMHYD*, the hydrate of hydrogen sulfide, a Type I former, at 10 °C has 93.83% of the small cages occupied and 98.12% of the large ones. Calculate the density of the hydrate. Compare this with the density if it was assumed that the hydrate was 100% saturated.

Answer: From Eqn (8.3):

$$\rho = \frac{\{46(18.015) + [(0.9383)2 + (0.9812)6]34.080\}}{(6.023 \times 10^{23})(1.728 \times 10^{-27})}$$

$$= 1.050 \times 10^6 \text{ g/m}^3$$

$$= 1.050 \text{ g/cm}^3$$

The density of this hydrate is slightly greater than that of pure water and, unlike ice, it will not float on water.

For comparison, assuming 100% saturation:

$$\rho = \frac{\{46(18.015) + [(1)2 + (1)6]34.080\}}{(6.023 \times 10^{23})(1.728 \times 10^{-27})}$$

$$= 1.058 \times 10^6 \text{ g/m}^3$$

$$= 1.058 \text{ g/cm}^3$$

This is an error of less than 1%.

Example 8.3

From *CSMHYD* a mixture of 50% methane, 30% ethane, and 20% propane forms a Type II hydrate at 10 °C and 1617 kPa. The saturation is given in the table below. Calculate the density of this hydrate.

	Feed (mol%)	Molar Mass (g/mol)	Cell 1 Satur'n (%)	Cell 2 Satur'n (%)
CH_4	50.0	16.043	53.06	0.68
C_2H_6	30.0	30.070	0.00	7.65
C_3H_8	20.0	44.094	0.00	91.52

Answer: Perform the inner summations first:

$$CH_4 : \quad [16(0.5306) + 8(0.0068)]16.043 = 137.07$$

$$C_2H_6 : \quad [16(0.0000) + 8(0.0765)]30.070 = 18.40$$

$$C_3H_8 : \quad [16(0.0000) + 8(0.9152)]44.094 = 322.84$$

Then, from Eqn (8.2):

$$\rho = \left[136(18.015) + (137.07 + 18.40 + 322.84\right]/6.023 \times 10^{23}/5.178 \times 10^{-27}$$

$$= 0.939 \times 10^6 \text{ g/m}^3$$

$$= 0.939 \text{ g/cm}^3$$

This hydrate is less dense than water.

Example 8.4

The hydrate locus for methane can be represented by the following equation:

$$\ln P = -146.1094 + 0.3165\ T + 16556.8/T$$

where P is in MPa and T is in K. Differentiate this equation to estimate the enthalpy of fusion for the hydrate at 0 °C. Assume that methane is an ideal gas ($z = 1$).

Answer: First, differentiate the expression for the hydrate locus:

$$\frac{\mathrm{d}\ln P}{\mathrm{d}T} = 0.3165 - 16556.8/T^2$$

Make a small transformation of Eqn (8.3)

$$\frac{\mathrm{d}\ln P}{\mathrm{d}1/T} = -\frac{\Delta H}{zR}$$

$$\frac{\mathrm{d}\ln P}{\mathrm{d}\ T} = \frac{\Delta H}{zRT^2}$$

$$\Delta H = zRT^2\left(\frac{\mathrm{d}\ln P}{\mathrm{d}T}\right)$$

Make the appropriate substitutions:

$$\begin{aligned}\Delta H &= zRT^2\left(0.3165 - 16556.8/T^2\right)\\ &= zR\left(0.3165\ T^2 - 16556.8\right)\\ &= (1)(8.314)(0.3165[273.15]^2 - 16556.8)\\ &= 58,676\ \mathrm{J/mol} = 58.7\ \mathrm{kJ/mol}\end{aligned}$$

This is comparable to the value of 54.2 kJ/mol from Table 8.3.

It was assumed that methane was an ideal gas. The reader should estimate the effect of this assumption on the calculated results. That is, determine the compressibility factor for methane at 0 °C and 2.60 MPa and substitute it into the above equation.

Example 8.5

Calculate the volume of propane released (at standard conditions, 60 °F and 1 atm) when 1 kg of propane hydrate is melted at 0 °F. Use the information provided in the tables in this chapter (or elsewhere in this book) and, where necessary, assume the properties are independent of temperature.

Answer: Calculate the moles of hydrate by using the molar mass from Table 8.1

$$1\ \mathrm{kg}/19.46\ \mathrm{kg/kmol} = 0.05139\ \mathrm{kmol} = 51.39\ \mathrm{mol}$$

Use the composition of the hydrate from Table 2.4 to determine what fraction of this is propane:

$$51.39 \times (5.55/100) = 2.85\ \mathrm{mol}$$

Use the ideal gas law to convert this to a volume at standard conditions (note 60 °F = 15.56 °C and 1 atm = 101.325 kPa) and $R = 8.314 \times 10^{-3}$ kPa m^3/mol K.

$$\begin{aligned}V &= nRT/P\\ &= (2.85)\left(8.314 \times 10^{-3}\right)(15.56 + 273.15)/101.325\\ &= 67.5 \times 10^{-3}\ \mathrm{m}^3 = 67.5\ \mathrm{L[std]}\end{aligned}$$

So 1 kg of propane hydrate releases almost 70 L[std] of gas.

REFERENCES

Handa, Y.P., Cook, J.G., 1987. Thermal conductivity of xenon hydrate. J. Phys. Chem. 91, 6327–6328.

Keenan, J.H., Keyes, F.G., Hill, P.G., Moore, J.G., 1978. Steam Tables. John Wiley & Sons, New York, NY.

Makogon, Y.F., 1997. Hydrates of Hydrocarbons. PennWell Publishing Co., Tulsa, OK.
Sloan, E.D., 1998. Clathrate Hydrates of Natural Gases, second ed. Marcel Dekker, New York, NY.
Wagner, W., de Reuck, K.M., 1996. Methane. International Thermodynamic Tables of the Fluid State, 13. Blackwell Science, Oxford, UK.

CHAPTER 9

Phase Diagrams

Phase diagrams are useful for both the theoretical discussion of gas hydrates and for engineering design. In this chapter, we will concentrate on fluid phase equilibria as they relate to hydrate formation. Thus, systems that do not form hydrates will not be discussed. This does not mean that they are not interesting or important, just beyond the scope of this book.

In this chapter, some rules are presented for constructing pressure-temperature (P-T) diagrams for single component systems and binary systems as well as pressure-composition (P-x) and temperature-composition (T-x) diagrams for binary systems. These rules can be found throughout this chapter.

9.1 PHASE RULE

Much of this material is based on the Gibbs phase rule. The phase rule is:

$$F = 2 + N - p \tag{9.1}$$

where F is the degrees of freedom, N is the number of components, and p is the number of phases.

An example of its use is as follows: a single component existing in two phases would have 1 degree of freedom. That means that there is one independent variable to "play with." Once one variable is fixed, all of the others are fixed as well. For example, a single component existing as a vapor and a liquid has one degree of freedom. If the temperature is specified, then there are zero degrees of freedom – the pressure is fixed. This pressure is called the vapor pressure.

9.2 COMMENTS ABOUT PHASES

There are some general observations about phases that are useful for developing phase diagrams. The first of these is that gases are always miscible. Therefore, a system, regardless of the number of components, cannot contain two vapor phases. Some may argue that at extreme conditions, two gas phases may exist, but these are dense phases and it may be more logical to treat one of these as a liquid. In addition, such phenomena usually occur beyond the range of temperature and pressure, which is of interest to those in the natural gas business.

Natural Gas Hydrates
ISBN 978-0-12-800074-8
http://dx.doi.org/10.1016/B978-0-12-800074-8.00009-0

A pure component has only one liquid phase. Thus, a single component cannot exhibit liquid phase immiscibility. Binary and multicomponent systems can and do exhibit liquid phase immiscibility. For example, it is common knowledge that oil and water do not mix.

A critical point is a point where the properties of the coexisting phases become the same. Critical points exist when a vapor and a liquid are in equilibrium and when two liquids are in equilibrium. The first critical point is the usual critical point, which should be familiar to natural gas engineers. The second type of critical point is usually called a consolute point.

Another common critical point occurs when two phases become critical while in the presence of a third phase. It may be a bit of a misnomer, but these points are called three-phase critical points.

It is also theoretically possible for three phases to simultaneously become critical (for example a gas and two liquids). Such a point is called a tricritical point.

On the other hand, a pure component can have more than one solid phase. For example, sulfur has two solid forms – rhombic and monoclinic. Carbon also exhibits three solid phases – the common diamond and graphite phases and the more recently discovered buckminsterfullerenes (the so-called "bucky balls"). Water has many different solid forms (denoted ice I, ice II, etc.), but most of these occur at extreme conditions (that is, at very high pressures).

9.3 SINGLE COMPONENT SYSTEMS

For a pure component system, we are only interested in the P-T diagram. Actually, we are primarily concerned with how the single component information relates to the two-component systems.

The first rule about phase diagrams pertains to a one-component system existing as a single phase. The phase rule ($N = 1$ and $p = 1$) says that there are two degrees of freedom. Therefore, we present the following:

1. A single component system in a single phase occupies a region in the temperature-pressure plane.

   This leads to the second rule. The phase rule for a single component system ($N = 1$) and two phases in equilibrium ($p = 2$) indicates that there is one degree of freedom.

2. For a single component system, the location where two phases are in equilibrium corresponds to a curve in the temperature-pressure plane.

   The curve where vapor and liquid are in equilibrium is called the vapor pressure curve. If the component does not decompose, then this curve ends in a critical point. The solid–liquid curve is called the melting curve and the gas–solid the sublimation curve.

These curves bound the various single-phase regions. For example, the vapor region lies at pressures less than and temperatures greater than the vapor pressure curve and at temperatures greater that and pressures less than the sublimation curve.

The next rule arises when there are three phases in equilibrium. In this case, the phase rule says there are zero degrees of freedom. Thus, this is a fixed point in the P-T plane and is called a triple point. The location of the triple point is at the intersection of the two-phase loci.

3. Three two-phase loci intersect at a triple point, a point where three-phases are in equilibrium.

As a corollary to rule 3, the vapor pressure, melting, and sublimation curves intersect at a vapor-liquid-solid triple point. This is the most common triple point, but because multiple solids can exist, there may be other triple points. However, there cannot be a single component liquid-liquid-vapor triple point, because, as was stated earlier, two liquid phases cannot exist for a pure component.

The critical points of components found in natural gas are listed in Table 9.1. This table also lists the vapor-liquid-solid triple points for these substances. Critical points for pure components are fairly well established and large tabulations are available. One reason for this is that the critical point is an important parameter in the correlation of fluid properties.

Table 9.1 Critical and Triple Points for Common Natural Gas Components

	Triple Point		Critical Point	
	Temperature (K)	**Pressure (MPa)**	**Temperature (K)**	**Pressure (MPa)**
Water	273.16	6.12×10^{-4}	647.3	22.06
Methane	90.7	0.0117	190.6	4.60
Ethane	91.7	1.1×10^{-6}	305.4	4.88
Propane	85.5	1.69×10^{-10}	369.8	4.25
n-butane	134.7	4×10^{-7}	425.2	3.80
Isobutane	113.5	1.95×10^{-6}	407.7	3.65
Carbon dioxide	216.7	0.518	304.2	7.383
Hydrogen sulfide	187.7	0.0232	373.4	8.963
Nitrogen	63.1	0.0125	126.2	3.39

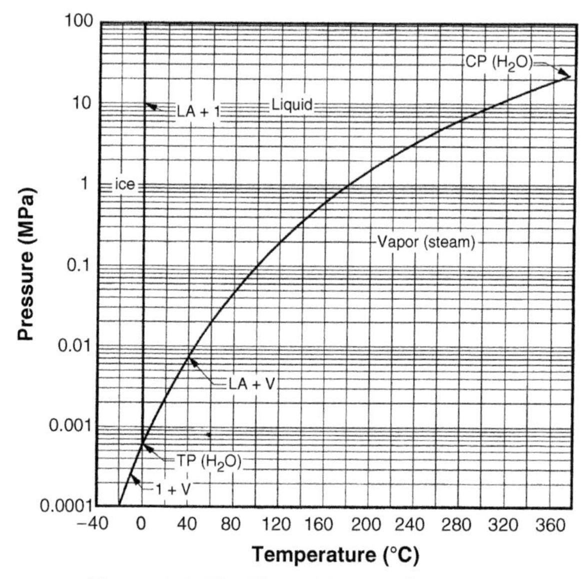

Figure 9.1 *The Phase Diagram for Water.*

9.3.1 Water

Figure 9.1 shows the P-T diagram for water. This plot is to scale and shows the vapor pressure, melting curve, and sublimation curve.

As was discussed earlier, note how the three loci intersect in a triple point. Also, note how the three two-phase loci map out the single-phase regions. For example, the single-phase vapor region is bounded by the vapor + liquid locus and the vapor + solid locus.

Similar diagrams could have been constructed for the other components commonly found in natural gas.

9.4 BINARY SYSTEMS

By adding a second component, we have added a degree of freedom. In Eqn (9.1) when N is increased by one, then the degree of freedom, F, is increased by one as well.

The P-T plane becomes a prism with composition being the third dimension. Two of the faces of the prism represent the pure components and are the P-T diagrams discussed earlier. Because it is difficult to interpret, the prism is often projected onto the P-T plane. In this manner, a binary P-T diagram is constructed.

A binary system existing in two-phases occupies a region in the P-T plane, as opposed to a curve for a pure component. On the other hand, for three-phase equilibrium, which is triple points for a one-component system, there are curves in the P-T plane for a binary system.

To construct a P-x diagram from the P-T diagram, the temperature is fixed and a plane is cut through the P-T projection. Similarly, for the T-x diagram the pressure is fixed.

4. The two-phase equilibrium for a pure component intersects the pure component axes on the P-x and T-x diagrams.
   For example, the vapor pressure is a single point on the pure component axis. If, for a given temperature or pressure, both of the vapor pressures exist, then both occur on the P-x or T-x diagram.
5. Binary critical loci extend from the pure component critical point. The critical locus does not always extend continuously between the two pure component critical points. Critical loci exist that do not end at a pure critical point.
   In a classic paper in the area of fluid phase equilibria, Scott and Van Konynenburg (1970) showed that there are six basic types of binary fluid phase equilibria based on the possible binary critical loci. These are shown in Fig. 9.2. One problem with this is that solid phases can interfere with the predicted fluid phase behavior. For example, it is difficult to categorize the system water + methane into this scheme because an aqueous liquid phase cannot exist at the temperatures required for a liquid methane-rich phase to form.
6. Three-phase surfaces are curves when projected into the P-T plane. That is, the compositional effect is apparently removed.
7. Pure component triple points are the endpoints from one of the many binary three-phase loci.
   The phase rule indicates that it is impossible for a pure component to simultaneously exist in four phases. This is not the case with a binary mixture. From the phase rule, for two components and four phases, there is zero degree of freedom. Thus, a point where four phases are in equilibrium, called a quadruple point, is a single point when projected into the P-T plane.
8. Quadruple points occur at the intersection of four three-phase loci.
   Next, consider constructing P-x and T-x diagrams. This involves taking an infinitely thin slice from the three-dimensional prism. For a T-x

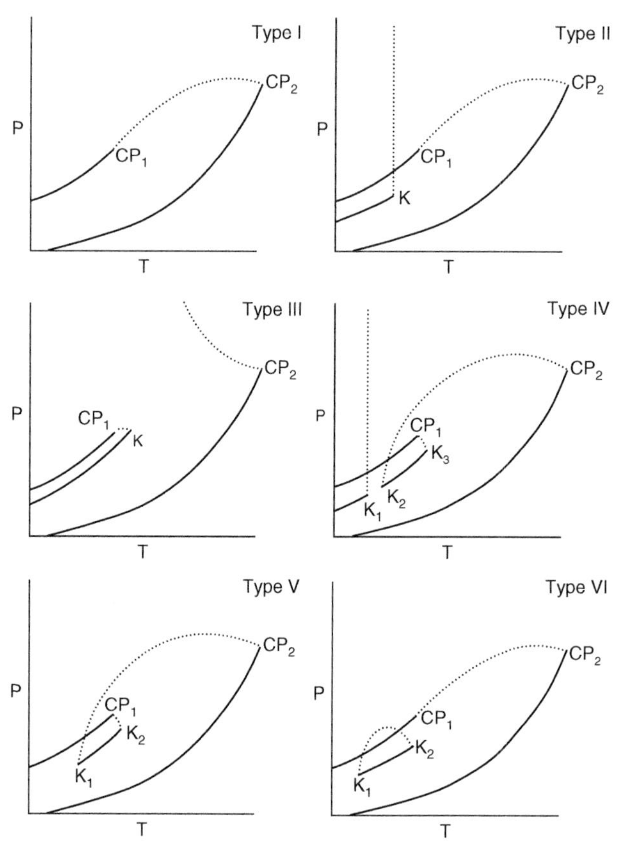

Figure 9.2 The classification of binary fluid phase behavior of binary systems *(Based on Scott and Van Konynenburg (1970).)*. The solid lines are either pure component vapor pressure curves or liquid–liquid equilibrium loci. Dotted lines are the binary critical loci.

diagram, the pressure is held constant and for a P-x diagram, the temperature is constant.

The next rules are useful for constructing P-x and T-x diagrams.

9. The two single-phase regions adjacent to a two-phase region must be the two phases that correspond to the two-phase region. For example, a vapor–liquid region is adjacent to a liquid region and a vapor region.
10. A three-phase point on a P-x or T-x diagram is a horizontal line. The two endpoints and a central point are the compositions of the three phases.
11. The three regions that connect to the three-phase line on a P-x or T-x diagram are two-phase regions, which make up the three-phase line. For example, a three-phase solid-liquid-vapor line is connected to three two-phase regions: vapor–liquid, vapor–solid, and solid–liquid.

A useful tool in the construction of these phase diagrams is the ability to calculate the water content of a gas. Several methods are presented in the next chapter for performing such calculations.

9.4.1 Constructing T-x and P-x Diagrams

The construction of either a T-x or a P-x diagram usually begins with a P-T diagram, or at least it requires knowledge of the various pure component two-phase loci and the binary three-phase loci. This information is best extracted from the P-T diagram.

The first step is to draw a line on the P-T diagram corresponding to the pressure of interest (for a T-x diagram) or the temperature of interest for a P-x diagram. For example, to construct a P-x diagram at 10 °C, draw the line at $t = 10$ °C. Determine which loci are intersected by this isotherm.

Plot the pure component two-phase points on the appropriate axis. Plot the horizontal line that corresponds to the various three-phase loci. The ends of this horizontal line correspond to the composition of two of the coexisting phases. The composition of the third phase is intermediate of these two compositions. That is, it lies on the line somewhere between the endpoints.

Use the rules presented earlier to join the points and lines to construct the various two-phase and single-phase regions. This quick construction results in a schematic P-x diagram, which, if carefully used, is sufficient for most applications. Phase equilibrium calculations are required in order to include the precise compositions of the various phases. Then these must be plotted to scale.

9.4.2 Methane + Water

Ironically, one system that does not fall into the classification of Scott and Van Konynenburg (1970) is methane + water. Because the critical point of methane is at such a low temperature, a methane-rich liquid cannot form in the presence of liquid water. The P-T diagram for this system is shown in Fig. 9.3. The vapor pressure curve of methane is at such a low temperature that it is not included.

From the P-T diagram, and some additional information regarding composition (given in a previous chapter), a P-x diagram at 10 °C (50 °F) can be constructed. As noted earlier, at 10 °C, methane does not liquefy and, thus, none of the loci intersects the pure methane axis. That is, there is no possibility of equilibria between two phases composed of pure methane.

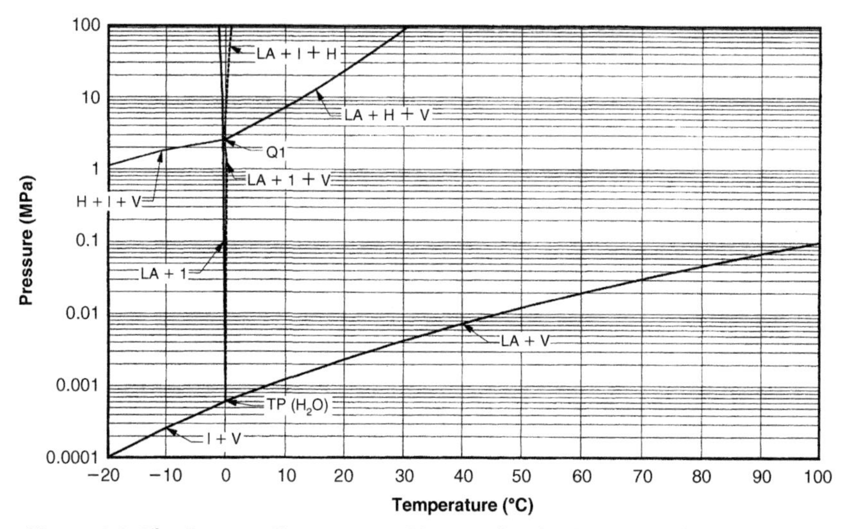

Figure 9.3 ***The Pressure-Temperature Diagram for the System Methane + Water.***

On the other hand, the vapor pressure of water at 10 °C is 1.23 kPa (0.18 psia).

In this and subsequent discussion, the reader is cautioned about kPa and MPa. The various pressures differ by orders of magnitude and the switch between kilo and mega makes the numbers more rational.

In addition, from information presented earlier, the hydrate pressure at this temperature is 7.25 MPa. From this information and the rules presented earlier, the diagram in Fig. 9.4 was constructed. Please note that this plot is not to scale, although several pressure and compositions are noted.

At the three-phase point (hydrate + vapor + aqueous liquid), the vapor is essentially pure methane and the aqueous phase is nearly pure water. The compositions of these phases are given in Table 2.2.

From the phase diagram, we can make a few observations. Consider an equimolar mixture of methane and water at 10 °C and 10 MPa (50 °F and 1450 psia). According to the phase diagram, this mixture exists in two phases: (1) a hydrate and (2) a vapor. In other words, a hydrate exists without free-water being present.

Next, consider a mixture that is very lean in methane, for example 0.1 mol%. At this concentration, there is not enough methane present to form a hydrate. All of the methane remains in an aqueous solution, regardless of the pressure.

Figure 9.4(a) is a magnification of a region of Fig. 9.4, but it is a mirror image and it is to scale so that qualitative analysis can be made based on this

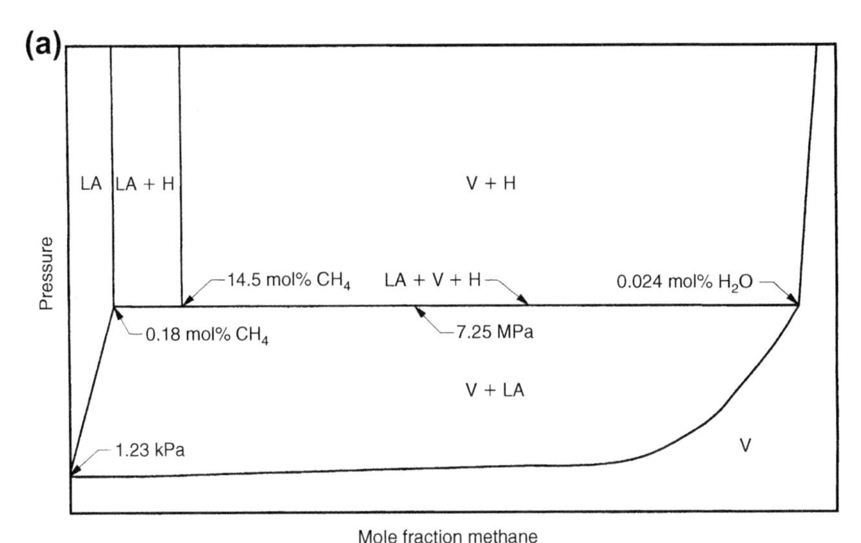

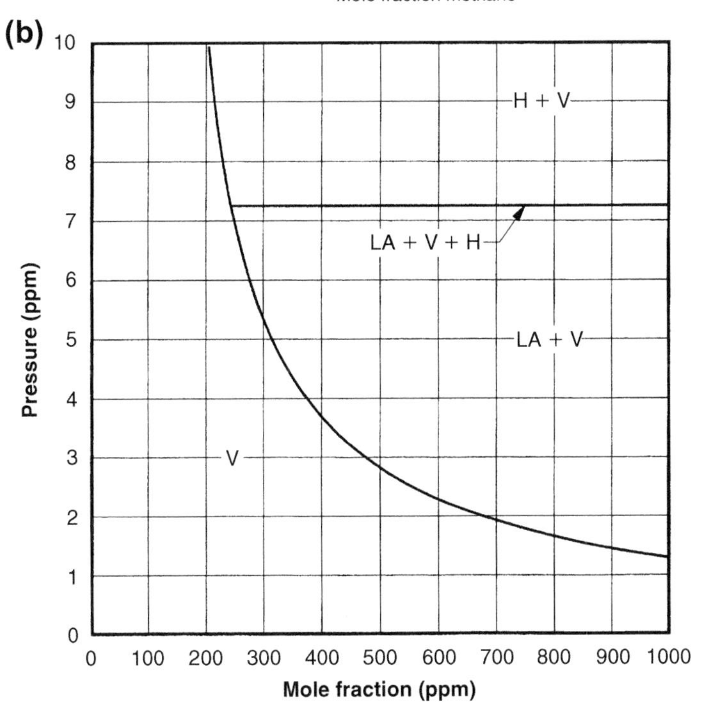

Figure 9.4 (a) Pressure-composition diagram for methane + water at 10 °C (not to scale). (b) Pressure-composition diagram for water + methane at 10 °C (magnified region to scale).

chart. The aqueous dew point portion of the curve is the prediction from AQUAlibrium and the water content in the hydrate region is an extrapolation, based on information presented in the next chapter.

Consider a mixture containing 400 ppm of water. From Fig. 9.4(a), this mixture has a water dew point of about 3800 kPa (550 psia). A hydrate will form once the pressure reaches 7250 kPa (1050 psia). At this point, the gas in equilibrium with the hydrate is estimated to be 245 ppm water. Therefore, any mixture of methane + water less than 245 ppm water will have a hydrate pressure greater than 7250 kPa (1050 psia). For example, a mixture containing 215 ppm water would not form a hydrate until 9000 kPa (1300 psia).

9.4.3 Free-Water

Figure 9.4, the P-x diagram for methane + water, demonstrates the so-called "phase diagram argument" against the need for free-water to be present in order to form a hydrate. For example, for an equimolar mixture of methane + water at 10 °C and 10 MPa, what phases exist? From the phase diagram, there is only hydrate and vapor – where is the free-water? This demonstrates clearly that it is not necessary to have free-water in order to have a hydrate.

The "frost argument" against the need for free-water was presented at the end of Chapter 1. However, we can revisit the frost argument using the phase diagram. This was demonstrated earlier with the mixture of methane + water with a water content of 215 ppm. When this mixture is compressed, an aqueous phase (i.e., free-water) is not encountered. The hydrate sublimes directly out of the gas phase, just like frost from the air.

9.4.4 Carbon Dioxide + Water

The P-T diagram for the system carbon dioxide + water is shown in Fig. 9.5. A quick comparison of Figs 9.3 and 9.5 shows that the carbon dioxide + water are significantly more complicated than that for methane + water.

In this case, a second liquid, rich in carbon dioxide, can form. Note the three-phase locus, $L_A + L_C + V$, ends in a three-phase critical point, K. At this point, the properties of the vapor and the CO_2-rich liquid become the same. A small critical locus extends from the K point to the critical point of pure CO_2.

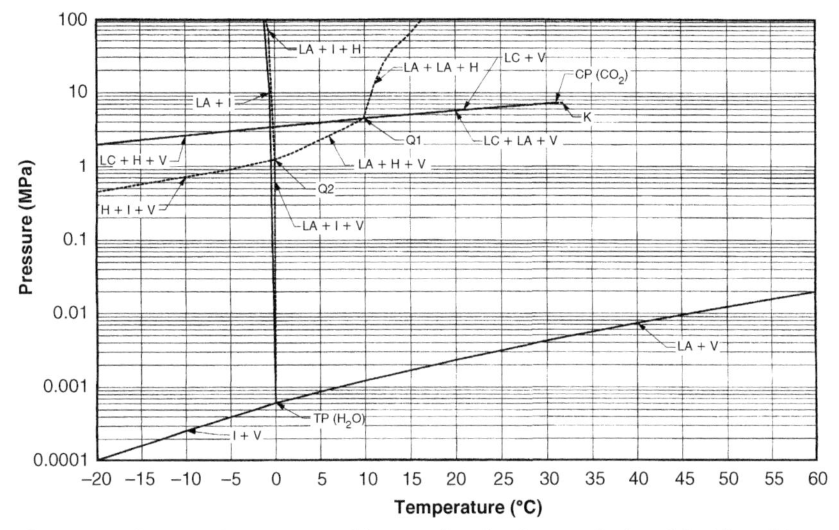

Figure 9.5 ***Pressure-Temperature Diagram for the System Carbon Dioxide + Water.***

The reader will also notice some loci on this diagram that have not been discussed up to this point. In particular, are the LC + H + V loci. This curve is almost coincident with the vapor pressure of pure CO_2.

The phase diagrams for ethane, propane, isobutane, and hydrogen sulfide with water are similar to carbon dioxide + water.

The P-x diagrams for the binary mixture carbon dioxide + water are also considerably more complicated than those for methane + water. As a first example, consider the P-x diagram at 5 °C, which is presented as Fig. 9.6. At this temperature, the vapor pressure of water is 0.873 kPa and that of CO_2 is 3.97 MPa. Interpolating Table 2.7 gives the pressure for the $L_A + H + V$ locus as 2.23 MPa. From the previous discussion, the $L_C + H + V$ locus is assumed to be at a pressure slightly less than the vapor pressure. The compositions noted on this plot are also from Table 2.7.

Next, consider the P-x diagram at 11.3 °C. At this temperature, the vapor pressure of water is 1.34 kPa and CO_2 is 4.65 MPa. At this temperature, the $L_A + L_C + V$ and $L_C + L_A + H$ loci are also crossed. The pressure and composition for the $L_C + L_A + H$ were taken from Table 2.7. The pressure at this point is 20.0 MPa. The pressure and compositions along the $L_A + L_C + V$ were calculated using AQUAlibrium. This information was used to construct the P-x diagram shown in Fig. 9.6.

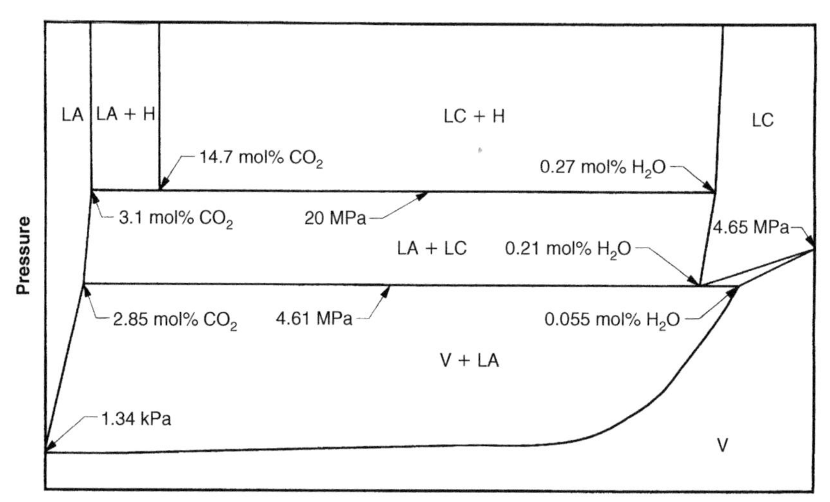

Figure 9.6 ***Pressure-Composition Diagram for Carbon Dioxide + Water at 11.3 °C.*** (not to scale).

9.4.5 Hydrogen Sulfide + Water

A detailed description of the phase diagrams for the binary system hydrogen sulfide + water was presented in the papers by Carroll (1998a,b). The phase equilibria in this system are analogous to those for water + carbon dioxide.

9.4.6 Propane + Water

A detailed review of the phase diagrams for the binary system propane + water was described in the paper by Harmens and Sloan (1990).

9.5 PHASE BEHAVIOR BELOW 0 °C

It is an interesting question to ask which solid phase will form at temperatures below 0 °C – ice or hydrate? To address this question, we examine the low temperature phase diagrams.

At this point, it is worth repeating that a discussion of water content of gases, including temperatures below 0 °C, is included in the next chapter.

9.5.1 Methane + Water

First, consider the binary mixture methane + water and construct a P-x diagram at −10 °C. From Fig. 9.3, we can see that two loci are crossed: (1) the sublimation curve for pure water (I + V) at 0.206 kPa and (2) a

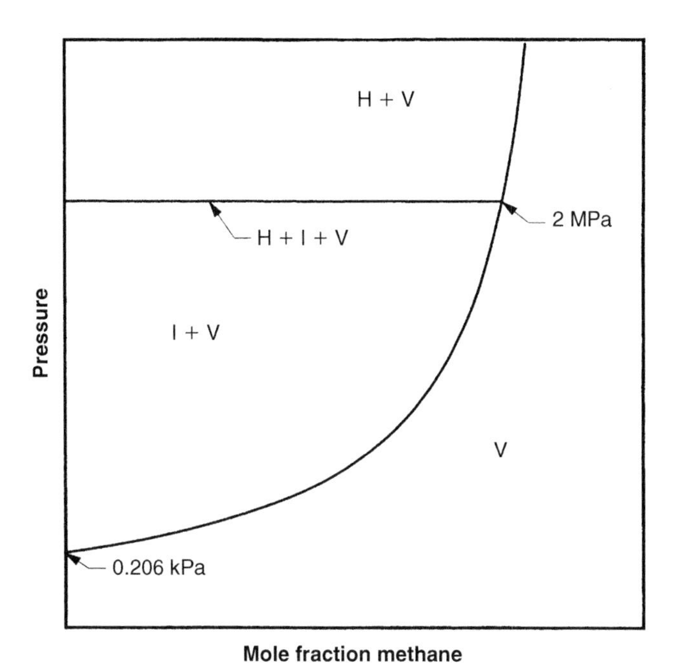

Figure 9.7 ***Pressure-Composition Diagram for Water + Methane at −10 °C.*** (not to scale).

binary three-phase locus (H + I + V) at about 2 MPa. Fig. 9.7 shows the P-x diagram at this temperature.

If you compress a mixture containing 10% methane from a low pressure, the frost point is reached at about 0.23 kPa. The frost point is where the first crystal of solid forms from a gas mixture. This solid is ice and contains no methane. As the compression continues, more ice is formed and the water from the mixture is consumed. The resultant vapor is richer in methane.

At a pressure of 1.827 MPa, the first crystal of hydrate forms. At this point, three phases are in equilibrium: ice (solid water), hydrate (containing about 14% methane and 86% water), and vapor (about 99.99% methane and a little bit more than 0.01% water (about 0.092 mg/Sm^3 or 5.7 lb/MMCF)).

As we continue to compress the mixture, more hydrate is formed and the vapor disappears. Finally, all of the vapor disappears and we enter a region where two solids are in equilibrium: ice and hydrate.

Next, consider a mixture that contains 24% methane. Again, at low pressure, the entire mixture is a gas. Once the mixture is compressed to about 0.275 kPa, the frost point is reached. Again, this solid is pure ice.

Upon further compression, this mixture behaves in a manner similar to the previous mixture – more solid is formed and vapor is consumed.

Again, like the previous mixture, at 1.827 MPa, a three-phase point is reached and hydrate begins to form. However, unlike the previous mixture, further compression results in the ice phase disappearing rather than the vapor. Finally, a point is reached where all of the ice is consumed and only hydrate and vapor remain.

As a final scenario, consider a mixture that is very rich in methane, say 99.999% (which is equivalent to 0.0076 mg/Sm^3 or 0.47 lb/MMSCF). With this mixture, the ice phase is never encountered. The first solid phase encountered is the hydrate. Because this mixture is lean in water, the hydrate is encountered at a pressure greater than the three-phase pressure.

Note that in none of these scenarios is liquid water ever encountered.

9.6 MULTICOMPONENT SYSTEMS

Beyond binary systems, the application of the rules becomes more difficult. With the addition of more components, the phase rule dictates that we now have more degrees of freedom.

With fluid phase equilibria, in multicomponent systems, the design engineer usually constructs a phase envelope, a map that shows the regions where the stream exists as a liquid, a vapor, or as two phases.

The construction of a phase envelope is virtually impossible without computer software. The number of calculations and the complexity of the calculations make hand calculations virtually impossible. Typically, the design engineer can handle perhaps one or two such hand calculations, but beyond that...

9.6.1 An Acid Gas Mixture

As an example of a phase diagram for a multicomponent mixture, consider an acid gas with the following composition (on a water-free basis):

Hydrogen Sulfide	47.20
Carbon dioxide	49.10
Methane	3.19
Ethane	0.51

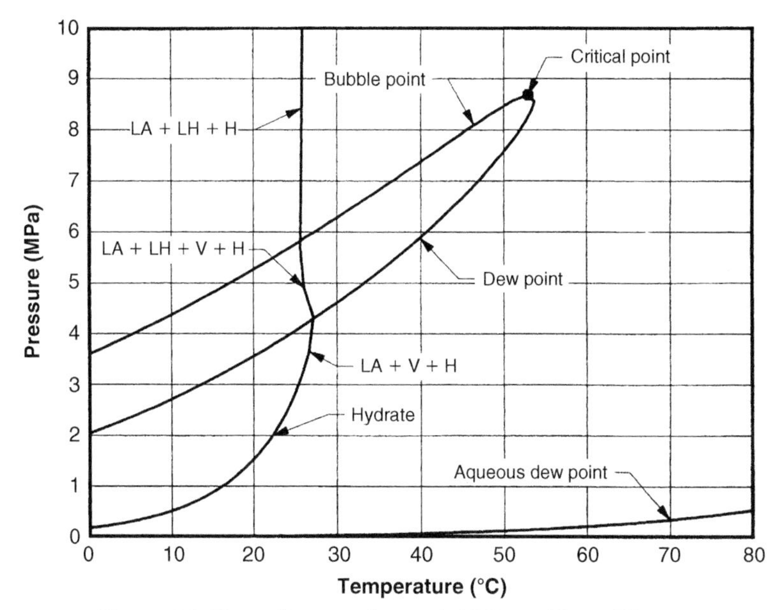

Figure 9.8 ***Phase Diagram for an Acid Gas + Water Mixture.***

Furthermore, the mixture is 90 mol% acid gas and 10 mol% water.

Figure 9.8 shows the P-T phase diagram for this system. The banana-shaped region is what is usually thought of as the "phase envelope." These are the nonaqueous phase dew- and bubble-points and they intersect at a multicomponent critical point.

The hydrate locus is also plotted on this figure. Along the hydrate locus, various phase combinations are encountered. At low pressure, the equilibrium is $L_A + V + H$. In this case, the hydrate locus intersects the phase envelope. As it traverses the phase envelope, there are four phases in equilibrium $L_A + L_H + V + H$, where L_H is used to designate the nonaqueous liquid phase. Note that this is a quadruple locus.

For systems containing more than two components, there can be a quadruple locus because we have additional degrees of freedom. Once the hydrate locus exits the phase envelope there is no more vapor and the equilibrium is $L_A + L_H + H$.

Finally, the curve near the bottom of the plot is the aqueous phase dew point locus. At pressures greater than this locus, an aqueous phase forms. It is not unusual to omit this locus because it is usually at very low pressure.

9.6.2 A Typical Natural Gas[1]

Consider a natural gas with a composition (on a water-free basis) as follows:

Methane	70.85 mol%
Ethane	11.34
Propane	6.99
Isobutane	3.56
n-butane	4.39
Carbon dioxide	2.87

Figure 9.9 shows the P-T diagram for such a system. The aqueous dew point locus has been omitted on this plot, but the reader should be aware of its existence.

The most obvious difference between this phase diagram and that presented previously for the acid gas is the large retrograde region.

Also shown on this plot is the hydrate locus. Compression from low pressure enables the hydrate locus to intersect the phase envelope and a second liquid begins to form. As the pressure raises, more of the second

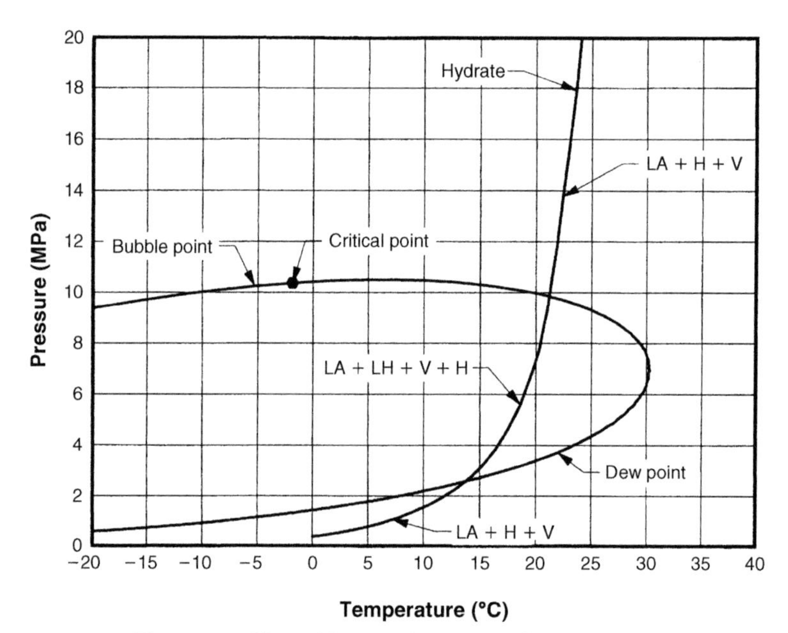

Figure 9.9 ***Phase Diagram for a Typical Natural Gas.***

[1] OK, there is no such thing as a "typical" natural gas, but this is a convenient title for this section.

liquid forms as expected, but a point is reached where the amount of the second liquid reaches a maximum. Beyond that point, the amount of the second liquid decreases until none remains – a retrograde dew point.

Examples

Example 9.1

Calculate the degrees of freedom for a single component system existing in four phases at equilibrium.

Answer: From the phase rule, we have $N = 1$ and $p = 4$. Thus:

$$F = 2 + 1 - 4$$
$$= -1.$$

This is an impossible situation – you cannot have a negative number of degrees of freedom. Therefore, a single component cannot exist in four equilibrium phases.

Example 9.2

Calculate the degrees of freedom for a two-component system existing in four phases at equilibrium.

Answer: From the phase rule, we have $N = 2$ and $p = 4$. Thus:

$$F = 2 + 1 - 4$$
$$= 0.$$

Zero degrees of freedom correspond to a fixed point in the P-T plane and is the quadruple point discussed earlier. Although a single component cannot exist in four phases, a two-component system can.

Example 9.3

At 15 °C and 30 MPa and a composition of 10% water and 90% methane (molar basis), what phases are present? Construct a phase diagram using the information provided in Chapter 2, to answer this question.

Answer: The construction of the phase diagram is left to the reader, but it will be similar to Fig. 9.4.

For the remainder of the question, from Table 2.2, the hydrate formation pressure for methane at 15 °C is 12.79 MPa. Therefore, at 30 MPa, a hydrate will form. Based on the overall composition of the mixture, the phases in equilibrium will be a vapor and a hydrate.

Example 9.4

At 10 °C, methane forms a hydrate at 7.25 MPa. How dry does methane have to be before a hydrate will form at a higher pressure at 10 °C? Use the information in Table 2.2 to answer this question. Express the results in ppm and in lb/MMCF.

Answer: From Table 2.2 the water content of the vapor in equilibrium with the hydrate is 0.025 mol%. Therefore, the gas must contain less than this amount of water.

Converting to ppm:

$$(0.025/100) \times 1,000,000 = 250 \text{ ppm}.$$

Converting to lb/MMCF (assuming standard conditions are 60 °F and 14.696 psi). The volume of 1 lb/mol of gas is:

$$V = nRT/P = (1)(10.73)(460 + 60)/14.696 = 379.7 \text{ ft}^3.$$

Then converting from mole fraction to lb/MMCF gives

$$\begin{aligned} &(0.025/100)\text{lb} - \text{mol water}/379.7 \text{ ft}^3 \times 18.015 \text{ lb/lb} - \text{mol} \\ &= 0.000\ 011\ 9 \text{ lb/ft}^3 \\ &= 12 \text{ lb/MMCF} \end{aligned}$$

Therefore, the gas will not form a hydrate if the water content is less than about 12 lb of water per million standard cubic feet of gas.

Example 9.5

A gas mixture made up of 0.01 mol% water (100 ppm) and the balance is methane. What is the water dew point pressure for this gas at 10 °C?

Answer: From Fig. 9.4, we can see that at these conditions the gas does not have a water dew point. For a mixture with this composition, at pressures below 44.1 MPa, an aqueous phase does not form. At pressures greater than 44.1 MPa, a hydrate may form, depending upon the slope of the boundary of the V + H region. However, an aqueous phase will never form.

The term "water dew point" is commonly encountered in the natural gas business. The term is used to describe the water content of a gas, even though that gas may not have a true water dew point, as was shown in this example.

Example 9.6

A gas mixture made up of 0.01 mol% water and the balance is carbon dioxide. What is the water dew point pressure for this gas at 5 °C?

Answer: This is similar to the previous example, but the phase behavior is different.

From Fig. 9.10, it can be seen that a mixture this lean in water does not have a water dew point. At pressures below about 3.95 MPa, the mixture is always a gas. Eventually, a point is reached where a liquid begins to condense, but this liquid is a carbon dioxide-rich liquid (L_C) and not an aqueous liquid. At higher pressures, the mixture becomes completely liquefied. The small amount of water will remain in solution and an aqueous phase will never form.

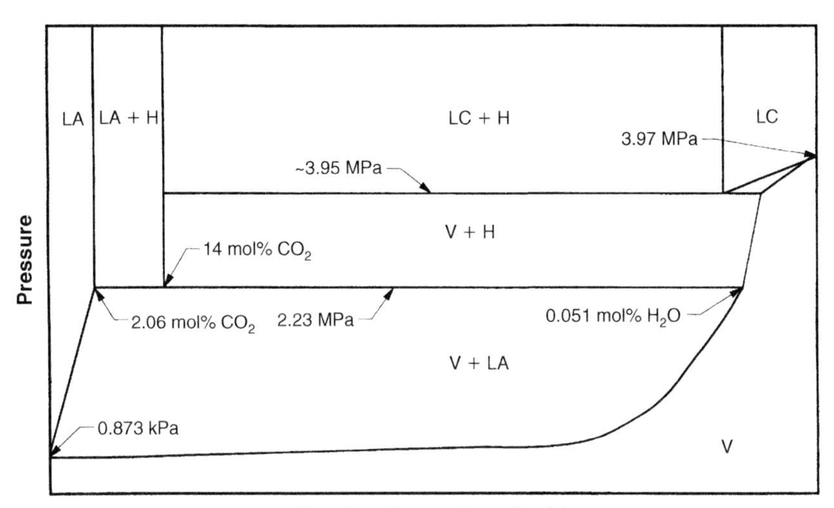

Figure 9.10 ***Pressure-composition diagram for carbon dioxide + water at 5 °C.***

Example 9.7

An equimolar mixture of methane and water is compressed at 10 °C from a very low pressure to a very high pressure. What phases will be encountered during this compression?

Answer: We will conduct this "thought" experiment in an imaginary cylinder-piston arrangement. The piston must be of immense size in order to conduct the actual experiment, so it is possible, but impractical.

To begin, at very low pressure the mixture is a gas.

From Fig. 9.4, as the gas is compressed, the first region entered is the $V + L_A$. That is how we reach an aqueous phase dew point. From AQUAlibrium, this dew point is estimated to be at 2.45 kPa.

Additional compression takes us further into the $V + L_A$ region. Once we reach a pressure of 7.25 MPa, the hydrate starts to form. Continued compression (i.e., reduction in volume) results in an increase in the amount of hydrate present and a reduction in the amount of aqueous liquid; however, the pressure remains unchanged. In addition, the compositions of the three phases are those given in Fig. 9.4 and remain unchanged during further compression.

Eventually, all of the aqueous liquid disappears and the V + H region is entered. For our purposes, we remain in the V + H region regardless of further compression. In reality, at very high pressures, other phases of solid water will be encountered.

Example 9.8

Return to Example 1.1 and we can now address the question of pressure and temperature. Will a hydrate form?

Answer: First, the VMGSim software was used to calculate the phase envelope for this mixture. The hydrate curve was estimated using: (1) VMGSim, (2) EQUI-PHASE Hydrate, and (3) CSMHYD. The three hydrate curves are plotted in Fig. 9.11. Finally, the operating conditions given are shown on the plot. The three hydrate predictions are in good agreement. The difference between the three of them is less than 1 °C on average from pressures up to 11 MPa.

From this plot, for this mixture, it is clear that pipeline conditions are in the region where a hydrate can be expected. The operating point lies to the left of the hydrate curve and, thus, is in the region where a hydrate can be expected.

3. The right combination of temperature and pressure:

However, there are some other interesting observations that can be made from this plot. Because the operating conditions are inside the phase envelope, the pipeline is operating in three phases: natural gas, condensate, and water, not just gas and water.

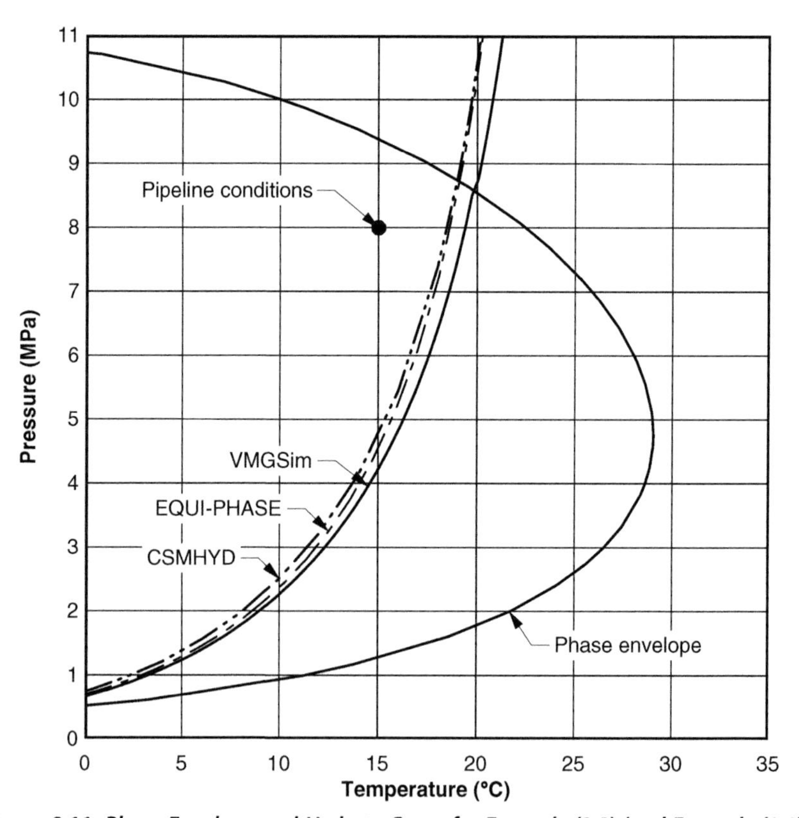

Figure 9.11 ***Phase Envelope and Hydrate Curve for Example (9.8) (and Example (1.1)).***

The critical point is not shown on this graph because it is at a temperature less than the scale of the graph. The critical point calculated by VMGSim is −53 °C and 7.64 MPa. The cricondentherm, the highest temperature at which two hydrocarbon phases can exist, is +29 °C and 4.76 MPa. The portion of the plot shown in Fig. 9.11 represents a portion of the retrograde region.

REFERENCES

Carroll, J.J., 1998a. Phase diagrams reveal acid-gas injection subtleties. Oil Gas. J. 96 (9), 92–97.

Carroll, J.J., 1998b. Acid gas injection encounters diverse H_2S, water phase changes. Oil Gas. J. 96 (10), 57–59.

Harmens, A., Sloan, E.D., 1990. The phase behaviour of the propane-water system: a review. Can. J. Chem. Eng. 68, 151–158.

Scott, R.L., Van Konynenburg, P.H., 1970. 2. Static properties of solutions. Van der Waals and related models for hydrocarbon mixtures. Discuss. Faraday Soc. 49, 89–97.

CHAPTER 10

Water Content of Natural Gas

The topic of the water content of natural gas is potentially a large subject. However, it will only be examined here in a cursory way. There are several models available for calculating the water content of natural gas. Only a few of them will be examined here.

In addition, it would be a valuable exercise for the reader to cross-check between the phase diagram in the previous chapter with the water content calculations presented in this chapter.

10.1 DEW POINT

It is common to express water content in terms of dew point and usually the dew point temperature, usually expressed in °F or °C. With ambient air, we have a feel for the dew point. As the air cools overnight, water will condense out and form a film of water on the grass, car, or any object that is outside. We refer to this film of water as "dew." Thus the dew point is the temperature to which the air must be cooled before this condensation occurs.

What happens if the ambient temperature is below the freezing point of pure water? In this case, the water condenses out as a solid phase called "frost" and the frost point is the temperature at which this occurs. As was discussed in Chapter 1, the process of forming frost can occur without the formation of free water.

The weather forecast will often report dew points even if the temperature is below the freezing point of pure water. This is done by extrapolating the high temperature dew points to lower temperatures. The value reported may have no physical meaning, but it is a sort of a standard in meteorology to do this.

A common method for measuring dew points is to use a chilled mirror. A mirror is placed in a process stream and it is cooled until the mirror "fogs"—until some kind of condensation occurs. This point is the called the dew point. Unfortunately, it is difficult to determine whether this is a water dew point or a hydrocarbon dew point. Water is a colorless liquid and the same is true for light hydrocarbons. From visual inspection, it is impossible to determine the type of liquid formed.

Natural Gas Hydrates
ISBN 978-0-12-800074-8
http://dx.doi.org/10.1016/B978-0-12-800074-8.00010-7

Next, it may not be possible to conclude whether the fog that forms on the mirror is a liquid or solid. This is particularly true if an automated system is used. Furthermore, it is impossible to tell whether the solid forming on the mirror is ice (pure water in the solid form) or hydrate.

The convention is to call the temperature at which the mirror fogs a dew point regardless of whether the phase that condenses is a solid or a liquid.

10.2 EQUILIBRIUM WITH LIQUID WATER

Let's begin with a few observations. The water content of sweet gas is a decreasing function of the pressure. That is, the amount of water in the gas continually decreases as the pressure increases. On the other hand, the water content of sweet gas is an increasing function of temperature—the higher the temperature, the more water in the gas.

For acid gases (hydrogen sulfide and carbon dioxide), this is not the case. Acid gas mixtures exhibit a minimum in the water content. Therefore, more rigorous methods, which usually require computer calculation, are required.

Sour gases (that is, natural gas with an appreciable amount of acid gas) behave in an intermediate fashion. If the mixture is lean in acid gas, then a minimum is not observed. If there is a sufficient amount of acid gas, then there will be a minimum.

In some of these models, the vapor pressure of pure water is required as an input. Poor estimates of the vapor pressure will lead to poor estimates of the water content. For calculations presented in this book, the correlation of Saul and Wagner (1987) is recommended. The correlation is:

$$\ln\left(\frac{P_{\text{water}}^{\text{sat}}}{P_C}\right) = \frac{T_C}{T}\left(-7.85823\tau + 1.83991\tau^{1.5} - 11.7811\tau^3 + 22.6705\tau^{3.5} - 15.9393\tau^4 + 1.77516\tau^{7.5}\right) \quad (10.1)$$

where:

$$\tau = 1 - \frac{T}{T_C} \quad (10.1a)$$

and the temperature is absolute temperature (either Kelvin or Rankine). In SI Units, $T_C = 647.14$ K and $P_C = 22.064$ MPa and in American Engineering Units (AEU) $T_C = 1164.85$ R and $P_C = 3200.12$ psia. The equation is applicable provided you use a consistent sent of units. That is, the critical pressure and the saturation pressure must have the same units. However, you can use psia and K, provided all pressures are in psia and all temperatures are in K.

10.2.1 Ideal Model

In this model, the water content of a gas is assumed to be equal to the ratio of the vapor pressure of pure water divided by the total pressure of the system. This yields the mole fraction of water in the gas. This quantity is converted to g of water per Sm^3 by multiplying by 760.4. Mathematically, this is:

$$w = 760.4 \frac{P_{\mathrm{water}}^{\mathrm{sat}}}{P_{\mathrm{total}}} \tag{10.2}$$

And to convert to lb water per MMCF, multiply by 47,484.

$$w = 47484 \frac{P_{\mathrm{water}}^{\mathrm{sat}}}{P_{\mathrm{total}}} \tag{10.3}$$

In these two equations, the units on the two pressure terms must be the same; $P_{\mathrm{water}}^{\mathrm{sat}}$ is the vapor pressure of pure water and P_{total} is the absolute pressure.

Clearly, this model is very simple and should not be expected to be highly accurate. However, it is reasonably good at very low pressures. For sweet natural gas, this equation can be used with reasonable accuracy ($\pm$15%) for pressures up to about 1400 kPa (200 psia).

A more thermodynamically correct model is to include the effect of gases dissolved in the water. Mathematically, this means:

$$y = \frac{x_{\mathrm{water}} P_{\mathrm{water}}^{\mathrm{sat}}}{P_{\mathrm{total}}} \tag{10.4}$$

In a typical application, the solubility of the gas is not known. Fortunately for hydrocarbons the solubility is so small that it is safe to assume that x_{water} equals unity. However, for acid gases, the solubility is significant, even at relatively low pressure.

10.2.2 McKetta–Wehe Chart

In 1958, McKetta and Wehe published a chart for estimating the water content of sweet natural gas. This chart has been reproduced in many publications, but most notably the *GPSA Engineering Data Book* (1997).

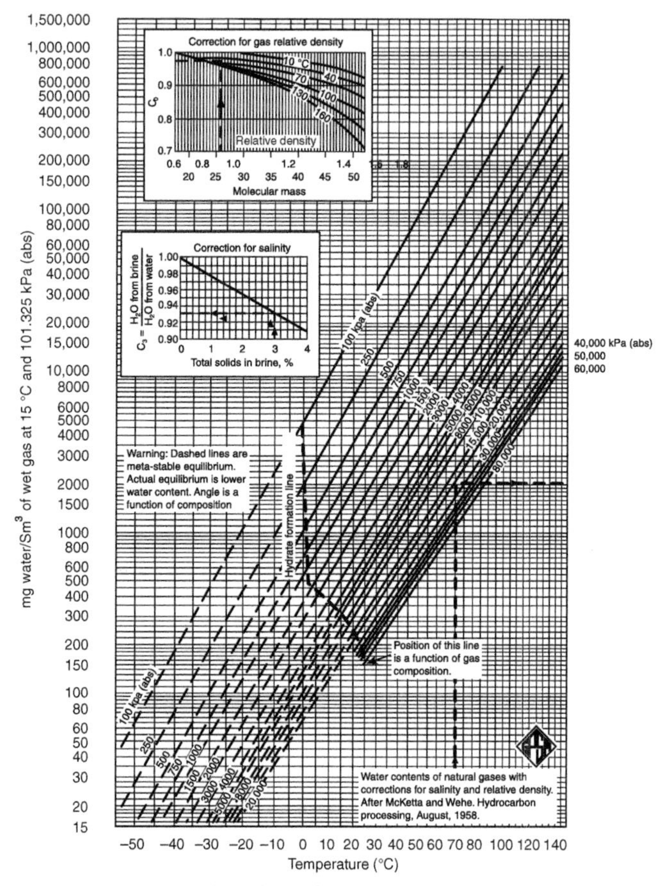

Figure 10.1 ***The McKetta-Wehe Chart for Estimating the Water Content of Sweet Natural Gas in SI Units.*** *Reprinted from the* GPSA Engineering Data Book, *11th ed.—reproduced with permission.*

The McKetta–Wehe chart is widely reproduced, including in this work as Fig. 10.1 in SI Units and Fig. 10.2 in Engineering Units. The chart shows the water content of a sweet gas as a function of pressure and temperature. In addition, simple corrections are provided to account for gas gravity and salinity of the water.

The main chart is for a relatively low-gravity gas. A smaller chart is provided such that the user can obtain a correction factor for higher gravity gas. The correction factor is applied as follows:

$$w = C_g w_{\text{light}} \tag{10.5}$$

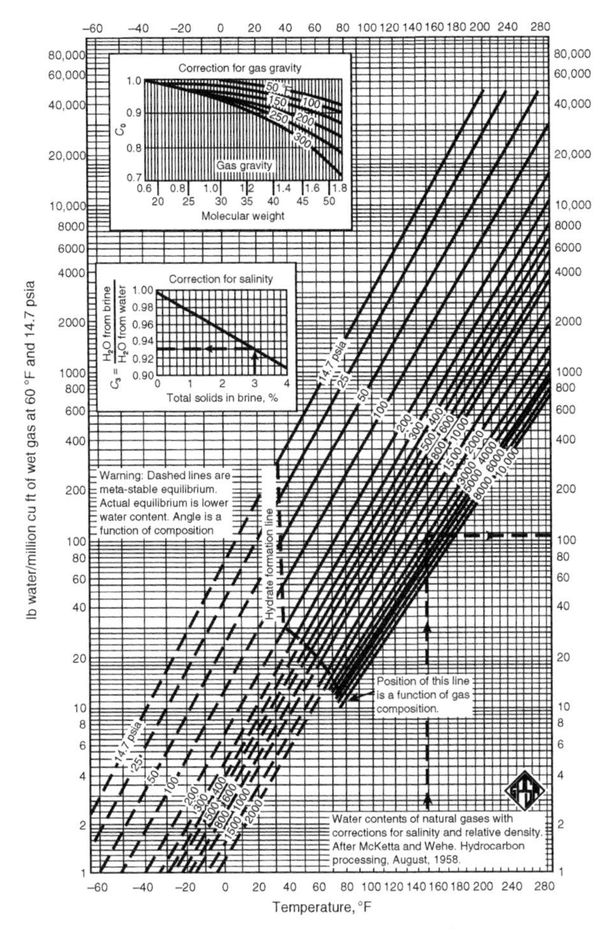

Figure 10.2 ***The McKetta–Wehe Chart for Estimating the Water Content of Sweet Natural Gas in American Engineering Units.*** *Reprinted from the* GPSA Engineering Data Book, *11th ed.—reproduced with permission.*

and C_g ranges from 0.7 to 1.0 and is for gases with gravities between 0.6 and 1.8.

A second correction factor is provided for the effect of brine versus pure water. This correction has been correlated with the following simple equation:

$$C_f = 1 - 0.023\,250 X_{solids} \tag{10.6}$$

where C_f is the ratio of the water content over pure water to the water content over brine and X_{solids} is the weight percent solids in the brine solution. Because of the set of units used, the same equation is applicable for both Engineering and SI Units.

The McKetta–Wehe chart is not applicable to sour gas. There have been corrections proposed to make the chart applicable to these systems and some of these will be discussed later in this chapter.

If used with care, and only for sweet gas, this chart is surprisingly accurate; errors of less than 5% can be obtained. On the other hand, the chart is a little difficult to read and this is probably the largest single source of error. Because of its ease of use and its high accuracy the McKetta–Wehe chart should be in the toolbox of all engineers who work in the natural gas industry.

10.2.3 Sharma–Campbell Method

Sharma and Campbell (1969) proposed a method for calculating the water content of natural gas, including sour gas. Although originally designed for hand calculations, this method is rather complicated. It is even rather complicated for computer applications.

Given the temperature and the pressure, the procedure is as follows. Determine the fugacity of water at the saturation conditions (T and $P_{\text{water}}^{\text{sat}}$), $f_{\text{water}}^{\text{sat}}$, and the fugacity at the system conditions (T and P_{total}), f_{water}. A chart is provided to estimate the fugacity of water at the system conditions. Then the correlation factor, k, is calculated from the following equation:

$$k = \left(\frac{P_{\text{water}}^{\text{sat}}}{P_{\text{total}}}\right)\left(\frac{f_{\text{water}}^{\text{sat}}/P_{\text{water}}^{\text{sat}}}{f_{\text{water}}/P_{\text{total}}}\right)\left(\frac{P_{\text{total}}}{P_{\text{water}}^{\text{sat}}}\right)^{0.0049} \tag{10.7}$$

In this equation, a consistent set of units is required for the pressure and fugacity terms and then k is dimensionless.

The next step is to obtain the compressibility factor, z, for the gas. Sharma and Campbell recommend using a generalized correlation for the compressibility.

Finally, the water content is calculated as:

$$w = 47484k\left(\frac{f_{\text{water}}^{\text{sat}}}{f_{\text{gas}}}\right)^{z} \tag{10.8}$$

where f_{gas} is the fugacity of the dry gas and z is the compressibility factor (z-factor) for the dry gas. Again, if a consistent set of units is used for the fugacity terms, then the calculated water content, w, is in lb/MMCF.

As was mentioned previously, this method is rather difficult to use for hand calculations. First, it requires the compressibility factor of the gas mixture. This is a nontrivial calculation. Next, it requires the fugacity of pure water at system conditions. The chart given to estimate this value is only valid for temperatures between 80 and 160 °F and for pressure less than 2000 psia. It is unclear how this method will behave if extrapolated beyond this range.

On the other hand, the effects of gas gravity and possibly the contribution of acid gas components are taken into account in this correlation.

10.2.4 Bukacek

Bukacek (quoted in McCain, 1990) suggested a relatively simple correlation for the water content of sweet gas. The water content is calculated using an ideal contribution and a deviation factor. In SI Units:

$$w = 760.4\frac{P_{\text{water}}^{\text{sat}}}{P_{\text{total}}} + 0.016016B \tag{10.9}$$

$$\log B = \frac{-1713.66}{273.15 + t} + 6.69449 \tag{10.10}$$

where w is in g/Sm^3 and t is in °C. In AEU, these equations become:

$$w = 47484\frac{P_{\text{water}}^{\text{sat}}}{P_{\text{total}}} + B \tag{10.11}$$

$$\log B = \frac{-3083.87}{459.6 + t} + 6.69449 \tag{10.12}$$

where w is in lb/MMCF and t is in °F. Note, the logarithm terms are common logs (i.e., base 10).

The pair of equations in this correlation is simple in appearance. The added complexity that is missing is that it requires an accurate estimate of the vapor pressure of pure water. The vapor pressure can be calculated using the Saul–Wagner correlation, which was presented earlier in this chapter.

This correlation is reported to be accurate for temperatures between 15 and 238 °C (60 and 460 °F) and for pressures from 0.1 to 69 MPa (15–10,000 psia). Again, it is only applicable to sweet gas. Some tests by the author indicate that within the stated range, the method of Bukacek is accurate to about ±5%, comparable to the McKetta–Wehe chart.

10.2.5 Ning et al

Ning et al. (2000) proposed a correlation based on the McKetta–Wehe chart. This correlation merits a brief discussion because it is a useful correlation and because it further reveals how difficult it can be to correlate something that is as seemingly simple as the water content of natural gas.

Their basic equation is quite simple in appearance:

$$\ln w = a_0 + a_1 T + a_2 T^2 \tag{10.13}$$

A table of values for the coefficients, a_0, a_1, and a_2 is given as a function of pressure, for pressures up to 100 MPa (14,500 psia). The values for a_0, a_1, and a_2 are tabulated in Table 10.1 and are plotted in Fig. 10.3. Unfortunately, as can be seen from Fig. 10.3, the coefficients are not smooth functions of the pressure. There appears to be no simple correlations for this

Table 10.1 Coefficients for Use with the Ning et al. (2000) Correlation for the Water Content of Natural Gas

Pressure (MPa)	a_0	a_1	a_2
0.1	−30.0672	0.1634	-1.7452×10^{-4}
0.2	−27.5786	0.1435	-1.4347×10^{-4}
0.3	−27.8357	0.1425	-1.4216×10^{-4}
0.4	−27.3193	0.1383	-1.3668×10^{-4}
0.5	−26.2146	0.1309	-1.2643×10^{-4}
0.6	−25.7488	0.1261	-1.1875×10^{-4}
0.8	−27.2133	0.1334	-1.2884×10^{-4}
1	−26.2406	0.1268	-1.1991×10^{-4}
1.5	−26.1290	0.1237	-1.1534×10^{-4}
2	−24.5786	0.1133	-1.0108×10^{-4}
3	−24.7653	0.1128	-1.0113×10^{-4}
4	−24.7175	0.1120	-1.0085×10^{-4}
5	−26.8976	0.1232	-1.1618×10^{-4}
6	−25.1163	0.1128	-1.0264×10^{-4}
8	−26.0341	0.1172	-1.0912×10^{-4}
10	−25.4407	0.1133	-1.0425×10^{-4}
15	−22.6263	0.0973	-8.4136×10^{-5}
20	−22.1364	0.0946	-8.1751×10^{-5}
30	−20.4434	0.0851	-7.0353×10^{-5}
40	−21.1259	0.0881	-7.4510×10^{-5}
50	−20.2527	0.0834	-6.9094×10^{-5}
60	−19.1174	0.0773	-6.1641×10^{-5}
70	−20.5002	0.0845	-7.1151×10^{-5}
100	−20.4974	0.0838	-7.0494×10^{-5}

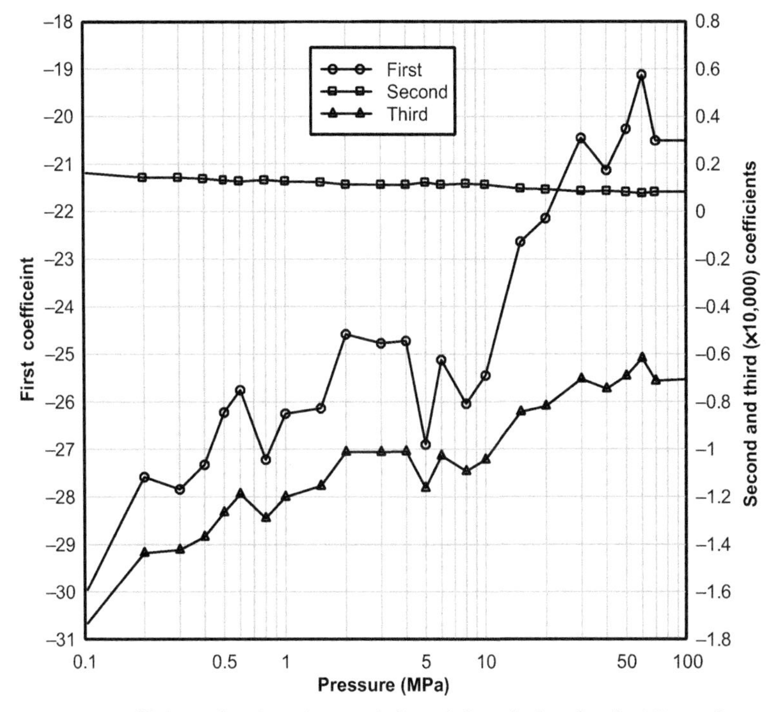

Figure 10.3 ***Coefficients for the Ning et al. (2000) Correlation for the Water Content of Natural Gas.***

pressure function and the authors recommend interpolating between the tabular values. They recommend calculating the water content at pressures that span the values in the tables and then linearly interpolate to the pressure of interest. However, this author found that a logarithmic interpolation results in slightly better predictions.

The article by Ning et al. (2000) includes the effect of gas gravity and salinity, but it does not include the effect of either H_2S or CO_2 and therefore is not applicable to sour gas mixtures. Their correlation of the gravity corrections is:

$$C_g = 1.01532 + 0.011t - 0.0182\gamma - 0.0142\gamma t \quad (10.14)$$

where t is the temperature in °C and γ is the gas gravity.

10.2.6 Maddox Correction

Maddox (1974) developed a method for estimating the water content of sour natural gas (Also see Maddox et al., 1988). This method assumes that

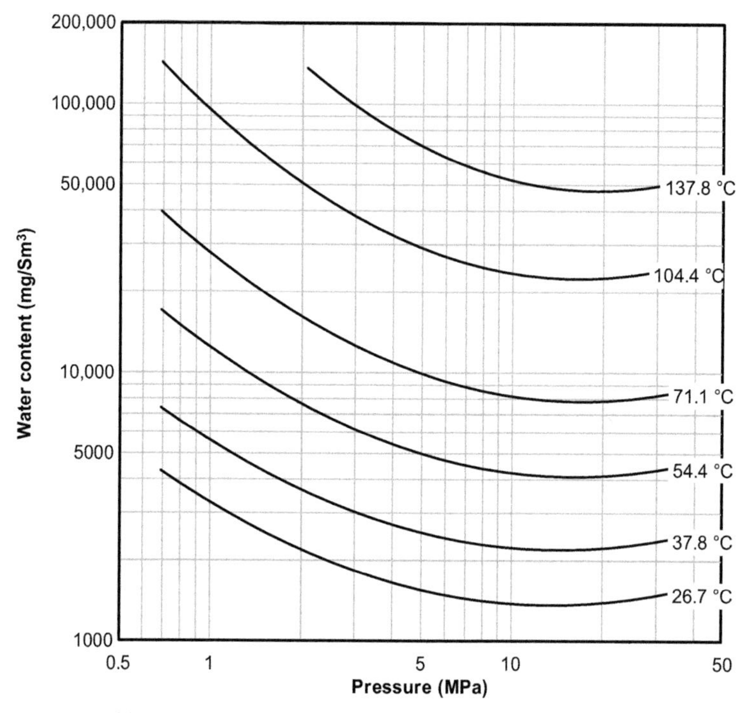

Figure 10.4 ***Maddox Correction for the Water Content of Sour Gas Hydrogen Sulfide Contribution (SI Units).***

the water content of sour gas is the sum of three terms: (1) a sweet gas contribution, (2) a contribution from CO_2, and (3) a contribution from H_2S. The water content of the gas is calculated as a mole fraction weighted average of the three contributions.

$$w = y_{HC}w_{HC} + y_{CO_2}w_{CO_2} + y_{H_2S}w_{H_2S} \tag{10.15}$$

where w is the water content, y is the mole fraction, the subscript HC refers to hydrocarbon, CO_2 is carbon dioxide, and H_2S is hydrogen sulfide. Charts are provided to estimate the contributions for CO_2 and H_2S. The chart for CO_2 is for temperatures between 80 and 160 °F (27–71 °C) and the chart for H_2S is for 80 and 280 °F (27–138 °C). Both charts are for pressures from 100 to 3000 psia (0.7–20.7 MPa). The correction charts are given here as Figs 10.4 and 10.5 in SI Units and Figs 10.6 and 10.7 in AEU.

To use this method, one finds the water content of sweet gas, typically from the McKetta–Wehe chart; the corrections for the acid gases are then obtained from their respective charts. The correction plots along with Eqn (10.15) are used to calculate the water content of the sour gas.

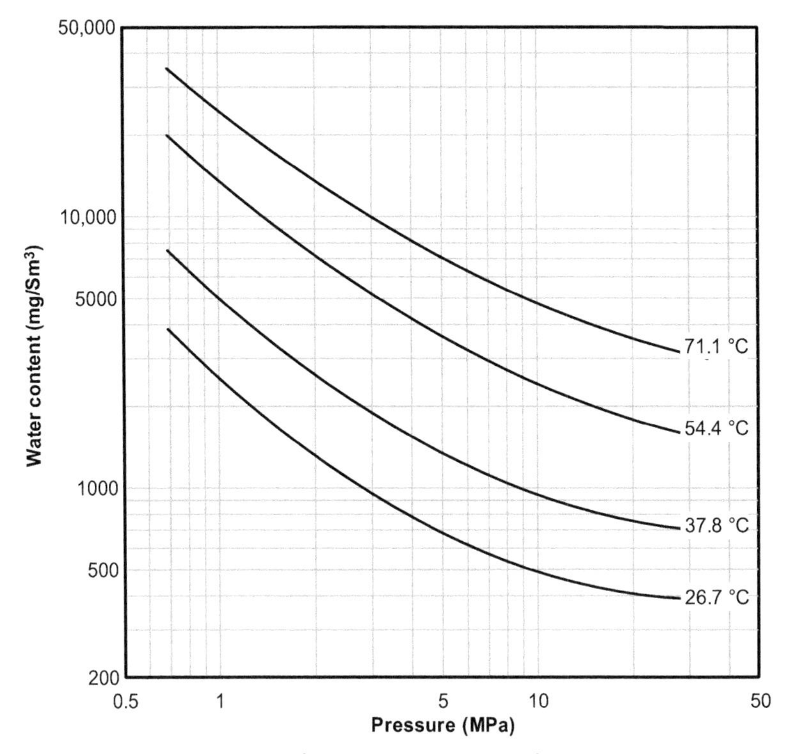

Figure 10.5 ***Maddox correction for the water content of sour carbon dioxide contribution (SI Units).***

Although these charts have the appearance of being useful for calculating the water content of pure H_2S and pure CO_2, Maddox advises that they should not be used for this purpose.

An attempt was made to correlate the correction factors as a function of both the pressure and the temperature, without much success. Therefore, the water content was correlated as a function of the pressure only using the following equation:

$$\log w = a_0 + a_1 \log P + a_2(\log P)^2 \tag{10.16}$$

where w is the water content in lb/MMCF, P is the total pressure in psia, and a set of coefficients, a_0, a_1, and a_2, was obtained for each isotherm. The coefficients are listed in Table 10.2. Note this equation uses common logarithms. Although these equations are not a perfect fit of the curves, they are probably as accurate as errors associated with reading the charts.

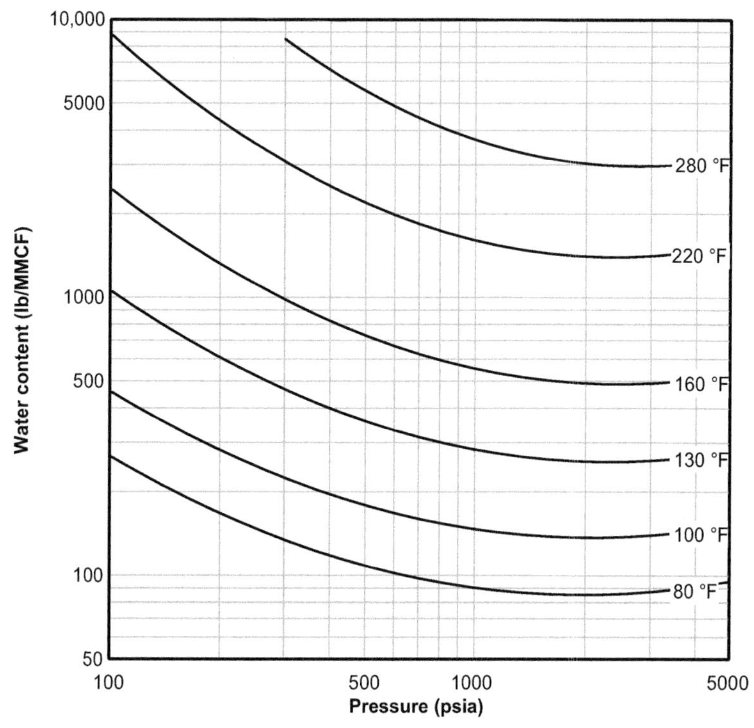

Figure 10.6 ***Maddox Correction for the Water Content of Sour Hydrogen Sulfide Contribution (AEU).***

10.2.7 Robinson et al. Charts

Robinson et al. (1978) used an equation of state method to calculate the water content of sour natural gases (Also see Robinson et al., 1980). Using their equation of state model, they generated a series of charts: one chart for 300, 1000, 2000, 3000, 6000, and 10,000 psia (2.07, 6.89, 13.79, 20.78, 41.37, and 68.95 MPa). The temperature range for the charts is 50–350 °F (10–177 °C), although it is slightly narrower at some pressures. A third parameter on the chart is the equivalent H_2S and it is calculated as follows:

$$y_{H_2S}^{equiv} = y_{H_2S} + 0.75 y_{CO_2} \tag{10.17}$$

where y is the mole fraction, the subscripts H_2S and CO_2 refer to hydrogen sulfide and carbon dioxide, and the superscript equiv is the equivalent.

The charts are applicable for H_2S equivalent up to 40 mol%.

These charts remain popular, but they require multiple interpolations making them a little difficult to use.

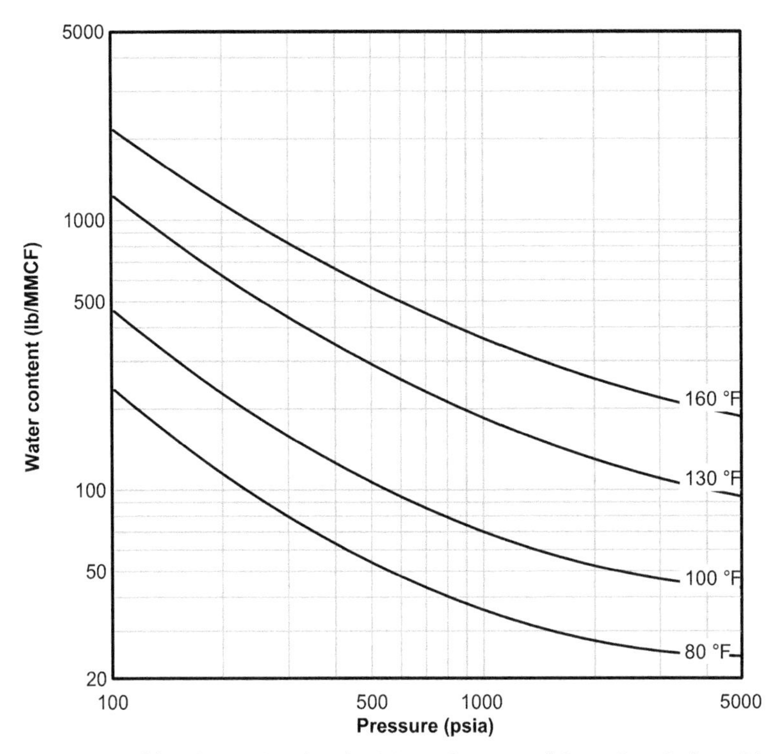

Figure 10.7 ***Maddox Correction for the Water Content of Sour Gas Carbon Dioxide Contribution (AEU).***

10.2.8 Wichert Correction

Wichert and Wichert (1993) proposed a relatively simple correction based on the equivalent H_2S content of the gas. The correlation was modified a decade later based on newer information (Wichert and Wichert, 2003). In the original article, they used Eqn (10.7) for the H_2S equivalent. However, they revised the definition to:

$$y_{H_2S}^{equiv} = y_{H_2S} + 0.70 y_{CO_2} \quad (10.18)$$

They presented a chart where given the temperature, pressure, and equivalent H_2S, one could obtain a correction factor, F_{corr}. Correction factors range from 0.9 to 5.0. Then the water content of the sour gas is calculated as follows:

$$w = F_{corr} w_{M-W} \quad (10.19)$$

Table 10.2 Correlation Coefficients for Calculating the Maddox Correction for the Water Content of Sour Natural Gas

Temperature (°F)	a_0	a_1	a_2
Carbon Dioxide			
80	6.0901	−2.5396	0.3427
100	6.1870	−2.3779	0.3103
130	6.1925	−2.0280	0.2400
160	6.1850	−1.8492	0.2139
Hydrogen Sulfide			
80	5.1847	−1.9772	0.3004
100	5.4896	−2.0210	0.3046
130	6.1694	−2.2342	0.3319
160	6.8834	−2.4731	0.3646
220	7.9773	−2.8597	0.4232
280	9.2783	−3.3723	0.4897

where w is the water content of the sour gas, F_{corr} is a correction factor, and w_{M-W} is the water content of sweet gas from the McKetta–Wehe chart. F_{corr} is dimensionless, so the two water content terms simply have the same units.

This method is limited to an H_2S equivalent of 55 mol% and is applicable for temperatures from 10 to 177 °C (50–350 °F) and pressure from 1.4 to 69 MPa (200–10,000 psia).

This method is much simpler to use than the charts of Robinson et al. (1978) because it does not require the interpolations of the earlier method (Also see Robinson et al., 1980).

10.2.9 AQUAlibrium

There are many computer programs available for calculating the equilibria in natural gas water systems. One such program is AQUAlibrium, and it can be used for acid gas.

AQUAlibrium uses a Henry's law approach for calculating the solubilities of gas in water. As a by-product of the solubility calculations, the water content of the gas, nonaqueous liquid, or combination of the two is also calculated.

AQUAlibrium is a software package developed specifically for calculating equilibrium in systems containing acid gas or sour gas and water. In addition to calculating the water content of both the gas and the liquid, the software can be used to calculate the solubility of gases and liquids in water.

Of the methods discussed previously, only AQUAlibrium was designed specifically to work with both a gas and a nonaqueous liquid phase. This is significant because often the sour gas is in a liquid state.

In addition, AQUAlibrium is based on a rigorous thermodynamic model, rather than mere empiricism. Thus it should be useful for extrapolating to conditions where no data exist.

AQUAlibrium 2 was a DOS-based program and AQUAlibrium 3 is an update and has a Windows interface. The calculation engine for the two versions is very similar, but there are some differences. Version 2 is no longer available and Version 3 is available from FlowPhase (www.flowphase.com).

10.3 EQUILIBRIUM WITH SOLIDS

There are much fewer data available for the water content of gas in equilibrium with solids and therefore there are few correlations available.

The McKetta–Wehe chart includes an extrapolation into a region where solid formation is expected. This is supposed to represent a water dew point and not the equilibrium with a solid. If the equilibrium is truly with a solid, then these values are for a metastable water phase.

There are more rigorous models that can be used for estimating the water content for a gas in equilibrium with a solid.

10.3.1 Ice

The water content of gases in equilibrium with ice can be estimate using fundamental thermodynamics. To begin with, it should be noted that the ice phase is pure water and does not contain any dissolved gas. Occasionally, gas may be trapped in the ice, but this is not an equilibrium effect.

For a pure component, the effect of pressure on the fugacity at constant pressure can be estimated from the following equation:

$$\ln\left[\frac{f(P_2)}{f(P_1)}\right] = \frac{1}{RT}\int_{P_1}^{P_2} v\mathrm{d}P \tag{10.20}$$

where f is the fugacity, P is the pressure, R is the universal gas constant, T is the absolute temperature, and v is the molar volume. The equation is useful for calculating the isothermal change in fugacity from state 1 to state 2. This equation is applicable to any pure component and is based in fundamental thermodynamics.

Now consider the case of ice subliming from a gas mixture. In this case, state 1 is at the saturation point. The saturation pressure along the ice-water vapor locus is so low that $f_{ice}(P^{sat}) = P^{sat}$. Therefore Eqn (10.20) becomes

$$\ln\left[\frac{f_{ice}(P)}{P^{sat}}\right] = \frac{1}{RT}\int_{P^{sat}}^{P} v_{ice}\,dP \tag{10.21}$$

This equation can be used to calculate the fugacity at any pressure and temperature.

Although the molar volume of ice is a function of the temperature, albeit a weak one, it is essentially independent of the pressure. Therefore the integration in Eqn (10.9) is straightforward and yields:

$$\ln\left[\frac{f_{ice}(P)}{P^{sat}}\right] = \frac{v_{ice}}{RT}\left(P - P^{sat}\right) \tag{10.22}$$

or

$$\ln[f_{ice}(P)] = \ln[P^{sat}] + \frac{v_{ice}}{RT}\left(P - P^{sat}\right) \tag{10.23}$$

The sublimation pressure for ice can be estimated from the following equation:

$$\ln P^{sat} = 22.5656 - 1.1172 \times 10^{-3}T - 6215.09/T \tag{10.24}$$

where P^{sat} is in kPa and T is in K. This equation is applicable from -40 to 0 °C (233–273 K). The molar volume of ice can be estimated from the following equation:

$$v = 58.018 + 9.270 \times 10^{-2}T \tag{10.25}$$

where v is in m^3/kmol and T is in K. This equation is also applicable from -40 to 0 °C (233–273 K). These correlations are derived from the tables of Keenan et al. (1978).

The procedure for using these equations to calculate the fugacity of ice is as follows. For the given T, obtain the saturation pressure from Eqn (10.24) and the molar volume from Eqn (10.23). Then calculate the fugacity at the pressure, P, using Eqn (10.21).

At equilibrium, the fugacity of the water in the vapor equals the fugacity of the ice. The previous equations can be used to calculate the fugacity of the ice and an equation of state can be used to calculate the fugacity of water in the gas. Mathematically, this can be expressed as:

$$f_{ice}(T,P) = \hat{f}_{water}(T,P,y) \tag{10.26}$$

Note in the vapor phase water is only one component in a multicomponent mixture. Hence the notation of the circumflex and the effect of composition on the vapor phase fugacity.

10.3.2 Hydrate

A cursory review of the literature reveals few simple correlations for the water content of gases in equilibrium with hydrates. One such correlation was presented by Kobayashi et al. (1987). In addition, a new chart method will be presented in the next section of this book.

On the other hand, the rigorous thermodynamic models presented in Chapter 4 are suitable for this type of calculation.

10.3.3 Methane

Figure 10.8 shows the water content of methane at temperatures below 0 °C (32 °F). Figure 10.9 is a similar plot except it is in AEU. At these temperatures, liquid water does not exist, except as a metastable form. The water content of the gas in equilibrium with ice was calculated using the procedure outlined previously. The water content of the gas in equilibrium with the hydrate was determined by an empirical correlation of the data presented by Aoyagi et al. (1979). The experimental data are plotted on this figure such that the reader can see the accuracy of the correlation. Finally, an adjustment was made to obtain a smooth transition between the two regions.

The broken line on the chart is the transition between hydrate and ice phases. At low pressure, the solid is ice and at higher pressure the solid is hydrate. Otherwise, the water content can be read directly from the chart.

Some interesting observations can be made based on this graph. First, it is not surprising that as the pressure increases the water content of the gas decreases. And this is true for both the ice and hydrate phase. Second, the water content increases with increasing temperature—again as expected.

This chart reveals both the saturation conditions (i.e., the temperature and pressure at which a solid will form) and the nature of the solid. For example, a gas mixture containing 100 ppm water is cooled at 0.8 MPa. The

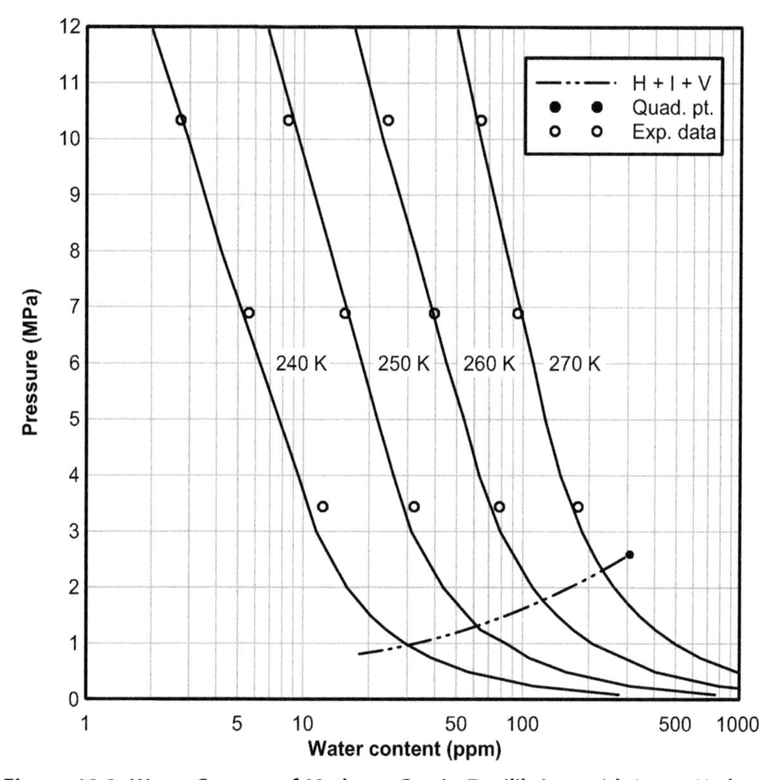

Figure 10.8 ***Water Content of Methane Gas in Equilibrium with Ice or Hydrate.***

solid that forms is ice (usually called "frost"), which forms at 250 K (−23 °C or −10 °F). On the other hand, if it is cooled at 2.4 MPa, the solid is a hydrate and it forms at 260 K (−13 °C or 8 °F).

10.3.4 Gas Gravity

It is interesting that if one examines the correction chart on the McKetta–Wehe chart, it is easy to conclude that the gravity correction is unity in the range of temperature encountered with hydrate formation. That is, there is no gravity correction. However, it has been observed that there is indeed an effect of gas gravity on the water content.

Song and Kobayashi (1982) measured the water content of a mixture of methane (94.69 mol%) and propane (5.31%), which has a gas gravity of 0.5851, in equilibrium with a hydrate. These data allow for some interpretation of the effect of gas gravity on the water content of a gas in equilibrium with a hydrate.

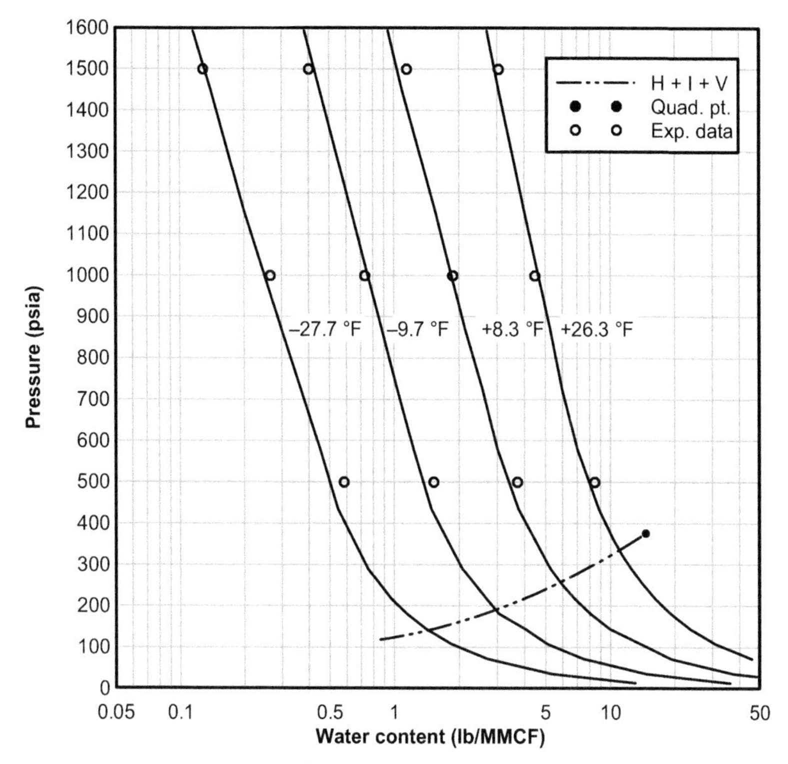

Figure 10.9 ***Water Content of Methane Gas in Equilibrium with Ice or Hydrate.***

Figure 10.10 shows a plot of the experimental data of Song and Kobayashi (1982) along with curves based on Fig. 10.8 (i.e., the water content of methane in equilibrium with solids). The mixture data are at consistently lower water content than those for pure methane. A close examination of this plot reveals that the difference in the logarithm of the water content is approximately independent of the temperature. A more detailed statistical analysis confirms this observation. However, there appears to be a pressure dependence, albeit a weak one. Regression produces the following expression:

$$\ln\left(\frac{w_{\mathrm{SK\ mixture}}}{w_{\mathrm{CH_4}}}\right) = 0.20284 + 0.02062P \tag{10.27}$$

where the two water content terms must have the same units (ppm, g/Sm^3, or lb/MMCF) and the pressure P must be in MPa.

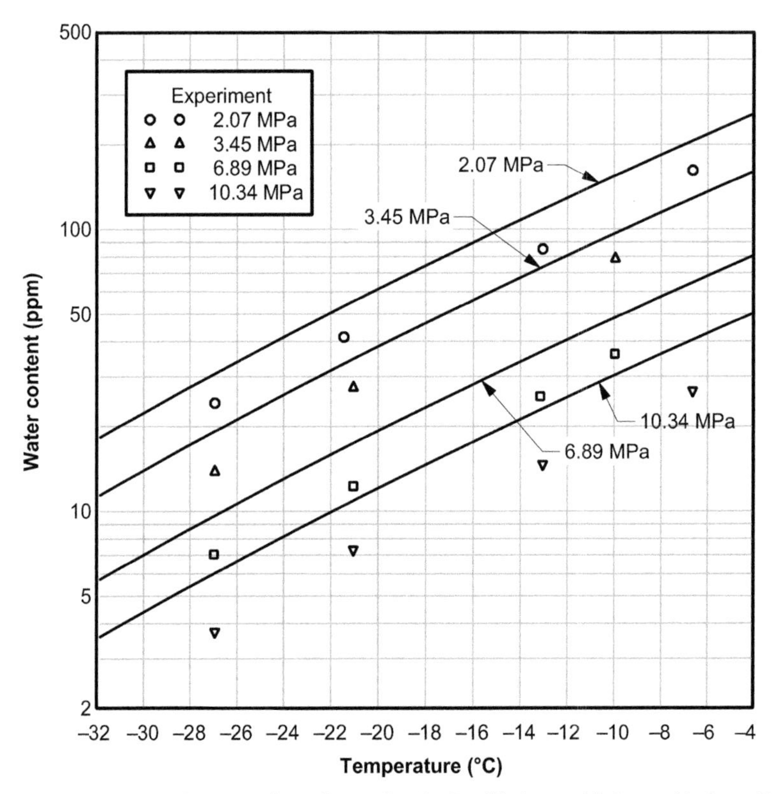

Figure 10.10 ***Water Content of Methane Gas in Equilibrium with Ice or Hydrate Data Points from Song and Kobayashi.***

If it is assumed that the water content is a linear function of the gas gravity, then the following equation can be used to estimate the water content of a gas mixture:

$$\ln\left(\frac{w(\gamma)}{w_{CH_4}}\right) = -2.1851 + 4.0813\gamma - 0.2221P + 0.4149P\gamma \qquad (10.28)$$

The two water content terms must have the same units, the pressure is in MPa, and the gas gravity is dimensionless. Because this equation is based on limited information, it should be used with some caution. In addition, it should not be applied to either sour gases or acid gases.

Another word of caution—because this analysis is at low temperature, one should be cautious of the formation of a hydrocarbon liquid. These simple correlations are only for hydrocarbons in the vapor phase. The design engineer would be wise to check the hydrocarbon dew point to ensure that liquid hydrocarbons do not form.

For temperatures between 270 K (the lowest temperature on Fig. 10.8) and the hydrate formation temperature (some temperature greater than 270 K), the method for calculating the water content is a little complicated.

First, given the temperature and pressure, it should be confirmed that a hydrate does indeed form. This is done using the methods given in Chapters 3 and 4. If a hydrate does not form, then simply use the correlations presented earlier in this chapter for the water content of gas in equilibrium with water.

If a hydrate does form, then it requires an interpolation. Calculate the water content at the given pressure and the hydrate formation temperature using the methods for calculating the water content in equilibrium with *liquid water*. The liquid methods are used because at this point the equilibrium is with both a liquid and a hydrate. Next calculate the water content at the given pressure and 270 K using Fig. 10.8 and correcting for the gravity using Eqn (10.26). Use the formula below to interpolate between these two values:

$$\ln w = \left(\frac{\ln w(T_{\text{hyd}}) - \ln w(270\text{ K})}{T_{\text{hyd}}(\text{in }^\circ\text{C}) + 3.15}\right)(T(\text{in }^\circ\text{C}) + 3.15) + \ln w(270\text{ K}) \tag{10.29}$$

This interpolation equation is based on T_{hyd} and T being in degrees Celsius. The various water content quantities can be in any units provided the same units are used for each occurrence.

The procedure for performing this type of calculation is given in the pseudo-code in Fig. 10.11.

10.3.5 Ethane

Figure 10.12 shows the water content of ethane at 500 psia (3.45 MPa) as a function of the temperature. The plot shows the experimental data of Song and Kobayashi (1994) and those of Song et al. (2004). The curve shown on the plot is simply a correlation of the experimental data and is:

$$\ln w = -0.309035 + 0.049845t - 8.2658 \times 10^{-5}t^2 \tag{10.30}$$

where w is in lb/MMCF and t is in °F.

Although these are at a single pressure, the water content over a liquid phase is a weak function of the pressure. As a first order of magnitude approximation, it can be assumed that the water contents are independent of

```
1. Input T and P

2. Estimate the hydrate temperature, Thyd, at P.

3. Is T > Thyd?
   3a. Yes (no hydrate) - goto step 4
   3b. No (hydrate) - goto step 6

4. Use the usual methods to calculate the water content at Thyd
   and P phase (for example the McKetta-Wehe chart).

5. Goto step 9

6. Calculate the water content at Thyd and P using the methods
   for the equilibrium with aqueous phase (for example the
   McKetta-Wehe chart). Call this w1

7. Calculate the water content at P and 270 K using Fig. 10-8 or
   10-9 and the gravity correction. Call this w2

8. Use the interpolation formula to calculate the water content
   at T and P

9. Stop
```

Figure 10.11 ***Pseudo-code for Estimating the Water Content of a Sweet Natural Gas in Equilibrium with Ice or Hydrate (1982).***

the pressure. This is only true for the hydrocarbon in the liquid phase and should not be used if the ethane vaporizes.

This chart can be used directly to determine water contents of ethane, but it can also be used to estimate the frost point temperature for a given water content. For example, at a water content of 1 lb/MMCF, the frost point temperature is about 6.5 °F (−14.2 °F). Thus, is the stream is warmer than 6.5 °F, then hydrate will not form, but if it is colder than 6.5 °F, a hydrate will form.

Figure 10.13 is a similar plot except it is in SI Units.

10.3.6 Propane

Figure 10.14 shows two data sets of data for the water content of liquid propane in AEU; the experimental data of Song and Kobayashi (1994) at 159 psia (1.10 MPa) and those of Song et al. (2004) at 125 psia (0.86 MPa). On the scale shown in this figure, it is clear that there is only a slight variation in the water content with the difference.

The curve on this plot is simply a correlation, and a single curve is used for both pressures.

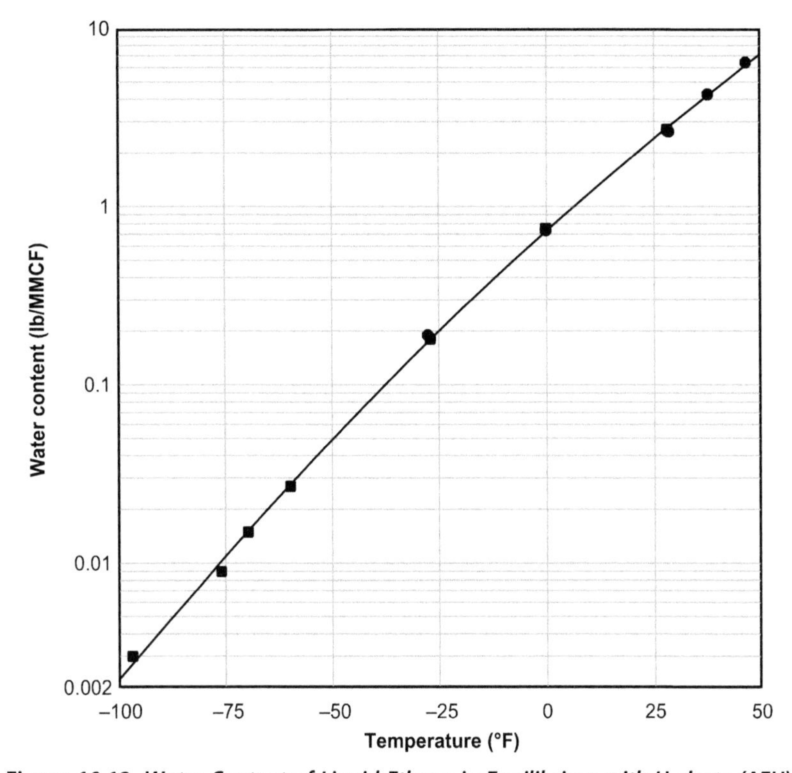

Figure 10.12 ***Water Content of Liquid Ethane in Equilibrium with Hydrate (AEU).***

$$\ln w = -0.008112 + 0.047461t - 3.0587 \times 10^{-5}t^2 \tag{10.31}$$

where w is in lb/MMCF and t is in °F.

When Fig. 10.12 is compared with Fig. 10.14, it can be seen that if the effect of pressure is ignored, the water content of propane is slightly larger than the water content of ethane.

Figure 10.15 is similar to Fig. 10.14 except it is in SI Units.

10.3.7 Carbon Dioxide

There are not a lot of experimental data for the water content of acid gas mixtures in equilibrium with hydrates. However, Song and Kobayashi (1987) published a few points for carbon dioxide. These data are plotted in Fig. 10.16.

Also plotted in Fig. 10.16 are the predictions from *CSMHYD* and *CSMGEM*. Although the predictions are not perfect fits (no model ever is),

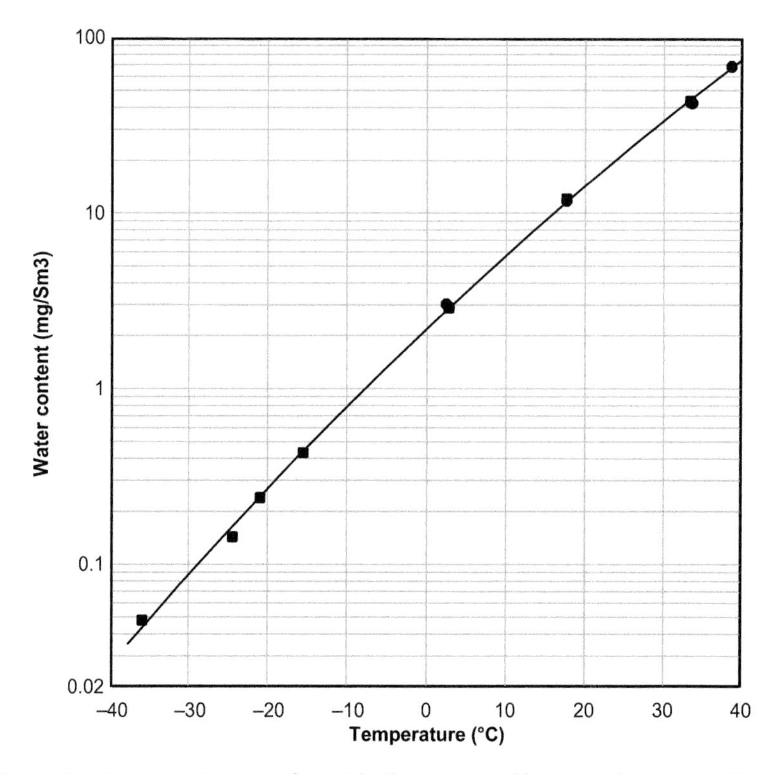

Figure 10.13 ***Water Content of Liquid Ethane in Equilibrium with Hydrate (SI Units).***

they are reasonably accurate. At higher pressure, the models fits better than at lower pressure. At 100 and 200 psia (689 and 1379 kPa), the models tend to under predict the water content.

At 500 psia (3447 kPa), CSMGEM predicts a first three-phase point (aqueous liquid + vapor + hydrate) at 46.5 °F (8.1 °C) and a second three-phase point (aqueous liquid + CO_2-rich liquid + hydrate) at 30.7 °F (−0.7 °C). Thus the curve shown for this isobar represents the water content of a vapor phase in equilibrium with a hydrate. For this reason, only a short curve is presented for his isobar.

At 100 psia, at 30.5 °F (689 kPa and −0.8 °C), CSMGEM predicts that the stable solid phase changes from hydrate to ice. At temperatures less than 30.5 °F (−0.8 °C), the stable solid is hydrate and the curve plotted in Fig. 10.16 for this isobar represents the water content of a vapor phase in equilibrium with a hydrate. Between 30.5 and 32 °F (−0.8 and 0 °C), the equilibrium is with ice and above 32 °F (0 °C), the stable water–rich phase is liquid.

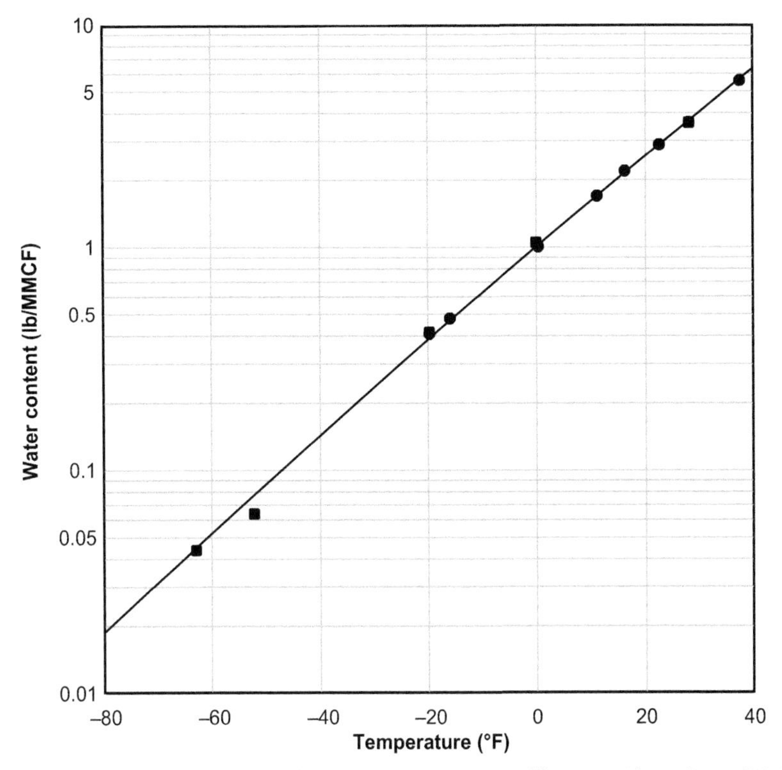

Figure 10.14 ***Water Content of Liquid Propane in Equilibrium with Hydrate (AEU).***

10.4 LOCAL WATER CONTENT MODEL

This model begins with the equality of the fugacity of water in the hydrate and water in the vapor:

$$\hat{f}_{w}^{hyd} = \hat{f}_{w}^{vap} \tag{10.32}$$

The fugacity in the solid phase, $\hat{f}_{w}^{hyd}$, can be estimated as:

$$\hat{f}_{w}^{hyd} = S_w f_w^{o} \exp\left[\frac{\overline{v}_{w}^{hyd}\left(P - P^{ref}\right)}{RT}\right] \tag{10.33}$$

where s_i is the mole fraction water, f_w^o is the standard state fugacity (much like the vapor press), $\overline{v}_{w}^{hyd}$ is the partial molal volume of water in the hydrate, P is the total pressure, P^{ref} is the reference pressure, R is the gas constant, and T is the absolute temperature

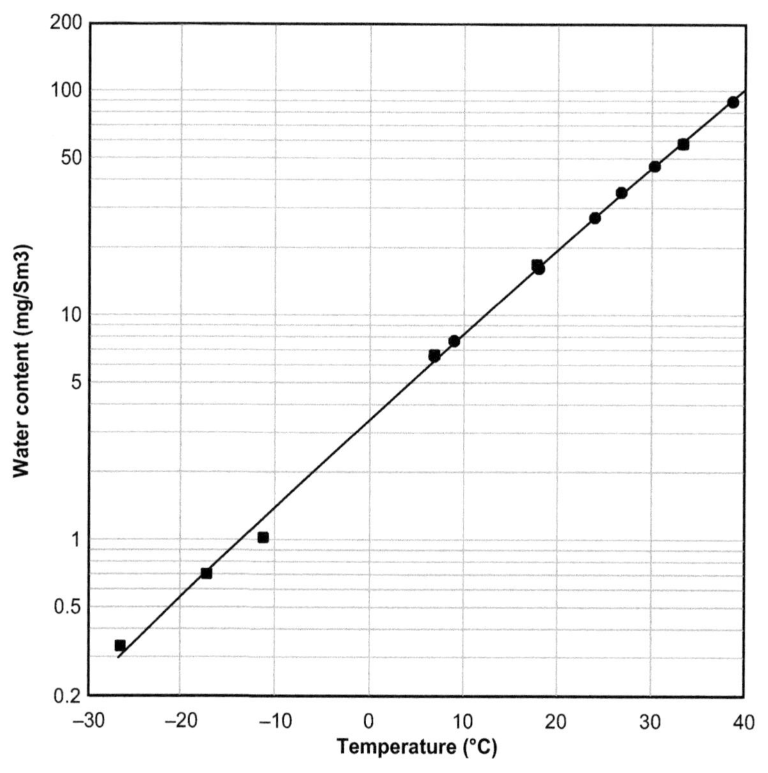

Figure 10.15 ***Water Content of Liquid Propane in Equilibrium with Hydrate (SI Units).***

The exponential term is the Poynting correction, the effect of pressure on the reference fugacity. Furthermore, the values of many of these quantities are not important. So for our purposes, the value of the partial molal volume is not important, but the observation that this quantity is approximately constant is. So:

$$\overline{v}_i^{hyd} \approx \text{constant} \tag{10.34}$$

The reference state is a hypothetical state and the exact pressure (or even where or not a hydrate exists at that pressure) is of little relevance here, so merely set this to a low pressure.

Since the reference fugacity is similar to the vapor pressure so it can be modeled using a Clausius–Clapeyron type equation:

$$f_1^0 = a + \frac{b}{T} \tag{10.35}$$

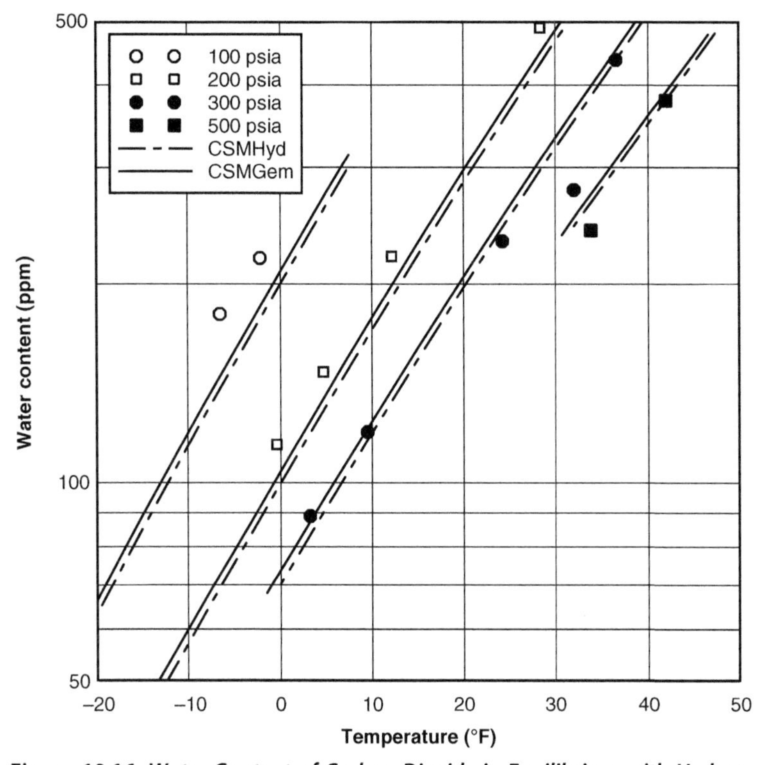

Figure 10.16 ***Water Content of Carbon Dioxide in Equilibrium with Hydrate.***

Finally, assume the concentration of water in the hydrate is approximately constant.

$$S_{\mathrm{W}} \approx \text{constant} \tag{10.36}$$

Although the concentrations of the various nonwater species (hydrate formers) in the hydrate phase are highly variable, the water concentration is not.

The fugacity of water in the vapor, $\widehat{f}_{\mathrm{W}}^{\mathrm{vap}}$, is calculated as

$$\widehat{f}_{\mathrm{W}}^{\mathrm{vap}} = y_{\mathrm{W}}\, P\widehat{\phi}_{\mathrm{W}}^{\mathrm{vap}} \tag{10.37}$$

where: y_{W} is the mole fraction water in the vapor phase and $\widehat{f}_{\mathrm{W}}^{\mathrm{vap}}$ is the fugacity coefficient for water in the vapor phase.

For our purposes, assume the vapor phase fugacity is the partial pressure:

$$\widehat{f}_{\mathrm{W}}^{\mathrm{vap}} \approx y_{\mathrm{W}}P \tag{10.38}$$

To convert the water concentration in the vapor from mole fraction to mass per unit standard volume, merely multiple by a conversion factor:

$$w = 760.4y_{\mathrm{w}} \tag{10.39}$$

where w is the water content of the gas in mg/Sm^3 (or in lb/MMSCF).

Finally, we have a single equation relating the pressure, temperature, and water content. The equation is:

$$\left(A + \frac{B}{T}\right)\exp\left[\frac{CP}{T}\right] = wP \tag{10.40}$$

where A, B, and C are empirical constants

The final equation, which relates the temperature, pressure, and the water content of the gas, is:

$$\left(A - 1 + \frac{B}{T}\right)\exp\left[\frac{CP}{T}\right] = wP \tag{10.41}$$

However, we desire an equation of the form, where the temperature is a function of the water content and the pressure. Taking the logarithms of both sides of the equation yields:

$$\ln\left(A + \frac{B}{T}\right) + \left[\frac{CP}{T}\right] = \ln w + \ln P \tag{10.42}$$

Again, because we desire an equation explicit in the temperature, we require one more approximation:

$$\ln(x) \approx x - 1 \tag{10.43}$$

Substituting that into Eqn (10.43) yields:

$$\left(A - 1 + \frac{B}{T}\right) + \left[\frac{CP}{T}\right] = \ln w + \ln P \tag{10.44}$$

After some final algebraic manipulation (no other approximations required), we obtain the equation:

$$\frac{1}{T} = a + b \ln w + c \ln P + \mathrm{d}P \tag{10.45}$$

From a set of P–T–w values, we can regress to get the constants a, b, c, and d. Note in this equation that the temperature should be in Kelvin or Rankine.

Although this equation is probably easy to work with when one has a set of data, it is probably more conventional to estimate the saturated water content of a gas mixture given the temperature and the pressure. Rearranging Eqn (10.43) gives:

$$\ln w = \left(\frac{1}{b}\right)\left(a - \frac{1}{T} + c \ln P + \mathrm{d}P\right) \qquad (10.46)$$

which can be used to solve for the water content directly given the temperature, the pressure, and a set of regression coefficients.

From Eqn (10.43), it can also be observed that there should be a roughly linear relation between the reciprocal temperature and the logarithm of the water content.

Figure 10.9 showed the smoothed data of Aoyagi et al. (1979) for the water content of methane. Here we will examine the raw data and see how well the correlation derived fits the data. The raw data set consists of 36 points ranging from 200 to 1500 psia. Performing least squares regression yields the following equation:

$$\frac{1000}{T} = 2.494189 - 0.090738 \ln w - 0.029784 \ln P - 9.36231 \times 10^{-5} P \qquad (10.47)$$

where T is in Rankine (°F + 459.67), w is in lb/MMSCF, and P is in psia. The r^2 for this correlation is 0.99117, which is indicative of a good fit. Note this correlation only applies to pure methane hydrates.

Figure 10.17 shows the raw data of Aoyagi et al. (1979) and the calculated water content from Eqn (10.45).

Examples

Example 10.1

Use the McKetta chart to estimate the water content of a sweet natural gas at reservoir conditions, 180 °F and 2200 psia.

Answer: Reading the chart (Fig. 10.2) gives approximately 300 lb/MMCF.

The small chart is difficult to read, so I make a large, poster-size copy of the chart, which is much easier to use. I get teased for this, but I believe the chart is sufficiently accurate and convenient that I continue to use it.

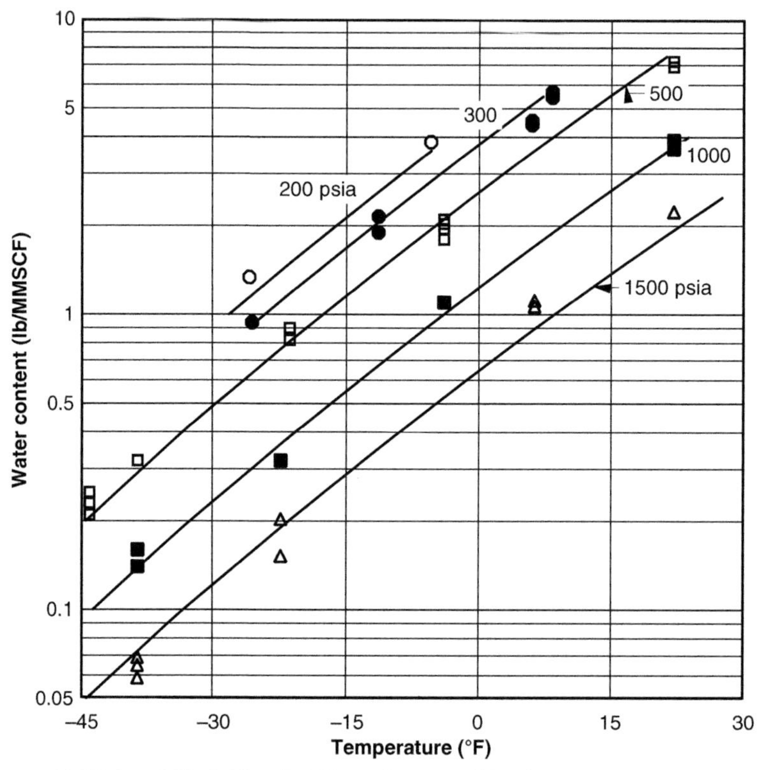

Figure 10.17 ***Local Model for the Water Content of Methane at Low Temperature.***

Example 10.2

Estimate the dew point temperature for a stream with 100 lb/MMSCF of water at 800 psia using the McKetta–Wehe chart.

Answer: Enter the chart (Fig. 10.2) at 100 lb/MMSCF the chart and go across to the 800 psia line. Go straight down and read the temperature, approximately 112 °F. This is the dew point temperature for this gas.

If the mixture is hotter than 112 °F, the stream is single phase and if it is colder, then water will condense.

Example 10.3

Assuming that the McKetta chart is applicable, estimate the water content of an acid gas containing 80% CO_2 and 20% H_2S at 120 °F and at: (1) 100 psi, (2) 250 psi, (3) 500 psi, and (4) 1000 psi. Comment on the correction factors.

Answer: Reading the chart (Fig. 10.2) gives approximately: (1) 800 lb/MMCF, (2) 375 lb/MMCF, (3) 180 lb/MMCF, and (4) 100 lb/MMCF.

Using the program for the Bukacek-Maddox method yields: (1) 883 lb/MMCF, (2) 375 lb/MMCF, (3) 225 lb/MMCF, and (4) 153 lb/MMCF.

This mixture contains too much acid gas to use the Wichert and Wichert (1993) method:

$$\begin{aligned} y_{H_2S}^{equiv} &= y_{H_2S} + 0.75 y_{CO_2} \\ &= 0.20 + (0.75)(0.80) = 0.80 \end{aligned}$$

This exceeds the 55% limit for the correlation.

Example 10.4

Redo Example 10.2 using AQUAlibrium 3.1.

Answer: The output from the AQUAlibrium 3.1 runs is appended to this chapter. From the output, we obtain: (1) 852 lb/MMCF, (2) 374 lb/MMCF, (3) 221 lb/MMCF, and (4) 170 lb/MMCF.

Note, previous editions of this book used AQUAlibrium 2.0, which gave the following results: (1) 841 lb/MMCF, (2) 368 lb/MMCF, (3) 217 lb/MMCF, and (4) 168 lb/MMCF.

Example 10.5

A low-pressure pipeline (200 psia, 1379 kPa) is transporting 10 MMSCFD of sweet natural gas a distance of 10 miles (16 km). The gas enters the line at 120 °F (48.9 °C) and cools to 50 °F (10.0 °C) when it reaches the field battery. For the purposes of this calculation, the pressure drop is negligibly small. There is no free water entering the line, only water-saturated gas. Calculate the amount of water that condenses in the line.

Answer: From the McKetta–Wehe chart (Figure 10.2):

At 200 psia and 120 °F the water content is approximately 420 lb/MMSCF.

At 200 psia and 50 °F the water content is approximately 47 lb/MMSCF.

Therefore the amount of water that condenses is:

10 MMSCFD (420 lb/MMSCF − 47 lb/MMSCF) = 3730 lb/day.

Given that the density of water is 8.34 lb/gal, this equals 447 gal/day or 10.6 bpd.

The fluid enters the pipeline single phase, but because of condensation, approximately 11 bpd of liquid water arrive at the battery.

Example 10.6

Use the McKetta–Wehe chart to estimate the water content of methane at 250 K (10 °F) and 580 psia (4.0 MPa). Compare this with the value obtained from Fig. 10.3.

Answer: Reading the chart McKetta–Wehe (Fig. 10.2) gives approximately 1.9 lb/MMCF. Converting this to ppm:

$$1.9/47484 \times 10^6 = 40 \text{ ppm}$$

From Fig. 10.3, the value is 25 ppm.

Thus if the design was based on the value obtained from the McKetta–Wehe chart, too much water would be present and solid formation would occur.

Example 10.7

A methane-water mixture is at 10 °C and 10 MPa. Estimate the minimum water content required to form a hydrate.

Answer: Interpolating from Table 2.2 methane forms a hydrate at 10 MPa and 12.8 °C and the water content of the gas is 240 ppm. From Fig. 10.8, the water content of methane in equilibrium with a hydrate at 270 K and 10 MPa is 63 ppm. Use Eqn (10.27) to interpolate between these values:

$$\ln w = \left(\frac{\ln w(T_{hyd}) - \ln w(270\ \mathrm{K})}{T_{hyd}(\mathrm{in}\ ^\circ\mathrm{C}) + 3.15}\right)(T(\mathrm{in}\ ^\circ\mathrm{C}) + 3.15) + \ln w(270\ \mathrm{K})$$

$$= \left(\frac{\ln(240) - \ln(63)}{12.8 + 3.15}\right)(10 + 3.15) + \ln(63)$$

$$= 5.2458$$

$$w = 190\ \mathrm{ppm}$$

Example 10.8

A sweet natural gas is flowing in a pipeline at 1000 psia (6.895 MPa). At this pressure, the gas forms a hydrate at 65 °F (18.3 °C), provided sufficient water is present. First, estimate the minimum amount of water required to form a hydrate at 65 °F and 1000 psia. If the gas has been dehydrated to 7 lb/MMCF (112 mg/Sm^3), at what temperature will this mixture form a hydrate? Finally, if the gas is dehydrated to 4 lb/MMCF (64 mg/Sm^3), at what temperature will a hydrate form?

Answer: The minimum water content can be estimated from the McKetta–Wehe chart. The reason why this chart can be used is because the gas is simultaneously in equilibrium with liquid water and hydrate. From the chart, one obtains 21 lb/MMCF (336 mg/Sm^3). This is the minimum water content.

To estimate the temperature, we must find the temperature at which a gas with a water content of 7 lb/MMCF forms a hydrate. Assume the water content of the gas is the same as methane. From Fig. 10.9 at 1000 psia and 26.3 °F, the required water content to form a hydrate is 4.5 lb/MMCF (72 mg/Sm^3). Linearly interpolating $\ln w$ vs T yields:

$$T = T_1 + \left(\frac{\ln w - \ln w_1}{\ln w_1 - \ln w_2}\right)(T_1 - T_2)$$

$$= 26.3 + \left(\frac{\ln 7 - \ln 4.5}{\ln 4.5 - \ln 21}\right)(26.3 - 65)$$

$$= 37.4\ ^\circ\mathrm{F}$$

Therefore if the gas is dehydrated to 7 lb/MMCF, then a hydrate forms at 37.4 °F (3.0 °C).

At 4 lb/MMCF, the temperature can be read directly from Fig. 10.9, although with some difficulty. Alternatively, from Fig. 10.9 at 26.3 °F, the water content is 4.5 lb/MMCF and at 8.3 °F, the water content is 1.8 lb/MMCF. Interpolating as previously yields:

$$T = 26.3 + \left(\frac{\ln 4 - \ln 4.5}{\ln 4.5 - \ln 1.8}\right)(26.3 - 8.3)$$

$$= 24.0\ ^{\circ}\mathrm{F}$$

Therefore if the gas has been dehydrated to 4 lb/MMCF, then the hydrate forms at 24.0 °F (−4.5 °C).

Example 10.9

Estimate the dew point temperature of liquid propane containing 1 lb/MMCF of water.

Answer: From Fig. 10.13, the dew point is approximately 0 °F (−17.8 °C); therefore, for temperatures greater than 0 °F a hydrate will not form.

Example 10.10

Using Eqn (10.45), estimate the hydrate point for a methane containing 5 lb/MMSCF of water at 750 psia.

Answer: Substituting the values in Eqn (10.45) yields:

$$\frac{1000}{T} = 2.494189 - 0.090738 \ln(5) - 0.029784 \ln(750) - 9.36231 \times 10^{-5}(750)$$

$$= 2.080761$$

$$T = 1000/2.08761 = 480.59\ \mathrm{R}$$

$$t = 480.59 - 459.67 = 20.9\ ^{\circ}\mathrm{F}$$

So the estimate frost point is about 21 °F (−6.2 °C).

APPENDIX 10A OUTPUT FROM AQUALIBRIUM

Prepared by:	AQUAlibrium 3.1 ©FlowPhase Inc. #330, 2749-39th Avenue N.E. Calgary, Alberta, Canada, T1Y 4T8 Phone: (403)250-7522; Fax: (403)291-9730 www.flowphase.com

HYDRATE BOOK EXAMPLE 10.4: 100 PSI

Water Content Calculation

Conditions

Temperature:	120.00	F
Pressure:	100.00	psia

Component Fractions

Components	Feed	Vapor	Aqueous	*K*-factor 1
Water	0	0.0179512	0.996582	0.0180127
H_2S	0.2	0.19641	0.0015234	128.929
CO_2	0.8	0.785639	0.00189445	414.706
Total	1	1	1	

Phase Properties

Properties	Units	Vapor	Aqueous
Mole percent		99.9999	0
Molecular weight	lb/lb·mol	41.593	18.0887
z-factor		0.967788	0.00470642
Density	lb/ft^3	0.690897	61.7861
Enthalpy	Btu/lb	15.731	−980.58
Heat capacity	Btu/lb·F	0.222628	0.999646
Viscosity	lb/ft·s	1.05149e-05	0.000374725
Thermal conductivity	Btu/h·ft·F	0.0107405	0.370174
Specific volume	ft^3/lb	1.44739	0.0161849

Water Content

Water content of gas	852.394	lb/MMSCF (60 °F and 14.696 psia)

Solubility

Solubility	25.3115	SCF/bbl water

Warnings

HYDRATE BOOK EXAMPLE 10.4: 250 PSI

Water Content Calculation

Conditions

Temperature:	120.00	F
Pressure:	250.00	psia

Component Fractions

Components	Feed	Vapor	Aqueous	*K*-factor 1
Water	0	0.00786641	0.991729	0.00793202
H_2S	0.2	0.198427	0.00375701	52.8151
CO_2	0.8	0.793707	0.00451437	175.818
Total	1	1	1	

Phase Properties

Properties	Units	Vapor	Aqueous
Mole percent		99.9999	0
Molecular weight	lb/lb·mol	41.8351	18.1927
z-factor		0.91837	0.0118127
Density	lb/ft^3	1.83078	61.8959
Enthalpy	Btu/lb	11.4093	−971.496
Heat capacity	Btu/lb·F	0.235762	1.00254
Viscosity	lb/ft·s	1.06259e-05	0.000375196
Thermal conductivity	Btu/h·ft·F	0.0110938	0.371386
Specific volume	ft^3/lb	0.546214	0.0161562

Water Content

Water content of gas	373.529	lb/MMSCF (60 °F and 14.696 psia)

Solubility

Solubility	61.5552	SCF/bbl water

Warnings

HYDRATE BOOK EXAMPLE 10.4: 500 PSI

Water Content Calculation

Conditions

Temperature:	120.00	F
Pressure:	500.00	psia

Component Fractions

Components	Feed	Vapor	Aqueous	*K*-factor 1
Water	0	0.0046449	0.984592	0.00471759
H_2S	0.2	0.199071	0.00720254	27.639
CO_2	0.8	0.796284	0.00820533	97.0447
Total	1	1	1	

Phase Properties

Properties	Units	Vapor	Aqueous
Mole percent		99.9999	0
Molecular weight	lb/lb·mol	41.9125	18.344
z-factor		0.830095	0.0237607
Density	lb/ft^3	4.05844	62.0553
Enthalpy	Btu/lb	3.51976	−958.588
Heat capacity	Btu/lb·F	0.270373	1.00666
Viscosity	lb/ft·s	1.0885e-05	0.000375917
Thermal conductivity	Btu/h·ft·F	0.0119371	0.373147
Specific volume	ft^3/lb	0.2464	0.0161147

Water Content

Water content of gas	220.559	lb/MMSCF (60 °F and 14.696 psia)

Solubility

Solubility	115.496	SCF/bbl water

Warnings

HYDRATE BOOK EXAMPLE 10.4: 1000 PSI

Water Content Calculation

Conditions

Temperature:	120.00	F
Pressure:	1000.00	psia

Component Fractions

Components	Feed	Vapor	Aqueous	*K*-factor 1
Water	0	0.00357928	0.975869	0.00366778
H_2S	0.2	0.199284	0.0107733	18.4979
CO_2	0.8	0.797137	0.0133577	59.6764
Total	1	1	1	

Phase Properties

Properties	Units	Vapor	Aqueous
Mole percent		100	0
Molecular weight	lb/lb·mol	41.9381	18.5353
z-factor		0.612781	0.047824
Density	lb/ft^3	11.0021	62.3056
Enthalpy	Btu/lb	−18.1138	−943.601
Heat capacity	Btu/lb·F	0.493225	1.01074
Viscosity	lb/ft·s	1.25317e-05	0.000377142
Thermal conductivity	Btu/h·ft·F	0.0154551	0.375917
Specific volume	ft^3/lb	0.0908915	0.0160499

Water Content

Water content of gas	169.958	lb/MMSCF (60 °F and 14.696 psia)

Solubility

Solubility	182.5	SCF/bbl water

Warnings

REFERENCES

Aoyagi, K., Song, K.Y., Sloan, E.D., Dharmawardhana, P.B., Kobayashi, R., March 1979. Improved Measurements and Correlation of the Water Content of Methane Gas in Equilibrium with Hydrate, 58th GPA Convention, Denver, CO.

Keenan, J.H., Keyes, F.G., Hill, P.G., Moore, J.G., 1978. Steam Tables. John Wiley & Sons, New York, NY.

Kobayashi, R., Song, K.Y., Sloan, E.D., 1987. Phase Behavior of hydrocarbon/water systems. In: Bradley, H.B. (Ed.), Petroleum Engineering Handbook, pp. 25-1–25-28.

Maddox, R.N., 1974. Gas and Liquid Sweetening, second ed. John M. Campbell Ltd. pp. 39–42.

Maddox, R.N., Lilly, L.L., Moshfeghian, M., Elizondo, E., March 1988. Estimating Water Content of Sour Natural Gas Mixtures. Laurance Reid Gas Conditioning Conference, Norman, OK.

McCain, W.D., 1990. The Properties of Petroleum Fluids, second ed. PennWell Books, Tulsa, OK.

Ning, Y., Zhang, H., Zhou, G., 2000. Mathematical simulation and program for water content chart of natural gas. Chem. Eng. Oil Gas 29, 75–77 (in Chinese).

Robinson, J.N., Moore, R.G., Heidemann, R.A., Wichert, E., February 6, 1978. Charts help estimate H_2O content of sour gases. Oil Gas J. 76 (5), 76–78.

Robinson, J.N., Moore, R.G., Heidemann, R.A., Wichert, E., March 1980. Estimation of the Water Content of Sour Natural Gas. Laurance Reid Gas Conditioning Conference, Norman, OK.

Saul, A., Wagner, W., 1987. International equations for the saturation properties of ordinary water substance. J. Phys. Chem. Ref. Data 16, 893–901.

Sharma, S., Campbell, J.M., August 4, 1969. Predict natural-gas water content with total gas usage. Oil Gas J. 67 (31), 136–137.

Song, K.Y., Kobayashi, R., 1982. Measurement and interpretation of the water content of a methane-propane mixture in the gaseous state in equilibrium with hydrate. Ind. Eng. Chem. Fundam. 21, 391–395.

Song, K.Y., Kobayashi, R., 1994. Water content of ethane propane, and their mixtures in equilibrium with water or hydrates. Fluid Phase Equilib. 95, 281–298.

Song, K.Y., Kobayshi, R., 1987. Water content of CO_2 in equilibrium with liquid and/or hydrates. SPE Form. Eval., 500–507.

Song, K.Y., Yarrison, M., Chapman, W., 2004. Experimental low temperature water content in gaseous methane, liquid ethane, and liquid propane in equilibrium with hydrate at cryogenic conditions. Fluid Phase Equilib. 224, 271–277.

Wichert, G.C., Wichert, E., 1993. Chart estimates water content of sour natural gas. Oil Gas. J. 90 (13), 61–64.

Wichert, G.C., Wichert, E., 2003. New charts provide accurate estimations for water content of sour natural gas. Oil Gas. J. 101 (41), 64–66.

CHAPTER 11

Additional Topics

In this final chapter, we will review some additional topics that do not fit into the earlier chapters, but are of importance to the topic of gas hydrates.

11.1 JOULE-THOMSON EXPANSION

When a fluid flows through a valve (or, as in the original experiments, a porous plug or any restriction to the flow), the fluid pressure drops. The process occurs adiabatically, that is without heat being transferred. This is because it happens quite quickly. The process is shown graphically in Fig. 11.1. The question is now what happens to the temperature of the fluid during such a process?

If you are unaware of the Joule–Thomson effect, but have some technical knowledge, you might conclude that the temperature is unchanged by the throttling process. There appears to be nothing happening to change the temperature of the fluid. This view is incorrect, except in a few relatively rare cases.

On the other hand, it should be clear to all that no work is done as the fluid crosses the valve. The potential work available from going from high pressure to low pressure is simply lost.

Those with some technical training might say that the fluid cools upon such an expansion and that the temperature leaving the valve is lower than that entering.

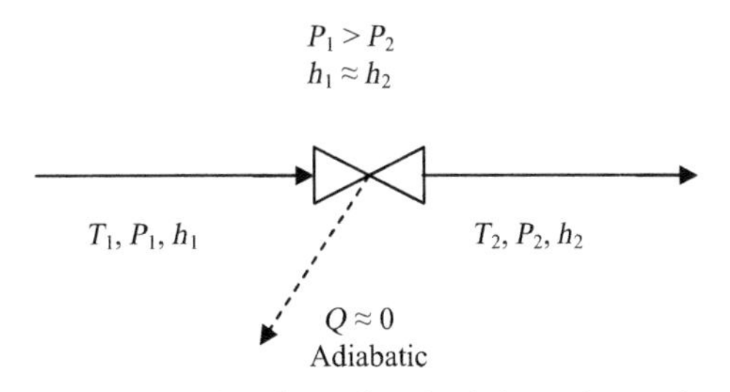

Figure 11.1 *The Flow of a Fluid through a Valve.*

Natural Gas Hydrates
ISBN 978-0-12-800074-8
http://dx.doi.org/10.1016/B978-0-12-800074-8.00011-9

11.2 THEORETICAL TREATMENT

The detailed thermodynamics of the Joule–Thomson expansion are well understood. Those interested can check almost any book on classical thermodynamics, but the details are outlined here briefly.

Mathematically this is given by:

$$\mu_{JT} = \left(\frac{\partial T}{\partial P}\right)_H. \tag{11.1}$$

When this quantity is positive, the fluid cools upon expansion, and when it is negative, the fluid warms.

From classical thermodynamics, we can derive the following expression for this quantity:

$$\mu_{JT} = \frac{T\left(\frac{\partial v}{\partial T}\right)_P - v}{C_P}. \tag{11.2}$$

The derivation of this expression is given in most textbooks on classical thermodynamics and will not be repeated here.

We have reduced the expression to one that contains only the pressure, temperature, molar volume (and derivatives of these quantities), and the heat capacity. An equation of state of the form $P = f(v, T)$ can be used to evaluate the numerator.

The detailed thermodynamics of the Joule–Thomson expansion are well understood and will not be repeated here. Those interested can check almost any book on classical thermodynamics or the paper by Carroll (1999).

11.3 IDEAL GAS

It is an easy matter to show from Eqn (8.2) that the Joule–Thomson coefficient for an ideal gas is exactly zero, regardless of the pressure and temperature. The ideal gas law is:

$$Pv = RT \tag{11.3}$$

then differentiating yields:

$$\left(\frac{\partial v}{\partial T}\right)_P = \frac{v}{T}. \tag{11.4}$$

Substituting this expression into Eqn (8.2) reveals that the Joule–Thomson coefficient is zero. Thus, for an ideal gas, the isenthalpic expansion does not affect the temperature. However, for real gases and for liquids, that is not the case.

11.4 REAL FLUIDS

For real fluids, the Joule–Thomson coefficient, as noted earlier, can be positive or negative. The boundary between the two regions is called the Joule–Thomson inversion curve. This is the temperature and pressure where the Joule–Thomson coefficient is zero.

There are three cases where negative Joule–Thomson values may be encountered in engineering practice. These are:

1. A gas at a relatively high temperature;
2. A low temperature liquid;
3. Very high-pressure fluids (both gases and liquids).

11.4.1 Compressibility Factor

A seemingly simple equation of state is obtained when the compressibility factor is introduced to the ideal gas law:

$$Pv = zRT \tag{11.5}$$

What is lost in the apparent simplicity of this equation is that the compressibility factor, z, is a function of the pressure and the temperature. On the other hand, with the appropriate values for z, this equation can be applied to both gases and liquids.

Differentiating this equation to obtain the expression for the Joule–Thomson equation, and after some manipulation, one obtains:

$$\left(\frac{\partial v}{\partial T}\right)_{\mathrm{P}} = \frac{R}{P}\left[T\left(\frac{\partial z}{\partial T}\right)_{\mathrm{P}} + z\right] \tag{11.6}$$

Substituting this into Eqn (11.2) yields:

$$\mu_{\mathrm{JT}} = \frac{\frac{RT^2}{P}\left(\frac{\partial z}{\partial T}\right)_{\mathrm{P}}}{C_{\mathrm{P}}} \tag{11.7}$$

All of the quantities in this expression are positive with the exception of the temperature derivative of the compressibility factor. This quantity can be either positive or negative.

Thus, if the compressibility factor is an increasing function of the temperature (i.e., z increases when T increases), then μ_{JT} is positive. If the compressibility factor is a decreasing function of the temperature, then μ_{JT} is negative.

A generalized compressibility chart can be used to roughly determine the location of these two regions. Although detailed calculation of Joule–Thomson coefficients from such a chart is not recommended.

11.4.2 The Miller Equation

Miller (1970) derived the following approximate equation for the Joule–Thomson inversion curve:

$$P_R = 24.21 - 18.54/T_R - 0.825\ T_R^2 \tag{11.8}$$

where T_R and P_R are the reduced temperature (the absolute temperature divided by the critical temperature) and reduced pressure (the pressure divided by the critical pressure), respectively. This function is plotted in Fig. 11.2.

We can make the following observations based on the Miller equation. First, at low temperatures, subcooled liquids, the Joule–Thomson coefficient is positive. In fact, according to the Miller equation, if T_R is less than approximately 0.8, then the Joule–Thomson is positive. At high temperatures, the Joule–Thomson is also positive. According to the Miller equation,

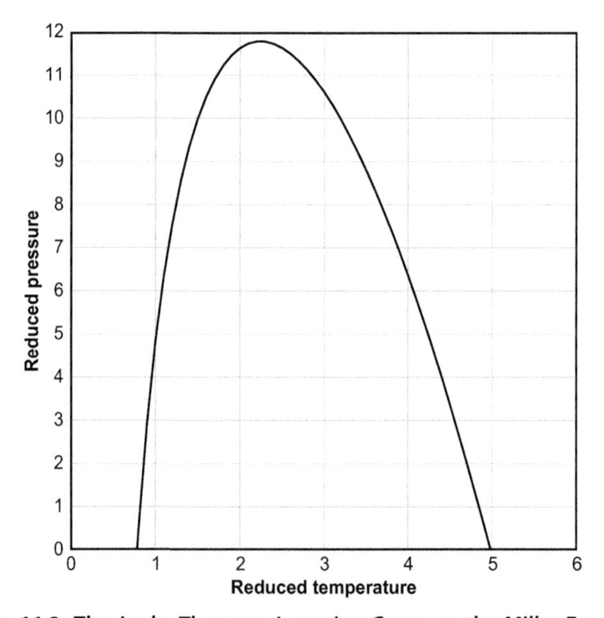

Figure 11.2 *The Joule–Thomson Inversion Curve on the Miller Equation.*

if the T_R is greater than approximately 5, then the Joule–Thomson is positive. Finally, at very high pressures, the Joule–Thomson is also positive. Thus, fluids under high pressure warm upon expansion.

Furthermore, it appears from this figure that, in the limit, as the pressure goes to zero, the ideal gas behavior is not exhibited. That is, at low pressure, the Joule–Thomson coefficient is not zero, which was demonstrated earlier, is the case for ideal gases. However, the Miller equation and the figure derived from it says nothing of the magnitude of the Joule–Thomson coefficient. Low-pressure gases have Joule–Thomson coefficients that are small in magnitude and, thus, for practical purposes, are zero.

11.5 SLURRY FLOW

A slurry is a heterogeneous mixture of an insoluble solid and a fluid (usually a liquid). Slurries are often used to transport a solid compound, such as coal, in a mixture with a liquid, water, or oil. Some discussion of the flow of hydrate slurries was presented earlier in this book, but additional comments are presented here.

In horizontal slurry flow, there are four flow regimes: (1) homogeneous flow, (2) heterogeneous flow, (3) saltation flow, and (4) flow with a stationary bed (Turian and Yuan, 1977).

In homogeneous flow, the slurry flows almost as a single phase with no separation into either phases or layers. This occurs when the settling velocity of the particles is small in comparison with the velocity of the fluid.

If the particles are somewhat larger or if the settling velocity of the particles is larger, then vertical gradients in the particle concentration can occur. This is the heterogeneous flow regime.

If the settling velocities become too large, then the particles with start to settle. The initial settling is the saltation regime. This regime is intermittent, with leaping movement of particles. Typically, the solids are transported by two mechanisms: bulk movement of the particle bed and movement with the fluid above the bed. To visualize this, imagine wind blowing sand off the top of a sand dune.

In the stationary bed regime, the particles have settled to the bottom of the pipeline. This is similar to the bedding of hydrates described below. In this case, the particles will only flow if they are swept along with the fluid flow.

The criteria for determining the flow regime is a function of the drag coefficient, the solids concentration, the velocity, the pipe diameter, the particle size, and the ratio of the densities. Different pressure drop correlations were developed for each flow regime.

In the slurries studied by Turian and Yuan (1977), there was no tendency for the particles to adhere. For example, in a coal-water or a coal-oil slurry, the coal will not agglomerate into a large mass. However, this is indeed the case with hydrates, thus, for hydrate slurries.

In addition, in typical slurries, the solid phase is denser than the liquid. For natural gas hydrates, if the liquid phase is water, then the hydrates are most likely less denser than the water, and will have some buoyancy and will not likely settle to the bottom of the pipe. If the fluid phase is a condensate or an oil, typical density less than 800 kg/m^3, then the hydrate will be of greater density, more like the traditional slurry.

11.6 HYDRATE FORMATION IN THE RESERVOIR DURING PRODUCTION

As a gas flow from the reservoir into the region of the wellbore, the gas expands. If there is no heat transfer with the surrounding formations, then this is an isenthalpic process.

The flow of the gas through the reservoir is governed by Darcy's law (or nearly so). For the radial flow of a gas, Darcy's law is (Smith, 1990):

$$Q = \frac{kh\left(P_2^2 - P_1^2\right)}{\mu z T_f \ln(r_2/r_1)} \tag{11.9}$$

where Q is the volumetric flow rate, k is the permeability, h is the thickness of the formation, P is the pressure, μ is the viscosity of the fluid, z is the compressibility factor (z-factor), T_f is the temperature of the fluid, and r is the radial distance. This equation is derived by integrating between two arbitrary points. One must be very careful with the units when using this equation. For American Engineering Units, this equation becomes:

$$Q = \frac{707.8kh\left(P_2^2 - P_1^2\right)}{\mu z T_f \ln(r_2/r_1)} \tag{11.10}$$

where Q is in ft^3[std]/d, k is in Darcy, h is in ft, P is in psia, μ is centipoise, z is unitless, T_f is in Rankin, and r is in ft.

Furthermore, when this equation was integrated, it was assumed that the properties of the fluid do not change significantly with the changes in the pressure and temperature of the fluid.

11.7 FLOW IN THE WELL

The gas from the reservoir enters the wellbore and travels upward. There are two significant factors in determining the pressure drop in the well flow: (1) friction pressure drop and (2) hydrostatic head. In the case of upward flow, both of these effects tend to reduce the pressure of the flowing gas. Smith (1990) provides a procedure for estimating the pressure and temperature in both shut-in and flowing gas wells.

In addition, moving from a depth to the surface, the surrounding temperature decreases the so-called geothermal gradient. Typically, the geothermal gradient is approximately 25 °C/km (1.5 °F/1000 ft).

Fig. 11.3 shows the pressure and temperature as a function of the depth for a hypothetical gas well. The well has a depth of 3000 m (9842 ft) and the pressure at the bottom of the well is 15 MPa (2176 psia), which includes some pressure drop because of the flow through the reservoir. The temperature at 3000 m is 75 °C (167 °F) and the geothermal gradient is 25 °C/km.

Fig. 11.4 shows the same data but on a pressure-temperature plot. Also shown on this graph is the hydrate curve for this gas mixture. It can be seen that the pressure-temperature curve intersects the hydrate curve, which means a hydrate can form in the wellbore. The intersection of the two

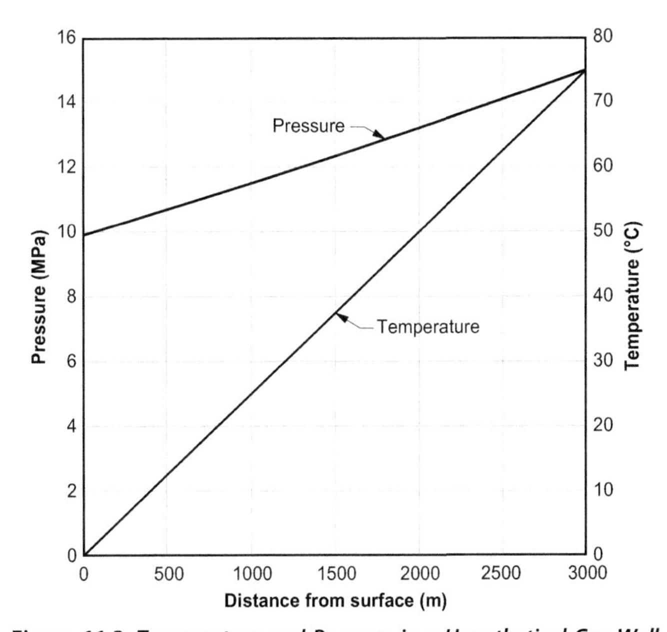

Figure 11.3 ***Temperature and Pressure in a Hypothetical Gas Well.***

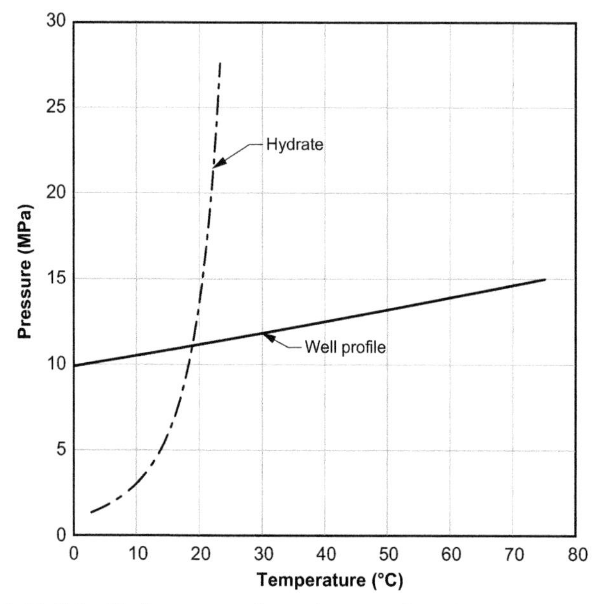

Figure 11.4 ***Well Profile for a Hypothetical Gas Well and Hydrate Curve Produced.***

curves is at approximately 19 °C (66 °F), which is equivalent to a depth of about 750 m (2460 ft). From the surface to about 750 m, hydrate formation is possible in this well. For depths greater than 750 m, the pressure-temperature of the gas is outside the hydrate region and freezing will not occur at greater depths.

In this situation, the injection of an inhibitor chemical is required. Some wells are completed with an injection string that allows for the injection of chemicals at the bottom of the well. This injection is typically continuously preventing production interruption because of plugging of the wellbore. Other wells have a pump at the surface for the injection of chemicals. Unlike pipelines, which are more or less horizontal, the injected chemical can flow down the well string, even if the flow is blocked, because of gravity.

11.8 CARBON STORAGE

Carbon dioxide has been implemented in global climate change. The capture of carbon dioxide from flue gas and its storage in porous subsurface reservoirs has been touted as a way to reduce anthropogenic CO_2 in the atmosphere.

A number of depleted gas reservoirs located in northern Alberta, Canada, and elsewhere, have pressure and temperature conditions that are identified as potential sites for CO_2 storage in gas hydrate form. These reservoirs are in the range of pressure from 2 to 5 MPa (290–725 psia) and temperatures from 1° to 10 °C (34°–50 °F). The CO_2 must flow into the reservoir for a sufficient distance and then solidify into a hydrate. If the gas solidifies in the wellbore region, the flow would be impeded.

This technology is very new and has only been studied in the laboratory and in simulation (Linga et al., 2009; Zatsepina and Pooladi-Darvish, 2012).

11.9 TRANSPORTATION

Another potential application of hydrates is for the transportation of natural gas. Currently, gas is transported across continents via pipelines (and there are a few intercontinental pipelines) and via liquid natural gas tankers over longer distances.

For long distance transmission lines, pressures are typically 1000 psia or about 7 MPa. In addition to the pipeline, this involves compression of the gas and, depending upon the distance of the line, may require intermediate recompression.

Transportation of liquid natural gas (LNG) is done at very low temperature but at near atmospheric pressures. It is expensive to liquefy natural gas and requires a vaporization facility at the delivery point.

The industrialization of the transportation of natural gas in the form of hydrates is in its infancy, with several problems that still must be worked out. However, this is not a new idea. Amongst the first to suggest this technology were Cahn et al. (1970).

One of the problems with this method is to produce a hydrate with minimal entrained water, ice (which contains no gas), and a high hydrate concentration. Gudmundsson (1996) describes one possible method for making hydrates on a large scale.

Table 11.1 gives a comparison of the amount of methane contained in 1 m^3 (35.3 ft^3) of hydrate, 1 m^3 of pipeline gas, and 1 m^3 (35.3 ft^3) of LNG. All of the values are for pure methane and the densities of methane are taken from Wagner and de Reuck (1996). One cubic meter of hydrate contains more than twice as much methane as the pipeline gas, but only about one-fourth of the methane in LNG.

Table 11.1 The Amount of Methane per Cubic Meter for Hydrate, Pipeline Gas, and Liquid

	Pressure (MPa)	Temperature (°C)	Amount (kg)	Comment
Hydrate	2.6	0	115	1 m^3 hydrate contains 170 Sm^3 of gas
Pipeline gas	7	27	50	Typical pipeline conditions
Liquid	0.101	−161	422	At the normal boiling point

	Pressure (psia)	Temperature (°F)	Amount (lb)	Comment
Hydrate	377	32	254	1 ft^3 hydrate contains 170 SCF of gas
Pipeline gas	1015	81	110	Typical pipeline conditions
Liquid	14.7	−258	930	At the normal boiling point

lb, pounds; SCF, standard cubic feet.

11.10 NATURAL OCCURRENCE OF HYDRATES

Combinations of water and hydrate formers are common in nature, so it should come as no surprise that there are natural occurrences of gas hydrates. This section reviews them briefly.

11.11 SEABED

There has been much discussion about the occurrence of methane hydrates in the seabed. As was shown in Example 2-7, hydrates of methane can form in the sea at depths of about 300 m (1000 ft). The methane required for the hydrate formation presumably comes from the anaerobic decay of organic material, but may come from other sources.

These hydrates are found throughout the world and not only in colder locations. For example, hydrates have been found in the Gulf of Mexico, the Caribbean Sea, off the cost of South America, India, and many other locations.

Estimates of how much hydrocarbon is locked in these resources are astronomical. For example, Dickens et al. (1997) estimated that there is about 15 GT7 equivalent of methane in the seabed tied up as natural gas hydrates. This is approximately equivalent to 3×10^{13} m^3[std] or 1×10^9 MMSCF of methane. However, such estimates range widely depending upon who is doing the estimate and what is their basis. For example, Kvenvolden (1999) estimates this amount of gas in seabed hydrates as between 1×10^{15} and 50×10^{15} m^3[std]. Lerche (2001) provides an interesting analysis of the estimates of the hydrate resource.

Others, such as Laherrere (1999), argue that the hydrates are of poor quality and may not be of commercial quality.

11.11.1 Nankai Trough

The Nankai Trough is on the southeast coast of Japan. It is a seismically active area and well known for being rich in seabed hydrates (Colwell et al., 2004). Because they do not have significant convention hydrocarbon reserves, the Japanese are keen to exploit these resources.

According to the Japan Oil, Gas, and Metals National Corporation (JOGMEC), there are 40 TCF of methane in the trough.

In March 2013, JOGMEC announced that it had completed a successful production test (http://www.jogmec.go.jp/english/news/release/) from the Nankai Trough. The test lasted approximately 6 days and resulted in a total of 120×10^3 m^3, which is an average of 20×10^3 m^3/d. They do not provide much detail other than to say that the production was achieved by depressurization. The gas produced was flared and no processing of the gas was attempted. There was no indication of the pressure or temperature at which the gas arrived at the surface. JOGMEC is hoping for commercial production by the year 2018.

11.12 NATURAL GAS FORMATIONS

In certain cold regions on earth, hydrates can be found in natural gas reservoirs. These include the arctic regions of Canada, Russia, and Alaska.

The production of the gas locked in the hydrates of such a reservoir is an interesting problem. Perhaps a thermal recover method, such as those used for heavy oil, could be used to melt the hydrate. Alternatively, perhaps a miscible flood of an inhibitor chemical could unlock the gas. There have been several papers discussing the production of natural gas from hydrate reservoirs by depressurization. One such paper is by Jia et al. (2001).

11.12.1 Messoyakha Field

An interesting example of a hydrate reservoir is the Messoyakha field in Siberia. This is a natural gas reservoir in a cold region of the world. The field was discovered in 1967 and has been exploited commercially with production from this field that began in 1970. The reservoir had about 850 BCF of natural gas, some of which is frozen in the hydrate.

The gas composition is 98.6% methane, 0.1% ethane, 0.1% C_3+, 0.5% carbon dioxide, and 0.7% nitrogen.

The top of the reservoir is at a depth of slightly more than 700 m and the porous formation extends to about 900 m. At approximately 800 m, the pressure and temperature intersect the hydrate curve for the gas in this reservoir. Therefore, there is a free gas zone on the top of the reservoir and a hydrate layer on the bottom.

Gas can be produced from the free gas zone in the top. As this gas is produced, the pressure in the reservoir falls. The reduction in the pressure melts the hydrate releasing additional gas and, thus, increasing the pressure.

More details about this field can be found in Tanahashi (1996) and Makogon et al. (2007).

11.12.2 Mallik

A joint research project lead by the Canadian Geological Survey has thoroughly studied a hydrate formation in the delta of the Mackenzie River in the Canadian Arctic (north of the Arctic Circle).

The drilled a well to a depth of about 1150 m (3770 ft) where hydrates were encountered in a sandstone formation. The formation is about 100 m (325 ft) thick. It is estimated that the structure contains about 100×10^9 m^3 of gas frozen in the hydrate. As of the writing of this book, this resource has not been exploited commercially.

There are many papers discussing this project, but a good summary is given by Hyndman and Dallimore (2001).

11.13 OUTER SPACE

The frigid conditions in outer space along with the possibility of water make for the potential of hydrates in outer space. Two examples are provided here.

11.13.1 Comets

Miller (1961) was among the first to speculate on the possibility of hydrates in outer space. It is ironic that, at the time of his paper, it was commonly believed that there were no natural hydrates on the Earth. In fact, Miller (1961) begins his paper by stating:

> *The gas hydrates... are not known to occur naturally on the earth because of the unfavorable combination of temperatures, pressures, and gases that are poor hydrate formers.*

It seems to be well known that comets are stellar "ice balls." However, comets are a witch's brew of water and organic compounds. These are the right combination for hydrate formation. Miller was among the first to speculate upon this possibility.

11.13.2 Mars

The Martian atmosphere is approximately 95.3% CO_2, 2.7% nitrogen, 1.6% argon, and the remainder (less than half of 1%) is oxygen, carbon monoxide, and others. The atmospheric pressure on the surface of Mars is 0.636 mbars. This data comes from the Website www.burro.astr.cwru.edu.

Water is also known to exist on Mars and there are prominent "ice caps" on both the south and north poles.

Much like the Earth, there are large variations in the temperature on the surface of Mars because of time of the Martian day, Martian seasons, and latitude. However, at the poles in the Martian winter, the temperature can be as low as 120 K (−153 °C).

Miller and Smythe (1970) speculated on the possibility of hydrates in the "ice" caps on Mars. They concluded that because Martian atmosphere contains carbon dioxide and if the "ice" caps contain water, there is a possibility that hydrates can be formed.

In January 1999, National Aeronautics and Space Administration launched the Mars Polar Lander (MPL) with the intention of exploring the south pole of Mars. It had the equipment to determine whether the ice cap was composed of hydrate or ice. Unfortunately, communication was lost with the spacecraft in December 1999 and it was never recovered. Therefore, the MPL is essentially "lost in space." Thus, an answer based on measurements from Mars will have to wait for another mission. For more details see: http://mars.jpl.nasa.gov/msp98/lander/.

In addition, the gas giant planets in our solar system: Saturn, Jupiter, and Neptune, and their moons, contain methane and water. Surely, there is a possibility that hydrates exist on those planets as well.

Examples

Example 11.1

Pure methane is throttled from 51.85 °C (325 K) and 5−3.5 MPa. The calculations for the exit temperature were made using the methane tables from Wagner and de Reuck (1996).

Answer: From the tables:

	5 MPa	320 K	91.416 J/mol
		330 K	501.42 J/mol
Interpolating enthalpy		325 K	296.42 J/mol

From the tables:

	3.5 MPa	320 K	301.82 J/mol
		310 K	−93.104 J/mol
Interpolating temperature		319.42 K	296.42 J/mol

Therefore, the outlet temperature is 319.42 K or 46.71 °C. Therefore, the temperature drop is 5.14 °C. From this, the Joule−Thomson coefficient can be approximated:

$$\mu_{JT} = \left(\frac{\partial T}{\partial P}\right)_H$$

$$= \left(\frac{T_o - T_i}{P_o - P_i}\right)_H = \frac{319.86 - 325.00}{3.5 - 5.0}$$

$$= 3.43 \text{ K/MPa}$$

Example 11.2

For the information given below, calculate the gas flow rate through the reservoir.

$k = 0.005$ Darcy	$h = 50$ ft
$\mu = 0.0175$ cp	$z = 0.925$
$T_f = 120$ °F = 580 R	
$r_2 = 2640$ ft	$r_1 = 0.25$ ft
$P_2 = 3000$ psia	$P_1 = 1000$ psia

Answer: Substituting into the Darcy equation and performing the arithmetic:

$$Q = \frac{707.8(0.005)(50)\left(3000^2 - 1000^2\right)}{(0.0175)(0.925)(580)\ln(2640/0.25)}$$

$$= 1.627 \times 10^7 \text{ft}^3/d = 16.27 \text{ MMCFD}$$

Example 11.3

Assuming the gas in the previous example is pure methane, estimate the temperature drop due to the expansion using the tables from Wagner and de Reuck (1996).

Answer: The tables are in SI units so the original pressures are converted from psia to MPa:

Convert from AEU to SI units:

$$2000 \text{ psia} = 13.789 \text{ MPa}$$

$$1000 \text{ psia} = 6.895 \text{ MPa}$$

$$120 \text{ F} = 322.04 \text{ K}$$

Double interpolate the table to get the value of the enthalpy at the initial conditions:

Pressure (MPa)	Temperature (K)	Enthalpy (J/mol)	Comments
13	320	−979.79	From tables
13	330	−493.33	From tables
13	322.04	−880.55	Interpolated
14	320	−604.60	From tables
14	330	−118.81	From tables
14	322.04	−505.50	Interpolated
13.789	322.04	−584.64	Interpolated

Because this is a constant enthalpy process, the enthalpy at the initial condition is equal to the enthalpy at the final conditions. Thus, find the temperature at 6.985 MPa and −584.64 J/mol.

Pressure (MPa)	Temperature (K)	Enthalpy (J/mol)	Comments
6.5	300	−974.38	From tables
6.5	310	−544.57	From tables
6.5	309.07	−584.64	Interpolated
7	310	−619.62	From tables
7	320	−187.80	From tables
7	310.81	−584.64	Interpolated
6.895	310.57	−584.64	Interpolated

Therefore, the estimated temperature is 310.57 K, which is equal to 37.4 °C or 99.4 °F. Thus, the expansion of the gas as it flows from the bulk reservoir (120 °F) to the wellbore has resulted in a cooling of about 20 °F.

This is an overly simple analysis and did not account for the changes in the properties because of the change in the temperature in the original Darcy's law calculation.

REFERENCES

Cahn, R.P., Johnston, R.H., Plumstead, J.A., 1970. Transportation of Natural Gas as a Hydrate, US Patent No. 3,514,274.

Carroll, J.J., 1999. Working with fluids that warm upon expansion. Chem. Eng. 106 (10), 108–114.

Colwell, F., Matsumoto, R., Reed, D., 2004. A review of the gas hydrates, geology, and biology of the Nankai Trough. Chem. Geol. 205, 319–404.

Dickens, G.R., Paull, C.K., Wallace, P., 1997. Direct measurements of *in situ* methane quantities in a large gas-hydrate reservoir. Nature 385, 426–428.

Gudmundsson, J.D., 1996. Method for Production of Gas Hydrates for Transportation and Storage, US Patent No 5,536,893.

Hyndman, R.D., Dallimore, S.R., 2001. Natural Gas hydrate studies in Canada. Recorder 26 (5), 11–20.

Jia, C., Ahmadia, G., Smith, D.H., 2001. Natural gas production from hydrate decomposition by depressurization. Chem. Eng. Sci. 56, 5801–5814.

Kvenvolden, K.A., 1999. Potential effects of gas hydrates on human welfare. Proc. Acad. Nat. Sci. 96, 3420–3426.

Laherrere, J.H., September, 1999. Data shows oceanic methane hydrate resource overestimated. Offshore, 156–157.

Lerche, I., April 30–May 3, 2001. Estimates of worldwide gas hydrate resources. In: Offshore Technology Conference, OTC Paper No. 13036, Houston, TX.

Linga, P., Haligva, C., Nam, S.C., Ripmeester, J.A., Englezos, P., 2009. Gas hydrate formation in a variable volume bed of silica sand particles. Energy Fuels 23, 5496–5507.

Makogon, Y., Holditch, S.A., Makogon, T.Y., February 7, 2007. Russian field illustrates gas-hydrate production. Oil Gas J. 103 (5).

Miller, S.L., 1961. The occurrence of Gas hydrates in the solar system. Proc. Nat. Acad. Sci. U.S.A 47, 1798–1808.

Miller, D.G., 1970. Joule-Thomson inversion curve, corresponding states, and simpler equations of state. Ind. Eng. Chem. Fundamen 9, 585–589.

Miller, S.L., Smythe, W.D., 1970. Carbon dioxide clathrate in the martian ice cap. Science 170, 53–54.

Smith, R.V., 1990. Practical Natural Gas Engineering, second ed. PennWell Pub. Co., Tulsa, OK.

Tanahashi, M., 1996. Massoyakha Gas Field. The First Commercial Hydrate Deposit?. http://www.aist.go.jp/GSJ/dMG/dMGold/hydrate/Messoyakha.html.

Turian, R.M., Yuan, T.-F., 1977. Flow in slurry pipelines. AIChE J. 23, 232–243.

Wagner, W., de Reuck, K.M., 1996. Methane. International Thermodynamic Tables of the Fluid State – 13. Blackwell Science (for IUPAC), Oxford, UK.

Zatsepina, O.Y., Pooladi-Darvish, M., 2012. CO2 storage as hydrate in depleted gas reservoirs. SPE Reserv. Eval. Eng. 15 (1), 98–108.

INDEX

Page references followed by "f" indicate figures, "t" indicate tables.

Printed and bound by CPI Group (UK) Ltd, Croydon, CR0 4YY
08/05/2025
01864799-0002